Bioactive Phytochemicals in By-products from Leaf, Stem, Root and Tuber Vegetables

Mohamed Fawzy Ramadan
Editor

Bioactive Phytochemicals in By-products from Leaf, Stem, Root and Tuber Vegetables

Springer

Editor
Mohamed Fawzy Ramadan
Faculty of Applied Medical Sciences, Clinical Nutrition Department
Umm Al-Qura University
Makkah, Saudi Arabia

ISBN 978-3-031-80699-5 ISBN 978-3-031-80700-8 (eBook)
https://doi.org/10.1007/978-3-031-80700-8

This Springer imprint is published by the registered company Springer Nature Switzerland AG
The registered company address is: Gewerbestrasse 11, 6330 Cham, Switzerland

Preface

The nutritional composition of leaf, stem, root, and tuber vegetable bio-waste is abundant in vitamins, minerals and a variety of health-promoting phytochemicals (phenolics, carotenoids, alkaloids, terpenoids, etc.). Recycling leaf, stem, root, and tuber vegetable by-products and utilizing those bio-wastes to produce added-value products are global trends that help accomplish sustainable development goals. As a result, it is anticipated that waste management in vegetable processing will grow significantly during the coming years.

Bioactive constituents from leaf, stem, root, and tuber vegetable processing bio-wastes are tested for their technological functions (such as coloring and texturing), nutritional values, or potential benefits to human and/or animal health. Vegetable processing by-products have different chemical compositions due to crop variations, harvest timing, and processing techniques. Meanwhile, recent research has focused on high-pressure green solvent extraction. On the other hand, by-products from leaf, stem, root, and tuber vegetable processing could be crushed, roasted, and utilized as animal feed or fertilizer.

Publications throughout the past few years have concentrated on using leaf, stem, root, and tuber vegetable processing wastes for both edible and non-edible purposes. With its comprehensive reference content and up-to-date knowledge, this book is a valuable source for educators, researchers, academics, food processors, industry professionals, and policy developers. The goal of this book project was to establish a multidisciplinary discussion forum on the phytochemistry, functional traits, and health-promoting effects of bioactive compounds found in by-products of leaf, stem, root, and tuber vegetables, as well as the applications of those by-products in edible and non-edible products.

The book contains 16 chapters that cover the following topics:

1. The economic value of by-products from the processing of leaf, stem, root, and tuber vegetables.
2. The active ingredients and chemical makeup of extracts obtained from by-products of leaf, stem, root, and tuber vegetables.

3. The biological and functional traits of the active compounds and extracts from by-products of leaf, stem, root, and tuber vegetables.
4. Uses of phyto-extracts and active compounds from leaf, stem, root, and tuber vegetable bio-wastes for food and non-food products.

The editor thanks all authors for their valuable contributions and collaboration during the book's stages. The Springer staff's help and support, especially *Sofia Costa* and *Neelofar Yasmeen*, was essential for completing the task and is appreciated.

Makkah, Saudi Arabia
October 2024

Mohamed Fawzy Ramadan

Description

Improving bio-waste management is one of the cornerstones to guaranteeing the industry's sustainability in terms of the environment, economy, and society. While vegetables are grown for fresh consumption, processing them produces a significant amount of by-products. The by-products of vegetable processing are rich in hydrophilic and lipophilic phytochemicals. These wastes could have substantial positive effects on the environment and economy if they are properly valorized. The idea of the "*Zero Waste Economy*," which transforms wastes into valuable products and finds practical uses by offering potential energy savings, may be used to value vegetable manufacturing by-products.

Phenolic compounds, carotenoids, pigments, tocols, vitamins, and phyto-extracts are examples of phytochemicals and secondary metabolites that may enhance the nutritional value of a diet. Due to their excellent nutritional value and quality, those bioactive constituents may be classified as functional food. In addition, because of their biological and antioxidant attributes, phytochemicals obtained from leaf, stem, root, and tuber vegetable by-products may be able to add value and prolong the shelf life of a variety of items, including food, cosmetics, and pharmaceuticals.

Research showed how to recover, characterize, and use beneficial chemicals from vegetable manufacturing by-products. Moreover, it has been demonstrated in a number of model systems that phytochemicals isolated from vegetable species, including leaf, stem, root, and tuber vegetables, exhibit biological traits that promote health. Thus, it seems promising to turn vegetable by-products into natural additives with added value.

Leaf, stem, root, and tuber vegetable by-products represent a major disposal problem for the industry concerned. However, they are promising sources of value-added phytochemicals, with particular emphasis on the retrieval of bioactive phytochemicals and technologically important secondary metabolites. Meanwhile, manufacturing conditions affect the by-products generated during processing. Therefore, these processes might affect the quantity and quality of phytochemicals, including antioxidants and vitamins.

The present book characterizes ingredients and phytochemicals in leaf, stem, root, and tuber vegetables and their health-enhancing traits. As a result of this book's findings, (1) the health-promoting effects of phyto-compounds from leaf, stem, root, and tuber vegetable production bio-wastes are discussed, (2) the chapters review the extraction techniques to develop novel applications from specific phyto-compounds, (3) the consumer comprehends the health-enhancing impact of phyto-compounds from leaf, stem, root, and tuber vegetable manufacturing bio-wastes, and (4) the economy will be strengthened by incorporating bioactive compounds from the bio-wastes obtained during the processing of leaf, stem, root, and tuber vegetables into added-value products.

Bioactive Phytochemicals in By-Products from Leaf, Stem, Root, and Tuber Vegetables aims to discuss the nutritional value, phytochemistry, and health-enhancing effects of phyto-compounds in leaf, stem, root, and tuber vegetable processing bio-wastes and their uses in food and non-food products. The benefits of using leaf, stem, root, and tuber vegetable processing by-products to boost health are outlined.

Key Features

- Includes in-depth discussions on chemistry and functional traits of phyto-constituents in leaf, stem, root, and tuber vegetable manufacturing by-products
- Authored by international scientists and academic experts
- Highlights literature and coverage of recent developments
- Covers a vast array of special topics illustrating the growing application areas, including pharmaceuticals, nutrition, novel food, and cosmetics.

Mohamed Fawzy Ramadan

Contents

About the Editor and Contributors

About the Editor

Mohamed Fawzy Ramadan is a Professor of Food Chemistry and Biochemistry at the Department of Clinical Nutrition, Faculty of Applied Medical Sciences, Umm Al-Qura University, Makkah, Saudi Arabia. Prof. Ramadan obtained his Ph.D. (*Dr. rer. nat.*) in Food Chemistry from the Berlin University of Technology (Germany, 2004). He continued his postdoctoral research at ranked universities such as the University of Helsinki (Finland), Max-Rubner Institute (Germany), Berlin University of Technology (Germany), and the University of Maryland (USA). In 2012, he was appointed as a Visiting Professor (100% teaching) in the School of Biomedicine, Far Eastern Federal University in Vladivostok, Russian Federation.Prof. Ramadan published over 400 research papers and reviews in international peer-reviewed journals. He has also edited and published several books and book chapters (Google Scholar *h*-index is 65, and there are more than 16000 citations). He was an invited speaker at several international conferences. Since 2003, Prof. Ramadan has been a reviewer and editor in several highly-cited international journals, such as Earth Systems and Environment, Journal of Medicinal Food, eFood, Journal of Umm Al-Qura University for Applied Sciences, and Journal of Advanced Research. He is the editor-in-chief of Journal of Umm Al-Qura University for Medical Science.Prof. Ramadan received several prizes, including Abdul Hamid Shoman Prize for Arab

Researcher in Agricultural Sciences (2006), the Egyptian State Prize for Encouragement in Agricultural Sciences (2009), European Young Lipid Scientist Award (2009), AU-TWAS Young Scientist National Awards (Egypt) in Basic Sciences, Technology and Innovation (2012), TWAS-ARO Young Arab Scientist (YAS) Prize in Scientific and Technological Achievement (2013), and Atta-ur-Rahman Prize in Chemistry (2014).

Contributors

Banzeer Ahsan Abbasi Department of Botany, Rawalpindi Women University, Rawalpindi, Pakistan

Turan Akdağ Necmettin Erbakan University, Meram Vocational School, Medicinal and Aromatic Plants Program, Medicinal and Cosmetic Plants Application and Research Center, Meram, Turkey

Medicinal and Aromatic Plants Program, Meram Vocational School, Necmettin Erbakan University, Konya, Turkey

Ahmad Ali Center for Plant Sciences and Biodiversity, University of Swat, Charbagh, Pakistan

Habib Ali Department of Agricultural Engineering, Khwaja Fareed University of Engineering and Information Technology, Rahim Yar Khan, Pakistan

Zeliha Üstün Argon Necmettin Erbakan University, Meram Vocational School, Medicinal and Aromatic Plants Program, Medicinal and Cosmetic Plants Application and Research Center, Meram, Turkey

Anna Rafaela Cavalcante Braga Department of Chemical Engineering, Federal University of São Paulo (UNIFESP), Diadema, São Paulo, Brazil

Ying Ping Chang Department of Agricultural and Food Science, Faculty of Science, Universiti Tunku Abdul Rahman, Jalan Universiti, Bandar Barat, Kampar, Perak, Malaysia

Shikha Chauhan Amity Institute of Pharmacy, Amity University UP, Noida, India

Ahmad Cheikhyoussef Science and Technology Division, Multidisciplinary Research Services, Centre for Research Services, University of Namibia, Windhoek, Namibia

Natascha Cheikhyoussef Ministry of Higher Education, Technology, and Innovation (MHETI), Windhoek, Namibia

Moncef Chouaibi Higher School of Food Industries of Tunisia, University of Carthage, Elkhadra City, Tunisia

Department of Chemical and Food Engineering (DICIV), Università degli Studi di Salerno | UNISA, Fisciano, Italy

Josemar Gonçalves de Oliveira Filho Brazilian Agricultural Research Corporation, Embrapa Instrumentation, São Carlos, São Paulo, Brazil

Süleyman Doğu Necmettin Erbakan University, Meram Vocational School, Medicinal and Aromatic Plants Program, Medicinal and Cosmetic Plants Application and Research Center, Meram, Turkey

Medicinal and Aromatic Plants Program, Meram Vocational School, Necmettin Erbakan University, Konya, Turkey

Medicinal and Cosmetic Plants Application and Research Center, Necmettin Erbakan University, Konya, Turkey

Mariana Buranelo Egea Goiano Federal Institute of Education, Science and Technology, Rio Verde, Goiás, Brazil

Sibele Santos Fernandes Federal University of Rio Grande, School of Chemistry and Food, Rio Grande, Brazil

Samy Mohamed Galal Food Science Department, Faculty of Agriculture, Cairo University, Giza, Egypt

Zinar Pınar Gümüş Central Research Testing and Analysis Laboratory Research and Application Center, Ege-Matal, Ege University, Izmir, Turkey

Charu Gupta Amity Institute of Herbal Research & Studies, Amity University UP, Noida, India

Ahmed A. Hussein Chemistry Department, Cape Peninsula University of Technology, Cape Town, South Africa

Shumaila Ijaz School of Biomedical Engineering, Shenzhen University Medical School, Shenzhen University, Shenzhen, China

Martha Ikechukwu Sam Nujoma Marine and Coastal Resources Research Centre, Sam Nujoma Campus University of Namibia, Henties Bay, Namibia

Madiha Imtiaz Department of Botany, Bacha Khan University, Charsadda, Khyber Pakhtunkhwa, Pakistan

Javed Iqbal Department of Botany, Bacha Khan University, Charsadda, Khyber Pakhtunkhwa, Pakistan

Muhammad Jamil Department of Biotechnology and Genetic Engineering, Kohat University of Science and Technology, Kohat, Pakistan

Sobia Kanwal Department of Biology and Environmental Sciences, Allama Iqbal Open University, Islamabad, Pakistan

Ailton Cesar Lemes Department of Biochemical Engineering, Federal University of Rio de Janeiro (UFRJ), School of Chemistry, Rio de Janeiro, Brazil

Tariq Mahmood Department of Plant Sciences, Faculty of Biological Sciences, Quaid-i-Azam University Islamabad, Islamabad, Pakistan

Masixole Makhaba Chemistry Department, Cape Peninsula University of Technology, Cape Town, South Africa

Nashwa Fathy Sayed Morsy Food Science Department, Faculty of Agriculture, Cairo University, Giza, Egypt

Mohamed Fawzy Ramadan Department of Clinical Nutrition, Faculty of Applied Medical Sciences, Umm Al-Qura University, Makkah, Saudi Arabia

Rajan Sharma Chemistry Department, Cape Peninsula University of Technology, Cape Town, South Africa

Hassan Sher Center for Plant Sciences and Biodiversity, University of Swat, Charbagh, Pakistan

Vanessa Situmbeko Department of Biochemistry, Microbiology & Biotechnology, Faculty of Agriculture, Engineering and Natural Sciences, University of Namibia, Windhoek, Namibia

Abdul Rauf Tareen Department of Microbiology, Quaid-I-Azam University, Islamabad, Pakistan

Siraj Uddin Department of Plant Sciences, Faculty of Biological Sciences, Quaid-i-Azam University Islamabad, Islamabad, Pakistan

Zahid Ullah Center for Plant Sciences and Biodiversity, University of Swat, Charbagh, Pakistan

Zakir Ullah Department of Plant Sciences, Faculty of Biological Sciences, Quaid-i-Azam University Islamabad, Islamabad, Pakistan

Zeliha Üstün Argon Medicinal and Aromatic Plants Program, Meram Vocational School, Necmettin Erbakan University, Konya, Turkey

Medicinal and Cosmetic Plants Application and Research Center, Necmettin Erbakan University, Konya, Turkey

Kwan Kit Woo Lee Kong Chian Faculty of Engineering and Science, Universiti Tunku Abdul Rahman, Bandar Sungai Long, Kajang, Malaysia

Tabassum Yaseen Department of Botany, Bacha Khan University, Charsadda, Khyber Pakhtunkhwa, Pakistan

Farishta Zarshan Department of Botany, Bacha Khan University, Charsadda, Khyber Pakhtunkhwa, Pakistan

Part I
General Aspects

Introduction to Bioactive Phytochemicals in By-products from Leaf, Stem, Root, and Tuber Vegetables

Mohamed Fawzy Ramadan

Contents

Abstract Worldwide, vegetables are grown and consumed as the main part of the human diet. A high volume of by-products is generated annually from leaf, stem, root, and tuber vegetables. Phytochemicals are organic compounds formed by plants in their metabolic operations. Phytochemicals enhance the functional traits of food and play a significant role in maintaining ecological balance. Meanwhile, bioactive phytochemicals contribute to economic sectors, such as nutraceuticals, pharmaceuticals, cosmetics, and novel food. This work highlights the nutritional value of by-products from leaf, stem, root, and tuber vegetables. This book discusses the health benefits and applications of phytochemicals derived from processing by-products of leaf, stem, root, and tuber vegetables. Besides, the book highlights the recovery issues and various techniques for developing new applications from these phytochemicals. In addition, the book boosts the circular economy *via* using phytochemi-

M. F. Ramadan (✉)
Department of Clinical Nutrition, Faculty of Applied Medical Sciences, Umm Al-Qura University, Makkah, Saudi Arabia
e-mail: mfhassanien@uqu.edu.sa

M. F. Ramadan (ed.), *Bioactive Phytochemicals in By-products from Leaf, Stem, Root and Tuber Vegetables*, https://doi.org/10.1007/978-3-031-80700-8_1

cals from leaf, stem, root, and tuber vegetable bio-wastes for industrial applications (new edible items, pharmaceuticals, and livestock).

Keywords Phytochemistry · SDGs · Bio-waste · Environment · Economy · Recycling · Novel food · Food manufacturing · Bio-waste valorization · Food technology

1 Introduction

Vegetable by-products are the tissue yields of vegetable processing that are deemed unfit for human consumption and are hence discarded as bio-waste in the food supply chain. Vegetable by-products are, however, potential sources of bioactive molecules. Worldwide, a large volume of various types of vegetable by-products is generated annually. Leaf, stem, root, and tuber vegetables make up a significant proportion of residue yields from vegetable processing. This book reports on the bioactive phytochemicals in the by-products of leaf, stem, root, and tuber vegetables in terms of their compound classes, occurrence, extraction, and recovery techniques, with an emphasis on their utilization in food, nutraceutical, and pharmaceutical products (Reguengo et al. 2022; Coman et al. 2020; Marcillo-Parra et al. 2021; Espinosa-Alonso et al. 2020; Jiménez-Moreno et al. 2020; Mateus et al. 2023; Majumder and Annegowda 2021; Lau et al. 2021; Galali et al. 2020; Aqilah et al. 2023).

Vegetables are an essential component of a healthy diet worldwide. They are rich sources of bioactive phytochemicals and small-molecular-weight secondary metabolites that impart health benefits beyond basic nutrition. Dietary phytochemicals include phenolics, carotenoids, phytosterols, organosulfur compounds, and vitamins. Due to their functions in the defense system of plants against biotic and abiotic stresses, phytochemicals are abundantly and preferentially accumulated in the unopened leaf buds, young leaves, sprouts, stems, tubers, and root of plants, which are the edible tissues of vegetables.

Vegetable by-products also referred to as vegetable residues or wastes are the tissue yields of vegetable processing—such as sorting, peeling, trimming, and juicing—that are discarded as waste and deemed unfit for human consumption. They are potentially rich sources of bioactive phytochemicals. Vegetable by-products often exhibit higher bioactive compounds compared to their edible counterparts, the edible tissues of vegetables being the targeted tissues in vegetable processing to supply salad, sides, boiled stir-fries, or vegetable juices. Globally, a large volume of various types of vegetable by-products is generated annually, a substantial portion of which is due to leaf, stem, root, and tuber vegetables. By incorporating vegetable by-products into food products, they could be utilized as potential sources of bioactive compounds that impart health benefits (Coman et al. 2020; Reguengo et al. 2022; Cano-Lamadrid and Artes-Hernandez 2022; Espinosa-Alonso et al. 2020;

Ramadan Hassanien 2023; Sorrenti et al. 2023; Hassanien 2023; Vazquez-Olivo et al. 2023; Mohammadnezhad et al. 2023; Majumder and Annegowda 2021).

2 Bioactive Phytochemicals: Classification and Importance

Phytochemicals are organic compounds formed by plants in their metabolic operations. Although phytochemicals are not required for the normal growth and development of plants, they aid in performing critical tasks. Phytochemicals enhance the health traits of food and play a significant role in maintaining ecological balance and the biodiversity of the planet. Bioactive phytochemicals contribute to economic sectors, such as pharmaceuticals, cosmetics, and novel food. Recent advances in science have opened up avenues for the agricultural industry, where bioactive compounds alleviate biotic and abiotic stresses in plants. Phytochemicals are present in various plant materials, including flowers, roots, seeds, leaves, and bark (Vaghela and Gohel 2023; Samynathan et al. 2021; Kaushik et al. 2021; Saleem et al. 2022; Jan et al. 2021; Shahid et al. 2022; Yeshi et al. 2022; Teklić et al. 2021; Rizvi et al. 2022; Shah and Smith 2020).

In recent years, contemporary interest has been granted to bioactive phytochemicals due to their therapeutic application against chronic diseases. In human health, bioactive phytochemicals possess anti-inflammatory, anticarcinogenic, antiviral, antioxidant, and antimicrobial traits. Recent studies provide insights into numerous bioactive phytochemicals isolated from a range of plant sources with potential protection or treatment against cardiovascular diseases, cancers of various types, diabetes, Alzheimer's disease, and microbial infections. Generally, phytochemicals are present as bioactive compounds in the common diet. Consumption of plant sources is linked to a low incidence of different chronic diseases. Epidemiological studies reported a low prevalence of chronic diseases in populations that adopt a vegetarian diet rich in fruits and vegetables (Kaushik et al. 2021; Ali et al. 2022; Bader Ul Ain et al. 2022; Ralte et al. 2022; Najeebullah et al. 2021; El-Beltagi et al. 2022; Esmeeta et al. 2022; Samtiya et al. 2021).

2.1 Definition and Classification

Phytochemicals harmoniously coexist in a myriad of fruits and vegetables, bestowing a multitude of health benefits for centuries. Originating from the Greek language, "*phyto*" is a term that encloses plant origin and is depicted as "phytochemical." Phytochemicals do not have any direct nutritional value; however, they perform a plethora of important functions in the medical physiologic systems of herbivores and humans. Nonetheless, the roles of phytochemicals as natural antioxidants by scavenging free radicals, inhibiting oxidants, DNA saturation, and chelating metals to limit the risk of diseases have been widely reported.

Vegetables are an important part of the human diet. They are rich sources of bioactive phytochemicals, vitamins, and minerals and are paired with a low caloric value. Depending on the cultivation area and part of a plant that is consumed, vegetables might be classified into six groups, comprising leaf, stem, flower, root, tuber, and hybrid vegetables (Sharma and Kaushik 2021; Behl et al. 2021a, b; Mani et al. 2020; Thakur et al. 2020; Bitwell et al. 2023; Ghildiyal et al. 2020; Saroj et al. 2020; Shahrajabian et al. 2022; Boukhatem and Setzer 2020; Agidew 2022).

Bioactive phytochemicals are neither nutrients essential for growth nor energy production nor maintenance of cellular functions. Instead, these compounds modulate the metabolic processes influencing human growth, development, and health (Chojnacka et al. 2020; Wink 2022; Esmeeta et al. 2022; Shrinet et al. 2021; Yu et al. 2021; Rizvi et al. 2022; Iqbal et al. 2020; Godlewska-Żyłkiewicz et al. 2020; Stéphane et al. 2021).

2.2 Importance and Functions

Bioactive phytochemicals refer to a wide variety of naturally occurring plant compounds that possess scientifically proven health-promoting properties. Bioactive phytochemical resources are abundant in fruits, vegetables, seeds, tubers, nuts, and mushrooms and play a crucial role in the health of humans and livestock. Phytochemicals include polyphenols, isothiocyanates, phytosterols, glucobrassicin, lignans, carotenoids, vitamins, hydroxylated polyunsaturated fatty acids, folates, organosulfur compounds, terpenes, flavonoids, phytoestrogens, and non-nutrients.

Bioactive phytochemicals have garnered considerable interest due to their proven chemical, biological, and pharmacokinetic traits. Moreover, vegetative parts of leaf, root, stem, tuber, and root vegetables are often underutilized food resources in Asia, with or without post-harvest processing. Processing of vegetables not only reduces the resource but also generates huge amounts of unusable organic waste. On the other hand, industrially derived agricultural waste is another cheap and sustainable source of phytochemicals (Sharma and Kaushik 2021; Murthy and Paek 2020; Nguyen et al. 2021; Osuntokun et al. 2020; Gamba et al. 2021; Amagloh et al. 2022).

Accordingly, vegetative wastes and by-products from root, stem, and leaf vegetables, with or without post-harvest processing, are being explored as sources of phytochemicals of diverse biological functions. Investigations focusing on the extracts of inedible parts of vegetables have successfully yielded phytochemicals with bioactive properties ranging from anticancer, antioxidant, anti-inflammatory, antimycotic, and antimicrobial to nematicidal and insecticidal. Moreover, bioactivity investigations on commercially available phytochemicals have unveiled novel functional traits. There has been a growing understanding of the significance and physiological functions of bioactive phytochemicals in producing beneficial biological responses in the host. Enhanced scientific understanding of the health-promoting properties of phytochemicals has motivated growers, food processors, and pharmaceutical and nutraceutical companies to utilize them in health-promoting food applications. Accumulation of bioactive phytochemicals in the edible system can be regulated by environmental, physiological, and genetic factors, leading to the potential for bioengineering target crops for enhanced food functionality (Banwo et al. 2021; Mondal et al. 2021; Yu et al. 2021; Maheshwari et al. 2022; Riaz et al. 2023; Rizvi et al. 2022; Behl et al. 2021a, b; Favela-González et al. 2020; Saxena 2023).

3 Vegetable By-products: Sources and Composition

Vegetable by-products are defined as wastes produced following vegetable processing and are rich in bioactive phytochemicals and fiber. The major types of leaf, stem, root, and tuber vegetable by-products are discussed. The major vegetable by-products produced during vegetable processing are illustrated. In addition, the nutritional composition and functional traits of the vegetable by-products are presented.

Stem vegetables could be divided into herbaceous and woody stem vegetables. Stem by-products such as Brussels sprouts, leaf cabbage, broccoli, knoll celery, and cauliflower stalks and by-products are produced during processing, transport, and storage of vegetables. Leafy vegetables possess shallow root systems and are grown on soils that can retain moisture. Leafy vegetables are excellent sources of vitamins A, C, E, and K and are also potential anticancer foods due to the presence of bioactive phytochemicals. Leaf by-products produced during the processing of lettuce, spinach, and cabbage are shown to contain good nutritional composition (Coman et al. 2020; Lau et al. 2021; Hussain et al. 2020; Espinosa-Alonso et al. 2020; Reguengo et al. 2022; Trigo et al. 2022; Galali et al. 2020; Cano-Lamadrid and Artes-Hernandez 2022; Santos et al. 2022).

Bioactive phytochemicals from tuber vegetables such as potatoes and sweet potatoes, generated by-products from processing, are underutilized. In addition to bioactive compounds, dietary fiber content and composition in by-products are important parameters to be researched. The effect of drying and conditions of processing on the dietary fiber composition of the vegetable by-products is also highlighted. The major tomato, carrot, mushroom, spinach, cabbage, potato, and sweet potato by-products were illustrated along with the production process and by-product composition (de los Ángeles Rosell et al. 2024; Oleszek et al. 2023; Calcio Gaudino et al. 2020; Lau et al. 2021; Popoola et al. 2023).

3.1 *Types of By-products*

Worldwide, vegetables are grown and consumed as an integral part of the human diet. Vegetables are edibles that can be harvested from different parts of plants, mostly comprising leaf, stem, root, and tuber. Certain green plants with stems, either whole or cut pieces, edible in raw/cooked form, are called stem vegetables.

Root vegetables are either herbs or root plants where unopened buds, edible in raw/cooked form, are called flower/bud vegetables. As per the FAO, Asian countries such as India, Bangladesh, Pakistan, and Myanmar account for a major share of the transmission and distribution of vegetables.

Vegetable processing industries may include ready-to-eat salads, pickles, frozen vegetables, dehydrated vegetables, canned vegetables, etc. As per estimates, around 20–60% of processed feedstocks are discarded as by-products during vegetable processing. The remaining 40–80% constitutes the final edible products of vegetable processing industries. Handling, transporting, and reproduction of food waste are costly. It is important to note that vegetable by-products, like other food wastes, should not be disposed of contiguously. It is advisable to convert agricultural by-products to added-value products by microbial bioprocessing (Joshi 2020; Rao et al. 2021; Zhu et al. 2023; Tlais et al. 2020; Tiwari and Khawas 2021; Ganesh et al. 2022; Manhongo et al. 2022; Esparza et al. 2020).

Vegetable by-products constitute leafy green parts of kale, cabbage, knol-khol, lettuce, and leaves and stalks of cucumbers, tomatoes, carrots, radishes, peas, okra, brinjal, beets, carrots, etc. By-products from flower vegetables constitute flower buds of cauliflower, broccoli, etc. Vegetable by-products may also contain peelings/cuts and shrivels of tuber vegetables. Tuber vegetables include potatoes, sweet potatoes, yams, taro, and cassava (Majumder and Annegowda 2021; Fraga-Corral et al. 2021; Mateos-Aparicio 2021; Espinosa-Alonso et al. 2020; Coman et al. 2020; Oleszek et al. 2023; Cano-Lamadrid and Artes-Hernandez 2022; El-Feky and eL Batanony 2022; Echegaray et al. 2023).

3.2 Nutritional Composition

Vegetables play an important role in a healthy diet. These commodities supply essential nutrients in a well-balanced human diet and are also rich sources of antioxidants. The by-products generated during the processing of vegetables are rich in bioactive components. Studies have demonstrated that bioactive phytochemicals from vegetables show no or little loss of biological activity when consumed after extraction and/or processing. The term "vegetables" denotes edible parts of the plants and usually includes edible parts of the leaf (leafy vegetables), stem (bulbs and stalks), root (tubers and root vegetables), and flower (flowering vegetables). This work highlights the nutritional composition of by-products from leaf, stem, root, and tuber vegetables (Pacifico et al. 2021; Oleszek et al. 2023; Reguengo et al. 2022; Lau et al. 2021; Jiménez-Moreno et al. 2020).

From an agricultural perspective, vegetable by-products could be further converted into vegetable protein concentrate and/or vegetable protein isolate, which in turn could be applied to functional food development by adding them into the food or beverage formulation at varying levels (2–8%) without adversely affecting the sensory properties (Gatahi 2020; Bilenka et al. 2023; Chuin Lim et al. 2022; Lim et al. 2021; Kováčik et al. 2022; Fujii et al. 2021; Husak et al. 2024).

4 Aims and Features of the Book

In view of increased environmental concerns, plant by-products generated during processing are being explored as alternative sources for bioactive compounds. Wide varieties of edible vegetables are cultivated throughout the globe. Among them, tubers and roots represent an important part of the human diet. The production of leaf, stem, root, and tuber vegetables is millions of tons, as per the FAOSTAT report (2018), which is the summation of global production. Apart from their edible parts, large quantities of untapped by-products ranging from peel to seeds, stem, leaves, and other unutilized parts are generated during processing. These by-products are a rich source of bioactive phytochemicals, including fibers, phenolics, carotenoids, and vitamins. The bioactive compounds of diverse fruits and vegetables possess antioxidant, antidiabetic, anticholesterol, anti-obesity, and other beneficial health-promoting traits. Bioactive phytochemicals present in crude extracts of different crops or their extracts after the appropriate processing have shown improved bioavailability. They have also imparted beneficial properties to the biopolymer-based edible films, enhancing the overall techno-functional performance (Scott 2021; Jahan 2020; Otálora et al. 2024; Otekunrin et al. 2021; Prasad and Shivay 2021; Nabeshima et al. 2020; Muthoni and Shimelis 2023).

Exploring the bioactivity of novel phytocompounds is crucial to unveil their antioxidant and other health-promoting traits. However, the majority of the research has focused only on the bioactivity of agricultural practices, focusing on the cultivation or the edible parts, as there is very little work with respect to exploring the bioactivity of the by-products generated from the vegetables above. In pursuance of filling this gap, it is aimed to investigate the bioactivity of different by-products generated from the processing of tuber, root, stem, and leaf fresh vegetables as a potential alternative source of unexplored bioactive compounds and to characterize the bioactive phytochemicals. To achieve these objectives, this work highlights extracts prepared from different processing by-products (peel, pulp, stem, leaves, etc.) of leaf, stem, root, and tuber vegetables and screened for their antioxidant, antidiabetic, and anti-obesity traits. Studies showed that the bioactivity varied with the nature of the processing by-products. The potent by-products exhibited specific bioactivity characterized for the bioactive phytochemicals by using state-of-the-art technology. The intricate mechanism(s) of action explaining the bioactivity of bioactive phytochemicals of interest were explored. Moreover, the extracts containing bioactive compounds employed in the designs of protein or polysaccharide-based active edible films, as well as the techno-functional performance of the films, were reported. In addition, reports mentioned that post-harvest and processing interventions such as polishing, steaming, and peel removal influenced the physical properties and bioactivity of the bioactive compounds (Coman et al. 2020; Marcillo-Parra et al. 2021;

Galali et al. 2020; Trigo et al. 2022; Jiménez-Moreno et al. 2020; Reguengo et al. 2022; Teruel-Andreu et al. 2021; Kumar et al. 2021; Görgüç et al. 2020).

This book unveils the scientific knowledge on leaf, stem, root, and tuber vegetable by-products' health-enhancing impacts and applications, mainly from a phytochemistry perspective. The book will support the economy by integrating leaf, stem, root, and tuber vegetable processing by-products into applied items (food, nutraceuticals, pharmaceuticals, livestock, and cosmetics). It is a valuable source for researchers investigating agro-industrial bio-waste recycling from leaf, stem, root, and tuber vegetable by-products.

References

Agidew MG (2022) Phytochemical analysis of some selected traditional medicinal plants in Ethiopia. Bull Natl Res Cent 46:87. https://doi.org/10.1186/s42269-022-00770-8

Ali SK, Mahmoud SM, El-Masry SS, Alkhalifah DHM, Hozzein WN, Aboel-Ainin MA (2022) Phytochemical screening and characterization of the antioxidant, anti-proliferative and anti-bacterial effects of different extracts of Opuntia ficus-indica peel. J King Saud Univ Sci 34(7):102216

Amagloh FC, Kaaya AN, Yada B, Chelangat DM, Katungisa A, Amagloh FK, Tumuhimbise GA (2022) Bioactive compounds and antioxidant activities in peeled and unpeeled sweetpotato roots of different varieties and clones in Uganda. Future Foods 6:100183

'Aqilah, N. M. N., Rovina, K., Felicia, W. X. L., & Vonnie, J. M. (2023). A review on the potential bioactive components in fruits and vegetable wastes as value-added products in the food industry. Molecules 28(6), 2631

Bader Ul Ain H, Tufail T, Javed M, Tufail T, Arshad MU, Hussain M et al (2022) Phytochemical profile and pro-healthy properties of berries. Int J Food Prop 25(1):1714–1735

Banwo K, Olojede AO, Adesulu-Dahunsi AT, Verma DK, Thakur M, Tripathy S et al (2021) Functional importance of bioactive compounds of foods with Potential Health Benefits: a review on recent trends. Food Biosci 43:101320

Behl T, Kumar K, Brisc C, Rus M, Nistor-Cseppento DC, Bustea C et al (2021a) Exploring the multifocal role of phytochemicals as immunomodulators. Biomed Pharmacother 133:110959

Behl T, Rocchetti G, Chadha S, Zengin G, Bungau S, Kumar A et al (2021b) Phytochemicals from plant foods as potential source of antiviral agents: an overview. Pharmaceuticals 14(4):381

Bilenka OM, Horhan TM, Pidlubenko IM, Ovchinnikova OP, Shtepa LY (2023) Evaluation of the adaptive potential of promising shallot accessions. Veget Melon Grow 73:6–12

Bitwell C, Indra SS, Luke C, Kakoma MK (2023) A review of modern and conventional extraction techniques and their applications for extracting phytochemicals from plants. Scientific African 19:e01585

Boukhatem MN, Setzer WN (2020) Aromatic herbs, medicinal plant-derived essential oils, and phytochemical extracts as potential therapies for coronaviruses: future perspectives. Plants 9(6):800

Calcio Gaudino E, Colletti A, Grillo G, Tabasso S, Cravotto G (2020) Emerging processing technologies for the recovery of valuable bioactive compounds from potato peels. Food Secur 9(11):1598

Cano-Lamadrid M, Artes-Hernandez F (2022) By-products revalorization with non-thermal treatments to enhance phytochemical compounds of fruit and vegetables derived products: a review. Foods 11(1):59

Chojnacka K, Witek-Krowiak A, Skrzypczak D, Mikula K, Młynarz P (2020) Phytochemicals containing biologically active polyphenols as an effective agent against Covid-19-inducing coronavirus. J Funct Foods 73:104146
Chuin Lim W, Nadzir MNHM, Hiew MWH, Mamat MS, Nazli MH, Shohaimi S (2022) Potential of open-air hydroponic system in producing highly nutritional composition maize fodder for goat farming. Pertanika J Trop Agric Sci 45(1):115–131
Coman V, Teleky BE, Mitrea L, Martău GA, Szabo K, Călinoiu LF, Vodnar DC (2020) Bioactive potential of fruit and vegetable wastes. Adv Food Nutr Res 91:157–225
Echegaray N, Gullón B, Pateiro M, Amarowicz R, Misihairabgwi JM, Lorenzo JM (2023) Date fruit and its by-products as promising source of bioactive components: a review. Food Rev Intl 39(3):1411–1432
El-Beltagi HS, Mohamed AA, Mohamed HI, Ramadan KM, Barqawi AA, Mansour AT (2022) Phytochemical and potential properties of seaweeds and their recent applications: a review. Mar Drugs 20(6):342
El-Feky AM, eL Batanony M (2022) Potential phytoconstituents of some fruit and vegetable peels against oxidative damage, inflammatory and cytotoxic diseases. Egypt J Chem 65:261–272
Esmeeta A, Adhikary S, Dharshnaa V, Swarnamughi P, Maqsummiya ZU, Banerjee A et al (2022) Plant-derived bioactive compounds in colon cancer treatment: an updated review. Biomed Pharmacother 153:113384
Esparza I, Jiménez-Moreno N, Bimbela F, Ancín-Azpilicueta C, Gandía LM (2020) Fruit and vegetable waste management: conventional and emerging approaches. J Environ Manag 265:110510
Espinosa-Alonso LG, Valdez-Morales M, Aparicio-Fernandez X, Medina-Godoy S, Guevara-Lara F (2020) Vegetable by-products. Food wastes and by-products: nutraceutical and health potential, 223–266
Favela-González KM, Hernández-Almanza AY, De la Fuente-Salcido NM (2020) The value of bioactive compounds of cruciferous vegetables (Brassica) as antimicrobials and antioxidants: a review. J Food Biochem 44(10):e13414
Fraga-Corral M, Otero P, Echave J, Garcia-Oliveira P, Carpena M, Jarboui A, Prieto MA (2021) By-products of agri-food industry as tannin-rich sources: a review of tannins' biological activities and their potential for valorization. Food Secur 10(1):137
Fujii H, Mori S, Matsumoto Y, Sasaki Y, Ito C, Nakagawa S, Ando K (2021) Tohoku region. The soils of Japan, 185–244.
Galali Y, Omar ZA, Sajadi SM (2020) Biologically active components in by-products of food processing. Food Sci Nutr 8:3004–3022
Gamba M, Asllanaj E, Raguindin PF, Glisic M, Franco OH, Minder B, Muka T (2021) Nutritional and phytochemical characterization of radish (Raphanus sativus): a systematic review. Trends Food Sci Technol 113:205–218
Ganesh KS, Sridhar A, Vishali S (2022) Utilization of fruit and vegetable waste to produce value-added products: conventional utilization and emerging opportunities-a review. Chemosphere 287:132221
Gatahi DM (2020) Challenges and opportunities in tomato production chain and sustainable standards. Int J Hortic Sci Technol 7(3):235–262
Ghildiyal R, Prakash V, Chaudhary VK, Gupta V, Gabrani R (2020) Phytochemicals as antiviral agents: recent updates. Plant-derived bioactives: production, properties and therapeutic applications, pp 279–295
Godlewska-Żyłkiewicz B, Świsłocka R, Kalinowska M, Golonko A, Świderski G, Arciszewska Ż et al (2020) Biologically active compounds of plants: structure-related antioxidant, microbiological and cytotoxic activity of selected carboxylic acids. Materials 13(19):4454
Görgüç A, Gençdağ E, Yılmaz FM (2020) Bioactive peptides derived from plant origin by-products: biological activities and techno-functional utilizations in food developments–a review. Food Res Int 136:109504

Hassanien MFR (2023) Bioactive phytochemicals from vegetable oil and oilseed processing by-products

Husak VV, Stambulska UY, Pitukh AM, Lushchak VI (2024) Exposure of Paulownia seedlings to silver nitrate improves growth parameters via stimulation of mild oxidative stress. Agric Conspec Sci 89(3):1–10

Hussain S, Jõudu I, Bhat R (2020) Dietary fiber from underutilized plant resources – a positive approach for valorization of fruit and vegetable wastes. Sustain For 12:5401

Iqbal A, Hamayun M, Shah F, Hussain A (2020) Role of plant bioactives in sustainable agriculture. Environment, climate, plant and vegetation growth, 591–605

Jahan I (2020) Post harvest diseases in some selected root and tuber crops and identification their causal agents. http://www.saulibrary.edu.bd/daatj/public/uploads/18-09097_11.pdf

Jan R, Asaf S, Numan ML, Kim KM (2021) Plant secondary metabolite biosynthesis and transcriptional regulation in response to biotic and abiotic stress conditions. Agronomy 11:968

Jiménez-Moreno N, Esparza I, Bimbela F, Gandía LM, Ancín-Azpilicueta C (2020) Valorization of selected fruit and vegetable wastes as bioactive compounds: opportunities and challenges. Crit Rev Environ Sci Technol 50(20):2061–2108

Joshi VK (2020) Fruit and vegetable processing waste management – an overview. Int J Food Ferment Technol 10(2):67–94

Kaushik B, Sharma J, Yadav K, Kumar P, Shourie A (2021) Phytochemical properties and pharmacological role of plants: secondary metabolites. Biosci Biotechnol Res Asia 18(1):23

Kováčik J, Husáková L, Graziani G, Patočka J, Vydra M, Rouphael Y (2022) Nickel uptake in hydroponics and elemental profile in relation to cultivation reveal variability in three Hypericum species. Plant Physiol Biochem 185:357–367

Kumar K, Srivastav S, Sharanagat VS (2021) Ultrasound assisted extraction (UAE) of bioactive compounds from fruit and vegetable processing by-products: a review. Ultrason Sonochem 70:105325

Lau KQ, Sabran MR, Shafie SR (2021) Utilization of vegetable and fruit by-products as functional ingredient and food. Front Nutr 8:661693

Lim WC, Nadzir MNHM, Hiew MWH, Mamat MS, Nazli MH, Shohaimi S (2021) Potential of open-air hydroponic system in producing highly nutritional composition maize fodder for goat farming

de los Ángeles Rosell M, Quizhpe J, Ayuso P, Peñalver R, Nieto G (2024) Proximate composition, health benefits, and food applications in bakery products of purple-fleshed sweet potato (*Ipomoea batatas* L.) and its by-products: a comprehensive review. Antioxidants 13(8):954

Maheshwari S, Kumar V, Bhadauria G, Mishra A (2022) Immunomodulatory potential of phytochemicals and other bioactive compounds of fruits: a review. Food Front 3(2):221–238

Majumder P, Annegowda HV (2021) Fruit and vegetable by-products: novel ingredients for a sustainable society. In: Valorization of agri-food wastes and by-products. Academic Press, pp 133–156

Manhongo TT, Chimphango AFA, Thornley P, Röder M (2022) Current status and opportunities for fruit processing waste biorefineries. Renew Sust Energ Rev 155:111823

Mani JS, Johnson JB, Steel JC, Broszczak DA, Neilsen PM, Walsh KB, Naiker M (2020) Natural product-derived phytochemicals as potential agents against coronaviruses: a review. Virus Res 284:197989

Marcillo-Parra V, Tupuna-Yerovi DS, González Z, Ruales J (2021) Encapsulation of bioactive compounds from fruit and vegetable by-products for food application–a review. Trends Food Sci Technol 116:11–23

Mateos-Aparicio I (2021) Plant-based by-products. Food Waste Recovery

Mateus ARS, Pena A, Sendón R, Almeida C, Nieto GA, Khwaldia K, Silva AS (2023) By-products of dates, cherries, plums and artichokes: a source of valuable bioactive compounds. Trends Food Sci Technol 131:220–243

Mohammadnezhad P, Valdés A, Álvarez-Rivera G (2023) Bioactivity of food by-products: an updated insight. Curr Opin Food Sci 52:101065

Mondal S, Soumya NPP, Mini S, Sivan SK (2021) Bioactive compounds in functional food and their role as therapeutics. Bioactive Compounds in Health and Disease-Online ISSN: 2574-0334; Print ISSN: 2769-2426, 4(3):24–39

Murthy HN, Paek KY (2020) Health benefits of underutilized vegetables and legumes. Bioactive compounds in underutilized vegetables and legumes, 1–37

Muthoni J, Shimelis H (2023) An overview of potato production in Africa. Potato production worldwide

Nabeshima EH, Moro TM, Campelo PH, Sant'Ana AS, Clerici MTP (2020) Tubers and roots as a source of prebiotic fibers. In: Advances in food and nutrition research, vol 94. Academic Press, pp 267–293

Najeebullah S, Shinwari ZK, Jan SA, Khan I, Ali M (2021) Ethno medicinal and phytochemical properties of genus Allium: a review of recent advances. Pak J Bot 53:135–144

Nguyen HC, Chen CC, Lin KH, Chao PY, Lin HH, Huang MY (2021) Bioactive compounds, antioxidants, and health benefits of sweet potato leaves. Molecules 26(7):1820

Oleszek M, Kowalska I, Bertuzzi T, Oleszek W (2023) Phytochemicals derived from agricultural residues and their valuable properties and applications. Molecules 28:342

Osuntokun OT, Yusuf-Babatunde MA, Fasila OO (2020) Components and bioactivity of Ipomoea batatas (L.; sweet potato) ethanolic leaf extract. Asian J Adv Res Rep 10(1):10–26

Otálora A, García-Quintero A, Mera-Erazo J, Lerma TA, Palencia M, Mercado T (2024) Sweet potato, batata or camote (*Ipomoea batatas*): an overview about its crop, economic aspects and nutritional relevance. J Sci Technol Appl 17:1–10

Otekunrin OA, Sawicka B, Adeyonu AG, Otekunrin OA, Rachoń L (2021) Cocoyam [Colocasia esculenta (L.) Schott]: exploring the production, health and trade potentials in sub-Saharan Africa. Sustain For 13(8):4483

Pacifico D, Lanzanova C, Pagnotta E, Bassolino L, Mastrangelo AM, Marone D, Balconi C (2021) Sustainable use of bioactive compounds from Solanum tuberosum and Brassicaceae wastes and by-products for crop protection – a review. Molecules 26(8):2174

Popoola JO, Ojuederie OB, Aworunse OS, Adelekan A, Oyelakin AS, Oyesola OL et al (2023) Nutritional, functional, and bioactive properties of African underutilized legumes. Front Plant Sci 14:1105364

Prasad R, Shivay YS (2021) Below ground food producing plants and their contribution to the world food kitty – a review. Indian J Agric Sci 91(11):1551–1556

Ralte L, Khiangte L, Thangjam NM, Kumar A, Singh YT (2022) GC–MS and molecular docking analyses of phytochemicals from the underutilized plant, Parkia timoriana revealed candidate anti-cancerous and anti-inflammatory agents. Sci Rep 12(1):3395

Ramadan Hassanien MF (2023) Introduction to bioactive phytochemicals from vegetable oil and oilseed processing by-products. In: Bioactive phytochemicals from vegetable oil and oilseed processing by-products. Springer, Cham, pp 3–10

Rao M, Bast A, De Boer A (2021) Valorized food processing by-products in the EU: Finding the balance between safety, nutrition, and sustainability. Sustain For 13:4428

Reguengo LM, Salgaço MK, Sivieri K, Júnior MRM (2022) Agro-industrial by-products: valuable sources of bioactive compounds. Food Res Int 152:110871

Riaz M, Khalid R, Afzal M, Anjum F, Fatima H, Zia S, Aslam MA (2023) Phytobioactive compounds as therapeutic agents for human diseases: a review. Food Sci Nutr 11(6):2500–2529

Rizvi A, Ahmed B, Khan MS, El-Beltagi HS, Umar S, Lee J (2022) Bioprospecting plant growth promoting rhizobacteria for enhancing the biological properties and phytochemical composition of medicinally important crops. Molecules 27(4):1407

Saleem S, Ul Mushtaq N, Shah WH, Rasool A, Hakeem KR, Ul Rehman R (2022) Beneficial role of phytochemicals in oxidative stress mitigation in plants. In: Antioxidant defense in plants: molecular basis of regulation. Singapore, Springer Nature Singapore, pp 435–451

Samtiya M, Aluko RE, Dhewa T, Moreno-Rojas JM (2021) Potential health benefits of plant food-derived bioactive components: an overview. Food Secur 10:839

Samynathan R, Shanmugam K, Nagarajan C, Murugasamy H, Ilango RVJ, Shanmugam A, Thiruvengadam M (2021) The effect of abiotic and biotic stresses on the production of bioactive compounds in tea (Camellia sinensis (L.) O. Kuntze). Plant Gene 27:100316

Santos D, da Silva JAL, Pintado M (2022) Fruit and vegetable by-products' flours as ingredients: a review on production process, health benefits and technological functionalities. LWT 154:112707

Saroj A, Oriyomi OV, Nayak AK, Haider SZ (2020) Phytochemicals of plant-derived essential oils: a novel green approach against pests. In: Natural remedies for pest, disease and weed control. Academic Press, pp 65–79

Saxena R (2023) Exploring approaches for investigating phytochemistry: methods and techniques. Medalion J Med Res Nurs Health Midwife Participat 4(2):65–73

Scott GJ (2021) A review of root, tuber and banana crops in developing countries: past, present and future. Int J Food Sci Technol 56:1093–1114

Shah A, Smith DL (2020) Flavonoids in agriculture: chemistry and roles in, biotic and abiotic stress responses, and microbial associations. Agronomy 10:1209

Shahid S, Shahid W, Ihsan A, Anjum F, Shahid M (2022) Phytochemical and antioxidant profiling of allium sativum germinated under biotic and abiotic stress. Pak J Med Health Sci 16(11):211–211

Shahrajabian MH, Sun W, Cheng Q (2022) The importance of flavonoids and phytochemicals of medicinal plants with antiviral activities. Mini-Rev Organ Chem 19(3):293–318

Sharma M, Kaushik P (2021) Vegetable phytochemicals: an update on extraction and analysis techniques. Biocatal Agric Biotechnol 36:102149

Shrinet K, Singh RK, Chaurasia AK, Tripathi A, Kumar A (2021) Bioactive compounds and their future therapeutic applications. In: Natural bioactive compounds. Academic Press, pp 337–362

Sorrenti V, Burò I, Consoli V, Vanella L (2023) Recent advances in health benefits of bioactive compounds from food wastes and by-products: biochemical aspects. Int J Mol Sci 24(3):2019

Stéphane FFY, Jules BKJ, Batiha GES, Ali I, Bruno LN (2021) Extraction of bioactive compounds from medicinal plants and herbs. Natural medicinal plants, 1–39

Teklić T, Parađiković N, Špoljarević M, Zeljković S, Lončarić Z, Lisjak M (2021) Linking abiotic stress, plant metabolites, biostimulants and functional food. Ann Appl Biol 178(2):169–191

Teruel-Andreu C, Andreu-Coll L, López-Lluch D, Sendra E, Hernández F, Cano-Lamadrid M (2021) Ficus carica fruits, by-products and based products as potential sources of bioactive compounds: a review. Agronomy 11(9):1834

Thakur M, Singh K, Khedkar R (2020) Phytochemicals: extraction process, safety assessment, toxicological evaluations, and regulatory issues. In: Functional and preservative properties of phytochemicals. Academic Press, pp 341–361

Tiwari A, Khawas R (2021) Food waste and agro by-products: a step towards food sustainability. In: Innovation in the food sector through the valorization of food and agro-food by-products. IntechOpen

Tlais AZA, Fiorino GM, Polo A, Filannino P, Di Cagno R (2020) High-value compounds in fruit, vegetable and cereal byproducts: an overview of potential sustainable reuse and exploitation. Molecules 25:2987

Trigo JP, Alexandre EM, Saraiva JA, Pintado ME (2022) High value-added compounds from fruit and vegetable by-products – characterization, bioactivities, and application in the development of novel food products. Crit Rev Food Sci Nutr 60(8):1388–1416

Vaghela N, Gohel S (2023) Medicinal plant-associated rhizobacteria enhance the production of pharmaceutically important bioactive compounds under abiotic stress conditions. J Basic Microb 63:308–325

Vazquez-Olivo G, Cota-Pérez JL, García-Carrasco M, Zamudio-Sosa VE, Heredia JB (2023) Antioxidant phenolics from vegetable by-products. Plant phenolics in abiotic stress management, 89–104

Wink M (2022) Current understanding of modes of action of multicomponent bioactive phytochemicals: potential for nutraceuticals and antimicrobials. Annu Rev Food Sci Technol 13:337–359

Yeshi K, Crayn D, Ritmejerytė E, Wangchuk P (2022) Plant secondary metabolites produced in response to abiotic stresses has potential application in pharmaceutical product development. Molecules 27:313

Yu M, Gouvinhas I, Rocha J, Barros AI (2021) Phytochemical and antioxidant analysis of medicinal and food plants towards bioactive food and pharmaceutical resources. Scientific Reports 11:10041

Zhu, Y., Luan, Y., Zhao, Y., Liu, J., Duan, Z., & Ruan, R. (2023). Current technologies and uses for fruit and vegetable wastes in a sustainable system: a review. Foods 12(10), 1949

Part II
Phytochemicals from Leaf Vegetables By-products

Bioactive Phytochemicals from Red Cabbage (*Brassica oleracea*) By-products

Mariana Buranelo Egea, Sibele Santos Fernandes, Anna Rafaela Cavalcante Braga, Ailton Cesar Lemes, and Josemar Gonçalves de Oliveira Filho

Contents

M. B. Egea
Goiano Federal Institute of Education, Science and Technology, Rio Verde, Goiás, Brazil
e-mail: mariana.egea@ifgoiano.edu.br

S. S. Fernandes
Federal University of Rio Grande, School of Chemistry and Food, Rio Grande, Brazil

A. R. C. Braga
Department of Chemical Engineering, Federal University of São Paulo (UNIFESP), Diadema, São Paulo, Brazil
e-mail: anna.braga@unifesp.br

A. C. Lemes
Department of Biochemical Engineering, Federal University of Rio de Janeiro (UFRJ), School of Chemistry, Rio de Janeiro, Brazil
e-mail: ailtonlemes@eq.ufrj.br

J. G. de Oliveira Filho (✉)
Brazilian Agricultural Research Corporation, Embrapa Instrumentation, São Carlos, São Paulo, Brazil

M. F. Ramadan (ed.), *Bioactive Phytochemicals in By-products from Leaf, Stem, Root and Tuber Vegetables*, https://doi.org/10.1007/978-3-031-80700-8_2

Abstract Red cabbage is extensively cultivated worldwide and is commonly used to prepare salads, generating a considerable volume of by-products, such as damaged leaves. These leaves, in turn, constitute a rich source of bioactive compounds, including anthocyanins, other phenolic compounds, and dietary fibers. These bioactive compounds have been recovered through various methods involving both organic solvents. Bioactive compounds derived from red cabbage by-products have been explored as natural food colorants to enhance these products' nutritional and functional value. Furthermore, anthocyanins extracted from red cabbage by-products have been employed in developing active and intelligent biodegradable food packaging, owing to their antioxidant and antimicrobial properties and ability to change color in response to pH variations. Therefore, red cabbage by-products are a valuable source of natural food ingredients, presenting significant potential for various applications in food technology and fostering the development of innovative food products.

Keywords Functional foods · Anthocyanins · Dietary fibers · Active films

1 Introduction and Economic Values of Red Cabbage By-products

Global agriculture ensures global food security, providing most of the food consumed by around 8 billion people (Pawlak and Kołodziejczak 2020). Furthermore, it has a relevant economic role, as it generated around 13.4 trillion dollars in 2023 alone, with expectations of continued growth to meet the population's nutritional demand (Globe-NewsWire 2023). Added to this, agriculture is responsible for generating employment and income, for the development of rural areas, boosting national and international trade and also through sustainable agricultural practices and the use of technological innovation, contributing to environmental preservation and climate change mitigation (Çakmakçı et al. 2023; Wijerathna-Yapa and Pathirana 2022).

Agriculture produces a wide diversity of foods that meet the population's nutritional needs through a unique combination of essential nutrients, such as carbohydrates, proteins, fats, vitamins, minerals, and various bioactive compounds, allowing a balanced and healthy diet. These foods include a variety of fruits, cereals, legumes, oilseeds, meat, dairy products, and derived products (Egea et al. 2023; Lemes et al. 2022; Lira et al. 2023).

Among the vegetables produced, red cabbage (*Brassica oleracea*) stands out, a species with diverse applications, which include its use in (i) cooking due to its sensory properties such as purple color, flavor, and characteristic odor; (ii) application as a natural pigment in various food and pharmaceutical products (Fernandes et al. 2019); (iii) application in smart food packaging and sensors (Abedi-Firoozjah et al. 2022); and (iv) indicator in pharmaceutical formulations (Chigurupati et al. 2002), among others.

Red cabbage stands out for its composition, including a low-calorie content (24 kcal), a good source of carbohydrates (0.7–5.3%), fiber (1.0–3.6%), proteins (0 0.8–2.0%), and minerals (0.3–0.7%) (Stoica et al. 2023; Zayed et al. 2023). In addition, it has several bioactive compounds, including flavonoids, anthocyanins, and phenolic acids, among others, which are constantly associated with various positive effects or improvements in the health characteristics of individuals, such as the ability to cause hypolipidemia, dyslipidemia, reducing the risk of heart attack myocardium, antidiabetic, and antiulcer potential, in addition to presenting potential antioxidant, antimicrobial (antibacterial and antifungal), anti-inflammatory, analgesic activity, among others (Fernandes et al. 2019; Zayed et al. 2023).

In 2023, the red cabbage market was estimated at US$100 billion and was projected to reach US$200 billion by 2030, registering a compound annual growth (CAGR) of 10.41% from 2024 to 2030 (VMR 2023). Among the largest cabbage producers, China, India, Pakistan, South Korea, Japan, the USA, and several European countries stand out, which have a large production area and, in addition, use this vegetable in typical culinary preparations such as salads, stews, preserves, fermented foods, and juices, for example (Mordor-Intelligence 2024; Zayed et al. 2023).

The main by-product of harvesting and processing red cabbage is made up of the outer leaves, usually cut by farmers in the field, and also the base of the stalk, which is often discarded due to the preference for softer and juicier inner parts (BHG 2023; Featherstone 2016; Tanongkankit et al. 2012). Depending on the harvesting process, the transformation technology adopted, the efficiency of the process, market demand, and other factors, the proportion of by-products generated during red cabbage processing can reach up to 63% (Lemes et al. 2023; Stoica et al. 2023).

Therefore, due to its rich nutritional composition and the large amount of by-products, it is essential to adopt methods of use to reduce the accumulation and potential negative impacts caused by this material. To date, the literature has reported the application of cabbage by-products in the extraction of natural pigments and functional components (Stoica et al. 2023), such as fertilizers (Stoica et al. 2023), pH indicators in food and packaging (Abedi-Firoozjah et al. 2022), and bioethanol production (Thenmozhi and Victoria 2013), among others.

Thus, this chapter aims to present the (i) economic values of red cabbage by-products, (ii) chemical composition and bioactive compounds of red cabbage by-products, (iii) biological and functional properties of extracts and bioactive compounds from red cabbage by-products, (iv) food and non-food applications of extracts and bioactive compounds from red cabbage by-products; and (v) the valorization of red cabbage by-products for non-health purposes.

Table 1 Proximal composition of red cabbage and its by-products

g/100 g	Zhang et al. (2023)	David et al. (2024)	Mejías et al. (2024)
Protein	11.85	13.8	19.19
Lipids	0.91	2.83	0.21
Ash	7.54	16.1	9.99
Crude fiber	ND	9.45	13.31
Total dietary fiber	31.27	ND	ND

*Fresh red cabbage. *ND* not determined

2 Chemical Composition and Bioactive Compounds of Red Cabbage By-products

Studies that evaluate the chemical composition of red cabbage by-products use waste from industrial processing, consisting of damaged leaves and stubs. Table 1 presents the composition of the red cabbage by-products and fresh red cabbage.

Protein content tends to be higher in the fresh sample, while lipid concentration increases in by-products. This phenomenon may be related to the rupture of the cell wall during the processing of the raw material, which can lead to the release of intracellular lipids (Mejías et al. 2024). Furthermore, the differences in parameters presented by different studies may be related to the geographical location of the raw materials and associated with the climate, season, and maturation stage.

The dietary fiber content is present at around 31%, of which around 32% are soluble fibers and 68% are insoluble fibers (Zhang et al. 2023). David et al. (2024) found a neutral detergent fiber content of 15.4 g/100 g, where protein, lipids, and soluble sugars are removed from the sample. This fiber is mainly composed of cellulose, hemicellulose, and lignin.

Furthermore, the by-products are rich in minerals, such as potassium (K, 8.56 mg/100 g), magnesium (Mg, 4.48 mg/100 g), phosphorus (P, 2.87 mg/100 g), sodium (Na, 1.30 mg/100 g), and calcium (Ca, 0.03 mg/100 g) (David et al. 2024). The mineral concentration of red cabbage by-products differs significantly from that in red cabbage leaves, particularly in the higher calcium content (125 mg/100 g) (Leahu et al. 2018).

Regarding the composition of amino acids, the predominant essential ones are arginine (74.1 g/kg), leucine (55.4 g/kg), and valine (42.7 g/kg). In contrast, the predominant non-essential amino acids are glutamic acid (97.5 g/kg), aspartic acid (89.9 g/kg), and glycine (38 g/kg) (David et al. 2024).

The by-products contain a high amount of reducing oligosaccharides, approximately 38%, which serve as precursors to the Maillard reaction and participate in lipid oxidation, thereby contributing to the undesired formation of amines and carbonyl compounds. Among these, the prominent reducing monosaccharides are glucose, accounting for 18.9%, and fructose, accounting for 15.5% (Zhang et al. 2023).

In red cabbage by-products, extractable phenolic compounds are present at approximately 1.83 g GAE (gallic acid equivalent) per 100 g, while non-extractable phenolic compounds are present at around 0.47 g GAE per 100 g. The extractable phenolic fractions from red cabbage by-products include anthocyanins and hydroxycinnamic acids (Zhang et al. 2023).

Anthocyanins from red cabbage are gaining popularity in the food industry as natural colorants and for their health-promoting effects due to their natural antioxidant properties. Red cabbage is one of the edible sources of natural pigments with a high amount of anthocyanins ($\geq$ 10 g/kg dry matter). More than 30 anthocyanin pigments have been identified in red cabbage extract, mainly containing derivatives of cyanidin-3-diglucoside-5-glucoside in non-acylated or acylated forms (Ghareaghajlou et al. 2021).

Most studies evaluate the extraction of anthocyanins from fresh red cabbage (Müller-Maatsch et al. 2019; Sendri et al. 2023). Furthermore, some studies have utilized extracted anthocyanins to develop smart active food packaging (Anugrah et al. 2023; Sadi and Ferfera-Harrar 2023).

However, the utilization of by-products as a source of anthocyanins remains limited, emphasizing a future area of research that focuses on the principles of biorefinery, circular economy, and value addition. Zhang et al. (2023) found that 70% of the total extractable polyphenols in red cabbage by-products were anthocyanins, with the cyanidin 3-*O*-diglucoside-5-*O*-glucoside anthocyanin being the most representative among the 15 identified, accounting for more than 17.6% of the total anthocyanins.

In red cabbage, the extraction of anthocyanins is generally carried out using organic solvents, such as methanol and ethanol (Zanoni et al. 2020). Stoica et al. (2023) developed a sustainable and cost-effective process for recovering anthocyanins from red cabbage by-products using an aqueous ethanolic extractant that includes apple cider vinegar. The results verified that the contaminants (Ni, Cr, Cd, Pb, and Co) presented values below the limits of quantification. Therefore, the extract obtained can enhance the color of food products and beverages, making it suitable for various applications in formulations in the food industry or printing.

Some of the anti-nutritional factors can be found in red cabbage by-products, including oxalate (1.87 g/kg), saponin (8.20 g/kg), and phytate (0.53 g/kg) (David et al. 2024). Joel et al. (2020) extracted the enzyme peroxidase from red cabbage (*Brassica oleracea*) by-products. The authors reported a purification of 87.65% with a yield of 34.92%. Furthermore, they verified that the peroxidase obtained efficiently decomposes and biodegrades phenol and some common synthetic azo dyes.

3 Biological and Functional Properties of Extracts and Bioactive Compounds from Red Cabbage By-products

It is well known that a significant amount of fruit and vegetable waste is generated in the food chain, and there is potential for utilizing these by-products, which are rich in bioactive compounds, to develop sustainable food products (Bas-Bellver et al. 2022).

The Brassicaceae family contains several bioactive compounds; among them, red cabbage boasts vibrant red and purple leaves, cultivated in spring and reaped in autumn (Özbakır Özer et al. 2021), and is commonly enjoyed in dishes like coleslaw, salads, and drinks (Ávila et al. 2023) due to its exceptional nutritional profile, packed with essential minerals, vitamins, oligosaccharides, anthocyanins, and flavonols (Drozdowska et al. 2020).

Among its bioactive components, red cabbage shines with anthocyanins, carotenoids, vitamins A, B, C, and K, and beneficial phenolics that promote various health benefits (Murador et al. 2016). Besides the mentioned bioactive compounds, glucosinolates (GSLs) are secondary metabolites found in these plants with bioactive properties, particularly in plant defense mechanisms and potential human health benefits. The same bioactives are also found in the by-products generated from the production chain of this vegetable, having incredible potential as ingredients for bioproducts with high added value (Bas-Bellver et al. 2022).

The anthocyanins found in red cabbage and its by-products are water-soluble pigments that exhibit colors ranging from red to blue to purple, depending on pH levels. Notably, cyanidin-3-diglycoside-5-glycoside stands out as the primary anthocyanin compound in red cabbage (Ghareaghajlou et al. 2022). Furthermore, red cabbage by-products are a good soluble and insoluble fiber source, contributing to its dietary value (Shinali et al. 2024).

Numerous research studies have explored the biological activities of red cabbage extract, revealing its potential as an antibacterial, anticancer, antidiabetic, anti-inflammatory, and analgesic agent. Additionally, these vegetables and their by-products have significantly modulated the innate immune response system, underlining their robust health-promoting properties (Zayed et al. 2023).

The literature encompasses various extraction techniques to recover bioactive compounds from cabbage and its by-products, focusing predominantly on its phytoconstituents to facilitate the production of value-added products. Given the sensitivity of GLSs to degradation and the pivotal role of the extraction method in natural product isolation, efforts have been directed toward optimizing extraction methods for active constituents in red cabbages. These techniques are typically categorized into classical solvent extraction methods and non-conventional or modern approaches. Innovative extraction methods such as pretreatment, enzyme-assisted extraction, ultrasound-assisted extraction, and microwave-assisted extraction have shown promise in enhancing the liberation of phytoconstituents and maximizing yield (Salem et al. 2022).

The extraction of phenolic compounds and flavonoids from red cabbages and their by-products has been commonly performed using 60–80% methanol and 70–80% ethanol, as reported in the literature. Recent research has delved into the total phenolic content of red cabbage using various solvent types, revealing significant differences ($p < 0.05$). The study found that the crude chloroform extract exhibited the highest phenolic content, followed by dichloromethane, ethyl ether, toluene, ethanol, and methanol extracts. Similarly, the crude chloroform extract showed the highest level of total flavonoid content, with chloroform yielding the most phenolics from cabbages. This surprising finding contrasts the common belief that polar solvents yield higher amounts, possibly due to the relatively lipophilic nature of the phenolic components in cabbages, which are predominantly acylated (Rubab et al. 2020).

Efforts to extract GLSs have commonly involved hot or boiling 70% methanol at 75 °C for 10–20 min. On the other hand, anthocyanins extraction has typically utilized a mixture of MeOH/H_2O/0.01% HCl (50:50:1, v/v/v), resulting in a yield of 12%. To prevent enzymatic browning during anthocyanins' extraction from red cabbage, adding 0.1% ascorbic acid is recommended to inhibit polyphenol oxidase activity and stabilize the anthocyanins at acidic pH levels. For optimal anthocyanins extraction, a hydromethanolic mixture containing methanol, water, formic acid, and trifluoroacetic acid (70:27:2:1, v/v/v/v) has proven advantageous. Acetone is commonly used for carotenoid extraction (Murador et al. 2016).

Solvent extraction methods, while familiar, have limitations due to the need for large solvent volumes, extended extraction times, and high energy consumption, leading to increased costs (Murador et al. 2019). Therefore, green and non-conventional methods are preferred for bioactive recovery to avoid solvent residues in the final products as using ionic liquids, pulsed electric field processing, microwave-assisted extraction, supercritical CO_2 extraction, and ultrasound-assisted extraction can be good alternatives for an eco-friendlier option of bioactive compounds from red cabbage by-products. Besides that, it is essential to develop new production systems that minimize waste and reintroduce residues into the food chain, contributing to more sustainable food systems. Focusing on the circular economy system, reducing food waste, and creating functional food products aligns with nutrition, health, sustainability, circularity, and innovation priorities.

4 Food and Non-food Applications of Extracts and Bioactive Compounds from Red Cabbage By-products

Red cabbage extract and bioactive compounds have been investigated for food and non-food applications. Food colorants derived from red cabbage have been used in soft drinks, candies, and other food products (Patras 2019). Esmailian and Hosseini (2021) applied red cabbage anthocyanins as a natural colorant for developing low-calorie gummy candies.

The anthocyanins in red cabbage exhibit remarkable sensitivity to pH variations, displaying a wide range of colors depending on pH (Zhang and Jing 2022). Due to this ability, they can be used as natural pH indicators (Mousazadeh et al. 2021), applicable in contexts where changes in the pH of food products are related to fermentation or deterioration rates (Rawdkuen et al. 2020). For instance, the anthocyanin-rich extract from red cabbage has been extensively studied for the development of pH-sensitive smart films capable of monitoring the quality of foods such as mushrooms (Zhan et al. 2024), shrimp (Ren et al. 2023), and fish (Silva-Pereira et al. 2015). Active packaging with antioxidant and antimicrobial properties has also been developed using anthocyanin-rich extracts from red cabbage to enhance the preservation of products like meat (Sanches et al. 2021).

The anthocyanin-rich extract of red cabbage has also been used to develop a pH-sensitive smart therapeutic dressing from chitosan hydrogel (Arafa et al. 2021). An environmentally friendly biochromic diagnostic biosensor for urea recognition has also been developed by immobilizing red cabbage extract on cellulose fibers (Al-Qahtani et al. 2021). Li et al. (2013) investigated the optimal photoelectrochemical conditions for red cabbage extract as a natural dye to develop dye-sensitized solar cells.

While red cabbage residue contains a rich source of bioactive compounds with vast potential for various applications, studies have focused on using fresh raw material, overlooking the utilization of the by-product. Consequently, the number of works investigating the application of bioactive compounds derived from red cabbage by-products is limited.

Among the available studies, Oliveira Filho et al. (2024) utilized the solution blow spinning (SBS) technique to swiftly produce smart nanofiber mats derived from polycaprolactone (PCL). These mats were infused with extracts from agricultural residues abundant in anthocyanins sourced from red cabbage (CE) to monitor food quality. The smart nanofibers exhibited the capability to detect spoilage in fish fillets through noticeable color alterations ($\Delta E \geq 3$) during storage.

Zhang et al. (2023) investigated the effect of the water-insoluble fraction of red cabbage by-product (WI-RCBP) on the formation of acrylamide and 5-hydroxymethylfurfural in potato chips. By replacing 20% of potato flakes with WI-RCBP, there was a significant reduction in the formation of these compounds. This was attributed to the inhibition of the Maillard reaction by plant polyphenols and the decrease in asparagine and reducing sugars resulting from the partial substitution of potato flakes. These findings suggest that removing water-soluble compounds may be an effective strategy for using vegetable by-products as ingredients in fried products, thereby limiting the formation of contaminants during processing.

Another study by Patras (2019) investigated the effects of 19 food additives/ingredients on the stability and color of hydroethanolic extract from red cabbage residue during 20 weeks of storage. Tartaric and citric acids caused the most significant changes, decreasing anthocyanins, brightness, and antioxidant activity, along with increased parameters related to dark color. Ascorbic and benzoic acids had similar but less pronounced influences. Untreated extracts showed high stability during storage, while treated extracts, especially with sugars, were also stable but with some differences.

5 Final Considerations

Red cabbage (Brassica oleracea) by-products are usually underused and undervalued when, in fact, they constitute an attractive raw material for the extraction of natural pigments such as anthocyanins, with the cyanidin 3-*O*-diglucoside-5-*O*--glucoside anthocyanin being the most representative.

These natural pigments have several applications, including correcting or giving color to various materials, and enabling color change when intended to function as a product quality indicator. In this way, it seems to be a circular economy strategy to take advantage of these by-products in extracting pigments for application in food.

References

Abedi-Firoozjah R, Yousefi S, Heydari M, Seyedfatehi F, Jafarzadeh S, Mohammadi R, Rouhi M, Garavand F (2022) Application of red cabbage anthocyanins as pH-sensitive pigments in smart food packaging and sensors. Polymers 14(8):1629

Al-Qahtani SD, Alzahrani HK, Azher OA, Owidah ZO, Abualnaja M, Habeebullah TM, El-Metwaly NM (2021) Immobilization of anthocyanin-based red-cabbage extract onto cellulose fibers toward environmentally friendly biochromic diagnostic biosensor for recognition of urea. J Environ Chem Eng 9(4):105493

Anugrah DSB, Darmalim LV, Sinanu JD, Pramitasari R, Subali D, Prasetyanto EA, Cao XT (2023) Development of alginate-based film incorporated with anthocyanins of red cabbage and zinc oxide nanoparticles as freshness indicator for prawns. Int J Biol Macromol 251:126203

Arafa AA, Nada AA, Ibrahim AY, Sajkiewicz P, Zahran MK, Hakeim OA (2021) Preparation and characterization of smart therapeutic pH-sensitive wound dressing from red cabbage extract and chitosan hydrogel. Int J Biol Macromol 182:1820–1831

Ávila S, Zalamanski S, Tanikawa LM, Kruger CCH, Ferreira SMR (2023) Influence of cooking methods on in vitro bioaccessibility of phenolics, flavonoids, and antioxidant activity of red cabbage. Plant Foods Hum Nutr 78(1):124–131

Bas-Bellver C, Barrera C, Betoret N, Seguí L (2022) Impact of disruption and drying conditions on physicochemical, functional and antioxidant properties of powdered ingredients obtained from brassica vegetable by-products. Food Secur 11(22):3663

BHG (2023) How and when to harvest cabbage for the best flavor. In: Vegetable gardening, vol 2024. Better Homes & Gardens. www.bhg.com

Çakmakçı R, Salık MA, Çakmakçı S (2023) Assessment and principles of environmentally sustainable food and agriculture systems. Agriculture 13(5):1073

Chigurupati N, Saiki L, Gayser Jr C, Dash AK (2002) Evaluation of red cabbage dye as a potential natural color for pharmaceutical use. Int J Pharm 241(2):293–299

David OO, John MO, Mary EL, Usman KDJ, Olayinka AV, John AO, Sikiru AB, Adewale OS, Emmanuel O, Olusoji AS (2024) Nutritive evaluation of dried cabbage (*Brassica oleracea* var. capitata) waste meal as feed for grower-finisher pigs. Res Vet Sci 168:105151

Drozdowska M, Leszczyńska T, Koronowicz A, Piasna-Słupecka E, Domagała D, Kusznierewicz B (2020) Young shoots of red cabbage are a better source of selected nutrients and glucosinolates in comparison to the vegetable at full maturity. Eur Food Res Technol 246(12):2505–2515

Egea MB, De Sousa TL, Dos Santos DC, De Oliveira Filho JG, Guimarães RM, Yoshiara LY, Lemes AC (2023) Application of soy, corn, and bean by-products in the gluten-free baking process: a review. Food Bioprocess Technol 16:1429–1450

Esmailian A, Hosseini F (2021) Optimization of anthocyanin extraction from red cabbage and its application in low calorie Gummy candy. J Food Sci Technol (Iran) 18(110):129–140

Featherstone S (2016) 1 – Canning of vegetables. In: Featherstone S (ed) A complete course in canning and related processes14th edn. Woodhead Publishing, pp 3–84

Fernandes DL, Hedge K, Shabaraya AR (2019) Scientific approaches on red cabbage: a review. Int J Pharma Chem Res 5(1):123–130

Ghareaghajlou N, Hallaj-Nezhadi S, Ghasempour Z (2021) Red cabbage anthocyanins: stability, extraction, biological activities and applications in food systems. Food Chem 365:130482

Ghareaghajlou N, Hallaj-Nezhadi S, Ghasempour Z (2022) Nano-liposomal system based on lyophilization of monophase solution technique for encapsulating anthocyanin-rich extract from red cabbage. Dyes Pigments 202:110263

Globe-NewsWire (2023) Agriculture global market report 2023, Vol 2023, Globe Newswire. www.globenewswire.com

Joel EB, Mafulul SG, Adamu HE, Goje LJ, Tijjani H, Igunnu A, Malomo SO (2020) Peroxidase from waste cabbage (Brassica oleracea capitata L.) exhibits the potential to biodegrade phenol and synthetic dyes from wastewater. Scientific African 10:e00608

Leahu A, Ghinea C, Oroian M-A, Damian C (2018) Determination of essential and toxic elements, ascorbic acid content and color of different leaves in two cabbage varieties. Ovidius Univ Ann Chem 29(2):110–116

Lemes AC, Egea MB, Oliveira Filho JGd, Gautério GV, Ribeiro BD, Coelho MAZ (2022) Biological approaches for extraction of bioactive compounds from agro-industrial by-products: a review. Front Bioeng Biotechnol 9:1–18

Lemes AC, Coelho MAZ, Gautério GV, Paula LC, Filho JGdO, Egea MB (2023) Chapter 11 – Industrial wastes and by-products: a source of functional foods, nutraceuticals, and biopolymers. In: Biopolymers in nutraceuticals and functional foods. The Royal Society of Chemistry, pp 329–360

Li Y, Ku S-H, Chen S-M, Ali MA, Al Hemaid FM (2013) Photoelectrochemistry for red cabbage extract as natural dye to develop a dye-sensitized solar cells. Int J Electrochem Sci 8(1):1237–1245

Lira MM, Oliveira Filho JGd, Sousa TLd, Costa NMd, Lemes AC, Fernandes SS, Egea MB (2023) Selected plants producing mucilage: overview, composition, and their potential as functional ingredients in the development of plant-based foods. Food Res Int 169:112822

Mejías N, Vega-Galvez A, Gomez-Perez LS, Pasten A, Uribe E, Cortés A, Valenzuela-Barra G, Camus J, Delporte C, Bernal G (2024) Health-promoting properties of processed red cabbage (Brassica oleracea var. capitata f. rubra): effects of drying methods on bio-compound retention. Food Secur 13(6):830

Mordor-Intelligence (2024) Cabbages and other brassicas market size & share analysis – growth trends & forecasts (2024–2029). In: Cabbages and other Brassicas Market, Vol 2024, Mordor Intelligence™ Industry Reports. https://www.mordorintelligence.com

Mousazadeh S, Ehsani A, Kia EM, Ghasempour Z (2021) Zinc oxide nanoparticles and periodate oxidation in developing pH-sensitive packaging film based on modified gelatin. Food Packag Shelf Life 28:100654

Müller-Maatsch J, Gurtner K, Carle R, Steingass CB (2019) Investigation into the removal of glucosinolates and volatiles from anthocyanin-rich extracts of red cabbage. Food Chem 278:406–414

Murador DC, Mercadante AZ, de Rosso VV (2016) Cooking techniques improve the levels of bioactive compounds and antioxidant activity in kale and red cabbage. Food Chem 196:1101–1107

Murador DC, de Souza Mesquita LM, Vannuchi N, Braga ARC, de Rosso VV (2019) Bioavailability and biological effects of bioactive compounds extracted with natural deep eutectic solvents and ionic liquids: advantages over conventional organic solvents. Curr Opin Food Sci 26:25–34

Oliveira Filho JG, de Oliveira Silva C, do Canto RA, Egea MB, Tonon RV, Paschoalin RT, and Mattoso LHC (2024) Fast and sustainable production of smart nanofibers mats by solution blow spinning for food quality monitoring: potential of polycaprolactone and agri-food residue-derived anthocyanins. Food Chemistry 140057

Özbakır Özer M, Kar H, Murat Doğru S, Kobal Bekar N (2021) Morphological characterization of some hybrid red head cabbage (Brassica oleracea L. var. capitata subvar. rubra) varieties. Tekirdağ Ziraat Fakültesi Dergisi 18(3):428–435

Patras A (2019) Stability and colour evaluation of red cabbage waste hydroethanolic extract in presence of different food additives or ingredients. Food Chem 275:539–548

Pawlak K, Kołodziejczak M (2020) The role of agriculture in ensuring food security in developing countries: considerations in the context of the problem of sustainable food production. Sustain For 12(13):5488

Rawdkuen S, Faseha A, Benjakul S, Kaewprachu P (2020) Application of anthocyanin as a color indicator in gelatin films. Food Biosci 36:100603

Ren G, He Y, Lv J, Zhu Y, Xue Z, Zhan Y, Sun Y, Luo X, Li T, Song Y (2023) Highly biologically active and pH-sensitive collagen hydrolysate-chitosan film loaded with red cabbage extracts realizing dynamic visualization and preservation of shrimp freshness. Int J Biol Macromol 233:123414

Rubab M, Chelliah R, Saravanakumar K, Barathikannan K, Wei S, Kim JR, Yoo D, Wang MH, Oh DH (2020) Bioactive potential of 2-methoxy-4-vinylphenol and Benzofuran from Brassica oleracea L. var. capitate f, rubra (red cabbage) on oxidative and microbiological stability of beef meat. Food Secur 9(5):568

Sadi A, Ferfera-Harrar H (2023) Cross-linked CMC/Gelatin bio-nanocomposite films with organoclay, red cabbage anthocyanins and pistacia leaves extract as active intelligent food packaging: colorimetric pH indication, antimicrobial/antioxidant properties, and shrimp spoilage tests. Int J Biol Macromol 242:124964

Salem MA, Zayed A, Beshay ME, Abdel Mesih MM, Ben Khayal RF, George FA, Ezzat SM (2022) Hibiscus sabdariffa L.: phytoconstituents, nutritive, and pharmacological applications. Adv Tradit Med 22(3):497–507

Sanches MAR, Camelo-Silva C, da Silva Carvalho C, de Mello JR, Barroso NG, da Silva Barros EL, Silva PP, Pertuzatti PB (2021) Active packaging with starch, red cabbage extract and sweet whey: characterization and application in meat. LWT 135:110275

Sendri N, Singh S, Bhatt S, Gupta M, Bhandari P (2023) Insight into the influence of oxygen, sunlight and temperature on the stability and color attributes of red cabbage anthocyanins and in vitro gastrointestinal behaviour. Food Chem Adv 3:100359

Shinali TS, Zhang Y, Altaf M, Nsabiyeze A, Han Z, Shi S, and Shang N (2024) The valorization of wastes and byproducts from cruciferous vegetables: a review on the potential utilization of cabbage, cauliflower, and broccoli byproducts. Foods 13(8):1163.

Silva-Pereira MC, Teixeira JA, Pereira-Júnior VA, Stefani R (2015) Chitosan/corn starch blend films with extract from *Brassica oleraceae* (red cabbage) as a visual indicator of fish deterioration. LWT–Food Sci Technol 61(1):258–262

Stoica R, Ganciarov M, Constantinescu-Aruxandei D, Capră L, Şuică-Bunghez I-R, Senin R-M, Pricope GD, Ivan G-R, Călin C, Oancea F (2023) Sustainable recovery of anthocyanins and other polyphenols from red cabbage byproducts. Food Secur 12(22):4157

Tanongkankit Y, Chiewchan N, Devahastin S (2012) Physicochemical property changes of cabbage outer leaves upon preparation into functional dietary fiber powder. Food Bioprod Process 90(3):541–548

Thenmozhi R, Victoria JI (2013) Optimization and improvement of ethanol production by the incorporation of organic wastes. Adv Appl Sci Res 4:119–123

VMR (2023) Global Cabbage Red Market by type (type I, type II), by application (application I, application II), by geographic scope and forecast. In: Cabbage Red Market, Vol 2023, Verified Market Reports. www.verifiedmarketreports.com
Wijerathna-Yapa A, Pathirana R (2022) Sustainable agro-food systems for addressing climate change and food security. Agriculture 12(10):1554
Zanoni F, Primiterra M, Angeli N, Zoccatelli G (2020) Microencapsulation by spray-drying of polyphenols extracted from red chicory and red cabbage: effects on stability and color properties. Food Chem 307:125535
Zayed A, Sheashea M, Kassem IAA, Farag MA (2023) Red and white cabbages: an updated comparative review of bioactives, extraction methods, processing practices, and health benefits. Crit Rev Food Sci Nutr 63(24):7025–7042
Zhan S, Yi F, Hou F, Song L, Chen X, Jiang H, Han X, Sun X, Liu Z (2024) Development of pH-freshness smart label based on gellan gum film incorporated with red cabbage anthocyanins extract and its application in postharvest mushroom. Colloids Surf B: Biointerfaces 236:113830
Zhang N, Jing P (2022) Anthocyanins in Brassicaceae: composition, stability, bioavailability, and potential health benefits. Crit Rev Food Sci Nutr 62(8):2205–2220
Zhang H, Troise AD, Sun S, Fogliano V (2023) The water insoluble fraction from red cabbage and black currant pomace reduces the formation of acrylamide, 5-hydroxymethylfurfural and reactive aldehydes in fried potato-based crisps. LWT 173:114238

Bioactive Phytochemicals from Cabbage (*Brassica oleracea*) By-products

Siraj Uddin, Javed Iqbal, Banzeer Ahsan Abbasi, Abdul Rauf Tareen, Shumaila Ijaz, Zakir Ullah, Tabassum Yaseen, Habib Ali, Muhammad Jamil, Hassan Sher, Zahid Ullah, Ahmad Ali, and Tariq Mahmood

Siraj Uddin and Javed Iqbal contributed equally with all other contributors.

S. Uddin · Z. Ullah · T. Mahmood
Department of Plant Sciences, Faculty of Biological Sciences, Quaid-i-Azam University Islamabad, Islamabad, Pakistan
e-mail: zakirullah@bs.qau.edu.pk; tmahmood@qau.edu.pk

J. Iqbal (✉) · T. Yaseen
Department of Botany, Bacha Khan University, Charsadda, Khyber Pakhtunkhwa, Pakistan
e-mail: tabassumyaseen@bkuc.edu.pk

B. A. Abbasi
Department of Botany, Rawalpindi Women University, Rawalpindi, Pakistan

A. R. Tareen
Department of Microbiology, Quaid-I-Azam University, Islamabad, Pakistan
e-mail: abdulrauf@bs.qau.edu.pk

S. Ijaz
School of Biomedical Engineering, Shenzhen University Medical School, Shenzhen University, Shenzhen, China

H. Ali
Department of Agricultural Engineering, Khwaja Fareed University of Engineering and Information Technology, Rahim Yar Khan, Pakistan

M. Jamil
Department of Biotechnology and Genetic Engineering, Kohat University of Science and Technology, Kohat, Pakistan
e-mail: dr.jamil@kust.edu.pk

H. Sher · Z. Ullah · A. Ali
Center for Plant Sciences and Biodiversity, University of Swat, Charbagh, Pakistan
e-mail: zahidtaxon@uswat.edu.pk; draali@uswat.edu.pk

M. F. Ramadan (ed.), *Bioactive Phytochemicals in By-products from Leaf, Stem, Root and Tuber Vegetables*, https://doi.org/10.1007/978-3-031-80700-8_3

Contents

Abstract This chapter unveils the hidden potential of cauliflower by-products and explores their economic importance, biochemical composition, and diverse biocompounds. Cauliflower, a healthy food source, also possesses several health-promoting characteristics. The by-product of cauliflower mainly consists of leaves, stalk, and florets and constitutes a remarkable portion of waste in the agriculture fields. These unexplored by-products of cauliflower are rich sources of bioactive phytocompounds (secondary metabolites) such as glucosinates, phenolic compounds, and flavonoid contents. Additionally, these secondary metabolites in by-products have significant antimicrobial, anticancer, antioxidant, and anti-inflammatory properties, making them a vital addition in nutraceuticals, pharmaceuticals, and food industries.

Furthermore, this chapter also focuses on the nutritional profile of the reported by-products of cauliflower, which is mainly composed of protein, dietary fiber, vitamins, cellulose, hemicellulose, and mineral elements. This chapter has also discussed cauliflower by-products in bioenergy production (biogas, biomass, bioethanol, and biochar production) along with their human health-promoting and nutritional applications. The valorization of by-products for non-edible purposes reduces agricultural waste and produces sustainable energy, aids in developing a circular economy, and reduces the cost of non-renewable energy sources. Conclusively, this chapter provides an inclusive exploration of bioactive compounds derived from by-products of cauliflower, transcending their edible and non-edible potential to understand agro-economic and environmental prospects. This concise overview offers valuable insights for researchers, policy makers, and health experts to understand cauliflower by-products' health and non-health potential.

Keywords Cauliflower by-products · Glucosinolates · Phenolic acids · Amino acids · Mineral elements · Valorization · Biological properties, food and non-food applications

1 Introduction

The world's population is increasing daily; approximately 70% more food sources will be required by mid-century to meet the nutritional ends of 9–10 billion people (Ijaz et al. 2023a). Globally, food production is increasing regularly, and consumers' awareness about well-being and health is also growing (Oriekhoe et al. 2024). However, according to the Food and Agricultural Organization (FAO), a third of the produced food (1.3 billion tons) is wasted annually, which results in a financial loss of US $750 billion (Khalid et al. 2023). The remaining agricultural waste not only pollutes the ecosystem but also limits the availability of food (Koul et al. 2022). In recent decades, global waste has created significant issues directly related to pollution and malnutrition. Food waste contributes to greenhouse gas emissions by contributing 3.3 billion tons of CO_2 annually to the atmosphere. The biggest problem associated with the biowaste of vegetables is their high moisture and organic contents, which are responsible for microbial contamination and cause an unhealthy environmental impact (Ali et al. 2022; Ramos-Andrés et al. 2021). Therefore, reducing food waste is one of the United Nations' primary goals to create a more sustainable world by 2030 (Lemaire and Limbourg 2019). An abundant supply of bioactive compounds could be found in the by-products of over a million tons of vegetables discarded annually by industries (Ali et al. 2023). Extensive research has indicated that utilizing vegetable by-products to develop food products has enormous potential (Ramos-Andrés et al. 2021).

Over one-third of the total diet waste accounts for vegetable by-products or biowastes produced from industrial processing (Helkar et al. 2016). Among them, Brassica species have a significantly higher waste rate, for instance, 45–60% for cauliflower and 60–75% for broccoli (Castelão-Baptista et al. 2021; Petkowicz and Williams 2020). According to the report published by the FAO in 2022, the global production of cruciferous crops reached 72,604 kilotons. This indicates a rise of 0.876% compared to the previous year and a noteworthy increase of 6.53% compared to the last ten years. Therefore, the cruciferous plant creates waste and low-value by-products on a global scale during various phases of the vegetable supply chain. Brassica crops belong to the family Brassicaceae, also known as cruciferous crops, which are cultivated globally as temperate vegetables and are considered one of the significant crops after soybeans in oil seed generation. The Brassica genus includes broccoli, cauliflower cabbage, kale, brussels sprouts, rape, pak choi, mustard, turnip, and cress (Artés-Hernández et al. 2023). Brassica crops were of rising interest to several researchers due to their health profits (Šamec et al. 2019). Cauliflower by-products are intended to hold a large share in this regard. Discarding cauliflower biowaste (stems, leaves, stalks) contributes more than 60% of the vegetable's entire weight, and its maximum waste index among all other vegetables is a matter of apprehension (Ribeiro et al. 2015). The cauliflower is cultivated worldwide, with China and India being the major international producers, followed by Spain and Italy in Europe (FAOSTAT 2019). The high consumption rate of this crop is directly linked to reducing the danger of cancer, cardiovascular diseases, and degenerative pathologies (Sanlier and Guler 2018). Cauliflower is reported to be one of the healthiest vegetables owing to its healthy possessions. These healthiness benefits are due to the presence of a significant amount of dietary fibers, minerals, vitamins, and bioactive compounds, such as antioxidants, phenolic acids, glucosinolates, isothiocyanates, carotenoids, ascorbic acid, and polyamines (Table 1) (Nartea et al. 2023). These bioactive phytochemicals and mineral elements have been reported with significant health-promoting potential against various chronic diseases such as cardiovascular disease, type II diabetes, obesity, immune dysfunction, dementia, macular degeneration, and some kinds of cancers (Shahidi and Ambigaipalan 2015). It is reported that protein hydrolysate extracted from by-products of cauliflower displays its antioxidative properties in a cell-free system by inhibiting angiotensin I-converting enzyme (ACE) and also aids in the regulation of glucose and glycogen content in HepG2 cells (Tlais et al. 2020). Additionally, it has been described that by-products of cauliflower are also used as food additives, flavor enhancers, and coloring agents. The biowaste of cauliflower is the best source of nutrients, which can be used as animal nutrition, biostimulator, biocompost, and a bioenergy (Wadhwa et al. 2016).

Cauliflower by-products are used in the food sector and within the sustainable energy domain, highlighting their flexibility and value within the context of a circular economy (Oliveira et al. 2023). In this outline, many research studies have been directed to find more effective, ecologically friendly, and cost-effective management options to reduce the impact of this issue. For this purpose, several strategies and commercial models have been developed to increase the valorization of vegetable biomaterial in the outline of "industrial symbiosis," facilitating a circular economy by using by-products as raw materials from one sector to another industry (Ijaz et al. 2023a). The circular economy is a broadly used theoretical framework for effectively supervising biowaste to simplify sustainable growth within the dominion of the bioeconomy. Therefore, the acceptance of this model by vegetable waste-producing industries bears significant economic and environmental advantages (Ubando et al. 2020).

The utilization of cauliflower is primarily limited to flowers (the main white part). However, the by-products or non-edible parts (stem and leaves) can also be a cherished source of bioactive phytochemicals. Typically, the outer layers of this

Table 1 Different crops of *Brassica oleracea* with their edible organs

Specie	Common name	Organ	Reference
Brassica oleracea	Kale, collards	Leaves	Cartea et al. (2010)
	Tronchuda cabbage	Terminal leaf buds	
	Cabbage	Terminal leaf buds	
	Kohlrabi	Stem	
	Cauliflower	Inflorescences	
	Broccoli	Inflorescences	
	Brussels sprouts	Vegetative buds	
	Tronchuda cabbage	Loose heads	

vegetable detached during consumption comprise of stalks, peels, and crashed pulp. These discarded parts contain significant bioactive compounds, resulting in a considerable loss of nutrition (Table 2) (Ferreira et al. 2015).

Sustainable diet consumption is a noteworthy characteristic of sustainable growth. Imperative measures are needed to address this issue. Environment-friendly strategies should be established to increase food production while minimizing bio-waste. Implementing sustainable food production systems and improving food processes will play an active role in reducing environmental impact, dropping production costs, and enhancing food quality and dietary value (Fig. 1). These efforts are crucial for attaining a more sustainable and resilient food structure. Sustainable food production systems by-products of vegetables can be utilized to

Table 2 Cauliflower florets and discarded parts (stems and leaves) with a rich bioactive compound and their applications

Vegetable	Composition of the edible part	Composition of discarded parts	Uses	Reference
Cauliflower	Floret: Dietary fiber, minerals, glucosinolates, phenolics, carotenoids, ascorbic acid	Leaves and stems: Dietary fiber, minerals, carbohydrates, fat, glucosinolates, protein, organic acids, phenolic acids	Edible: Fortifying pizza, quinoa salad, beef sausage, wheat biscuits, noodles, chicken soup. Medicinal: antioxidant, anti-inflammatory, preservative, and anticancer Bioenergy: biomass, biofuel, biocompost, etc.	Chauhan (2015), Press (2015), Tukassar et al. (2023)

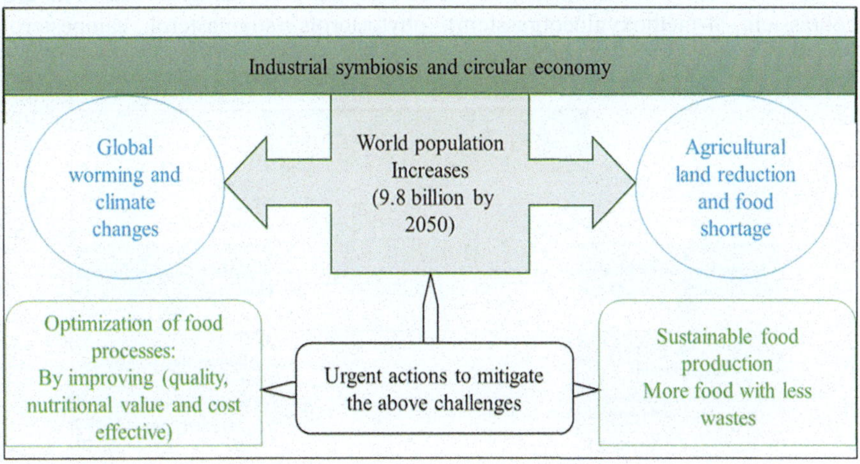

Fig. 1 Sustainable food production systems to produce more food for the next generation

gain novel bioingredients with functional, preservative, nutraceutical, and pharmaceutical characteristics. Additionally, the biowastes can obtain biofertilizers, biofuels, and enzymes that can develop commercial worth and decrease the environmental impact (Kumar et al. 2017; Magama et al. 2022).

This chapter focuses on valorizing and exploring bioactive compounds derived from cauliflower by-products. The main objective is to gather information about converting biowaste into valuable resources that can be used in various sectors, including food and non-food industries. This chapter is divided into four parts to deliver an inclusive understanding of the topic. The first part highlighted the economic value of the by-products of cauliflower and its phytocompounds. The second part elaborated on the biological properties of these by-products and their phytochemicals. The third and fourth parts of the chapter focus on the applications of cauliflower by-products in the food, non-food, and health sectors. By comprehensive examination of the objectives of this chapter, valuable resources and diverse applications of by-products obtained from cauliflower will be understood deeply.

2 Chemical Composition and Bioactive Compounds of Extracts from By-products

By-products, particularly vegetable wastes, comprise biocompounds that are the best choice for food and non-food sources. Vegetables are rich sources of primary metabolites, including dietary fibers, amino acids, lipids, cellulose, hemicellulose, lignin, carbohydrates, and secondary metabolites such as flavonoids, phenolic acids, alkaloids, glucosinolates, carotenoids, and organic acids (Fig. 2 and Table 3). Nartea et al. (2023) examined the bioactive compounds in the stalk and leaves of cauliflower and found that they mainly comprise glucosinolates (neoglucobrassicin, glucobrassicin, 4-methoxyglucobrassicin), phytosterols (stigmasterol, campesterol,

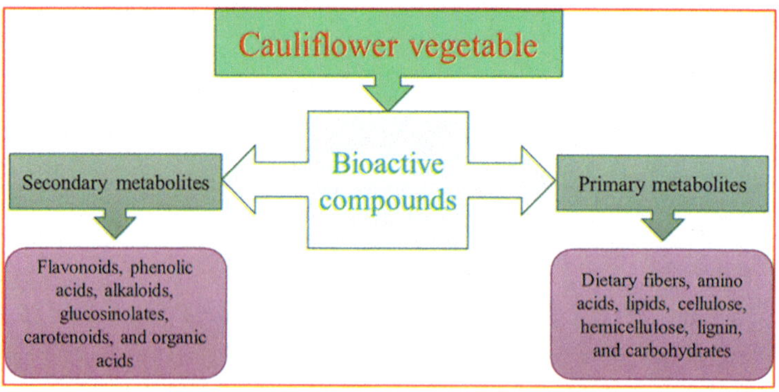

Fig. 2 Schematic representation of the bioactive compounds in cauliflower vegetable

Table 3 Major phytochemicals in cauliflower by-products

Biocompound	Major phytochemicals	Reference
Phenolic compounds	Kaempferol 3-diglucoside-7-diglucoside, kaempferol 3-triglucoside-7-diglucoside, kaempferol 3-disinapoyltriglucoside-7-diglucoside, kaempferol 3-sinapoyltriglucoside-7-glucoside, kaempferol 3-sinapoyltriglucoside-7-diglucoside, kaempferol 3-disinapoyltriglucoside-7-glucoside, and kaempferol 3-feruloyldiglucoside-7-diglucoside	Gul et al. (2022), Nilsson et al. (2006)
Glucosinolates	Glucoraphanin, progoitrin, glucoiberin, neoglucobrassicin, glucoerucin, sinigrin, 4-methoxy-glucobrassicin glucobrassicin, glucoiberverin, and 4-hydroxy-glucobrassicin	Gul et al. (2022), Nilsson et al. (2006)
Carotenoids	Lutein, cryptoxanthin, violaxanthin, zeaxanthin, antheraxanthin, neoxanthin, and β-carotene	Ijaz et al. (2024b), Picchi et al. (2020)
Amino acid	Alanine, asparagine, aspartate, γ-aminobutyric acid, arginine, glycine, glutamine, methiin, histidine, leucine, isoleucine, tryptophan, glutamate, valine lysine, phenylalanine, pyroglutamic acid, threonine, and tyrosine	Gupta and Prakash (2022), Helkar et al. (2016)
Organic acids	Acetic acid, formic acid, quinic acid, fumaric acid, malic acid, pyruvic acid, gallic acid, citric acid, succinic acid, ferulic acid, trans-cinnamic acid, and lactic acid	Lucarini et al. (2020), Ingallina et al. (2023)
Carbohydrates, polyols	α-glucose, myo-Inositol, β-glucose, β-fructofuranose, β-fructopyranose, sucrose, α-fructofuranose	Bastings et al. (2019)

and β-sitosterol) and carotenoids (lutein and β-carotene). It is crucial to notice that the accumulation of different biocompounds is diverse in the aerial portions of cauliflower. Furthermore, genetic and environmental factors generally influence the variation of these biocompounds in aerial parts.

Similarly, Drabińska and Nogueira studied that the cauliflower aerial organs have a rich source of biocompounds (glucosinolates, phenolics, and flavonoids). However, leaves showed the highest phytochemicals with potent antioxidant activities (Ijaz et al. 2024a). Table 4 represents leaves with the highest level of all reported phytochemicals compared to other parts. These studies reported that the distribution of the nutritional and bioactive compounds fluctuates significantly among the different parts of cauliflower. These studies propose that, along with the edible part, cauliflower's non-edible organs (leaves) could be utilized to produce novel functional diets (Shinali et al. 2024).

2.1 *Phenolic Acid*

Generally, vegetables have been studied as remarkable sources of phenolic compounds, amino phenolic compounds, and flavonoids. Furthermore, the variation of these phenolic compounds in aerial parts of vegetables is generally influenced by

Table 4 Bioactive compounds of cauliflower by-products

Bioactive compound	Florets	Leaves	Stems	Reference
Total glucosinolates (mmol/g DM)	94.02 ± 20.71	225.2 ± 48.88	125.05 ± 25.39	Magama et al. (2022), Ijaz et al. (2024a), Gul et al. (2022)
TPC (mg GAE/g DM)	2.76 ± 0.34	4.40 ± 0.22	2.65 ± 0.52	Magama et al. (2022), Ijaz et al. (2024a), Gul et al. (2022)
TFC (mg RUT/g DM)	0.29 ± 0.02	0.64 ± 0.06	0.22 ± 0.02	Magama et al. (2022), Ijaz et al. (2024a), Gul et al. (2022)
TCH (μg/g DM)	1.04 ± 0.14	22.43 ± 2.10	1.00 ± 0.13	Magama et al. (2022), Ijaz et al. (2024a), Gul et al. (2022)
Protein (%)	23.28 ± 1.46	20.74 ± 1.68	19.12 ± 1.57	Magama et al. (2022), Ijaz et al. (2024a), Gul et al. (2022)
Total amino acid (μmol/g DM)	1968.58 ± 364.87	3610.08 ± 544.06	2422.64 ± 517.67	Magama et al. (2022), Ijaz et al. (2024a), Gul et al. (2022)
Total carotenoids (μg/kg DW)	762.4 ± 135.3	817641.1 ± 71,400	–	Magama et al. (2022), Ijaz et al. (2024a), Gul et al. (2022)

climatic conditions, variety, agricultural practices, storage conditions, and genetic factors. Among phytochemicals bearing antioxidant potential in vegetables, phenolic compounds are one of the substantial groups (Llorach et al. 2003). In phenolic compounds, more than 8000 biocompounds are reported throughout the plants. Plant phenolics belong to secondary metabolites with an aromatic ring with one or more hydroxyl groups and other side groups. Phenolics are classified based on the number and arrangement of carbon atoms into non-flavonoids (phenolic acids, hydroxycinnamates, stilbenes, and others) and flavonoids (flavan-3-ols, isoflavones, flavanones, flavones, flavonols, anthocyanidins, and others) (Tsimogiannis and Oreopoulou 2019; Naczk and Shahidi 2006). Flavanols are among the main flavonoids, while kaempferol, quercetin, and isorhamnetin are the most plentiful flavonols found in vegetables (Ijaz et al. 2023b, 2024a). Furthermore, the variation of these phenolic compounds in aerial parts of vegetables is generally influenced by climatic conditions, variety, agricultural practices, storage conditions, and genetic factors.

In cauliflower by-products, Llorach et al. (2003) described the succeeding phenolic compounds: kaempferol 3-disinapoyltriglucoside-7-glucoside, kaempferol

3-feruloyldiglucoside-7-diglucoside, kaempferol 3-triglucoside-7-diglucoside, kaempferol 3-sinapoyltriglucoside-7-glucoside, kaempferol 3-diglucoside-7-diglucoside, kaempferol 3-sinapoyltriglucoside-7-diglucoside, and kaempferol 3-disinapoyltriglucoside-7-diglucoside (Llorach et al. 2003). Phenolic compounds are effective in neutralizing the reactive oxygen species by their ability to donate electrons and hydrogen atoms. Reactive oxygen species (ROS) are molecules generated from oxygen metabolism, including superoxide, singlet oxygen, hydroxyl radicals, and hydrogen peroxide. ROS have damaging properties at higher concentrations and are lethal for biomolecules such as proteins, lipids, and nucleic acids. To cope with such lethal concentrations of ROS, plants have antioxidant systems consisting of enzymatic components (superoxide dismutase, catalase, guaiacol peroxidase, and ascorbate peroxidase) and nonenzymatic components (polyphenols, vitamins, and carotenoid) (Ijaz et al. 2024c).

2.2 Glucosinolates

Glucosinolates are sulfur-containing glycosides, bioactive secondary metabolites found in plants, particularly cruciferous vegetables such as cauliflower, cabbage, broccoli, and kale (Mithen et al. 2000). More than 120 glucosinolates have been described in the plant kingdom, among which 96 have been found in *Brassicaceae* (Anjum et al. 2012). In these plants, the glucosinolates quantities are 0.1–2.5 g/kg, fluctuating due to several genetic and environmental factors. Structurally, all glucosinolates have the same basic structure comprising a sulfonated oxime group, β-D-thioglucose group, and a side chain. On the basis of the side chain, the glucosinolates are divided into the following groups: aliphatic, aromatic, x-methylalkyl, and heterocyclic (indole) glucosinolates (Śmiechowska et al. 2010).

Generally, vegetables, particularly cruciferous plants, are rich sources of glucosinolates (Ijaz et al. 2024d). Glucosinolates are mainly available in the leaves, stems, and florescences of cauliflower plants. Glucosinolates are mostly stable inside tissue due to their conjugation with sugar. However, through cutting, chewing, or crushing and enzymatic hydrolysis, they release many volatile and bioactive compounds, including isothiocyanates, oxazolidinethiones nitriles, thiocyanates indoles, and organic cyanides. The release of these derivatives mainly depends on many factors, such as the pH, glucosinolate substrates, ferrous ion concentration, the level of specific coenzymes and protein factors, and the handling of plant material (Ali et al. 2023; Arumugam and Razis 2018). In cauliflower, the main glucosinolates are glucoraphanin, progoitrin, glucoiberin, 4-methoxy-glucobrassicin, glucoerucin, sinigrin, glucobrassicin, glucoiberverin, neoglucobrassicin, and 4-hydroxy-glucobrassicin. In cauliflower by-products, the major glucosinolates are glucobrassicin, neoglucobrassicin, and 4-methoxy glucobrassicin. Glucosinolates and their corresponding derivatives (isothiocyantes) are important biological compounds with antimutagenic and anticarcinogenic potential (Abbasi et al. 2018; Arumugam and Razis 2018).

2.3 Carotenoid

Carotenoids are one of the most extensive pigments in microorganisms, plants, and animals. They are produced in all green photosynthetic organisms (plants and algae) and are responsible for characteristic colors, including orange, yellow, and red. However, in animals, carotenoids are supposed to originate from plant materials through colorful diets. Plants contain more than 600 carotenoids in the form of carotenes (beta-carotene and lycopene) and xanthophylls (lutein and zeaxanthin) (Singh and Devi 2015).

Cauliflower is recognized for its white color florets (edible parts), and the by-products (stem leaves and other parts) have different colors due to various pigments, such as carotenoids. Compared to florets, cauliflower's by-products have varying levels of carotenoids (Izadpanah et al. 2024). Izadpanah et al. (2024) reported that pigment concentrations such as carotenoid and total chlorophyll are higher in leaves than in florets. In cauliflower by-products, the main carotenoids are β-carotene, zeaxanthin, lutein, violaxanthin, antheraxanthin, neoxanthin, and cryptoxanthin (Ijaz et al. 2024b; Picchi et al. 2020).

2.4 Amino Acid

Cauliflower is a rich source of amino acids, particularly their by-products (leaves and stem), which showed remarkable amounts of amino acid content. Notably, leaves displayed many amino acids among the analyzed by-product components. This phenomenon is mainly due to the nitrogen assimilation process in the leaf (Yoneyama and Suzuki 2020). The key amino acids reported in cauliflower leaves are glutamine, glutamic acid, asparagine, and alanine, primary products of nitrogen assimilation (Zhang et al. 2010). By-products showed both essential (valine, phenylalanine, threonine, leucine, tryptophan, and isoleucine) and non-essential amino acids (glutamine, methiin, asparagine, arginine, aspartic acid, and glutamic acid), respectively (Zhang et al. 2010).

Green plants are rich sources of phytonutrients, including amino acids, essential for functional food and non-food sources. Amino acids have a vital role in plant physiological processes such as growth, development, transportation, and storage of nutrients (Ijaz et al. 2024a). Along with this, amino acids are vital molecules for the suitable metabolic activities of the human physique (Table 5). In the human body, amino acids are essential for metabolic activities. The perturbations in amino acids in the human body cause several illnesses and health-related issues, from simple pain to deadly cancer (Bastings et al. 2019). Human beings must consume amino acids to avoid such complexities or health issues. Therefore, the safe and rich reservoir of amino acids is green plants; particularly these amino acids have a significant role in the biosynthesis of other secondary metabolites against biotic and abiotic stresses. For example, tryptophan is used as a precursor for synthesizing indole, glucobrassicin, and other derivatives, whereas these defensive compounds are

Table 5 Biological applications of amino acids

Amino acids	Biological properties	Reference
Methiin, methionine	Antimicrobial and antidiabetic	Lucarini et al. (2020), Ingallina et al. (2023)
Isoleucine, leucine, lysine, threonine, valine, arginine	Muscle synthesis and repair, glycogen production, immune function, regulation of blood sugar levels	Lucarini et al. (2020), Ingallina et al. (2023)
Glutamic acid, aspartic acid, phenylalanine, tyrosine	Neurotransmitter synthesis	Lucarini et al. (2020), Ingallina et al. (2023)
Tryptophan	Precursor for serotonin, indole, glucobrassicin	Lucarini et al. (2020), Ingallina et al. (2023), Abul (2012)

related to antioxidant, antimicrobial, and other biological activities (Ijaz et al. 2024a; Sikorska-Zimny and Beneduce 2021).

2.5 *The Proximate Composition*

Cauliflower by-products have been studied to have rich reserves of nutrients such as high moisture, protein, carbohydrates, fat, crude fiber, and mineral contents. It is reported that products with low-fat values mainly reflect high moisture content and are one of the foremost parameters for food screening and processing (Baloch et al. 2015). Similarly, remarkable amounts of other nutrients such as protein, carbohydrates, and crude fiber have been reported in by-products of cauliflower. The cauliflower by-products are a rich source of macro- and micro-elements, making them suitable for fortifying food products. The mineral content is similar both in by-products and in floret. The by-products have been found to have higher amounts of macro-elements (calcium, sodium, magnesium, potassium, and phosphorus) and micro-elements (iron and zinc). These findings signify that this enriched proximate nature of by-products makes them an ideal candidate to use as food additives to enhance the nutritional value and nutraceutical ability of food stuff.

3 Biological Properties of Bioactive Compounds from By-products

It is reported that out of total plant weight, florets gain 35.42% as edible portions, while the remaining parts (stem, leaves, and midribs) secure more than 64% of the total weight as by-products. The wasted parts of cauliflower could be used as a source of dietary fiber and nutrients in food processing (Abul 2012). Cruciferous vegetables (cauliflower, broccoli, and cabbage) and their resulting harvests were documented as rich sources of bioactive compounds. The advancement in extraction and characterization techniques makes cruciferous plants an attractive topic for researchers due to their biochemical nature and rich resources of bioactive compounds (Fig. 3). Additionally, the by-products of these plants have gained attention in pharmaceutical and nutraceutical fields due to their biological activities such as antimicrobial, anti-inflammatory, anti-obesity, hepatoprotective, antioxidant, anticancer activity, and preservative properties (Table 6) (Shinali et al. 2024).

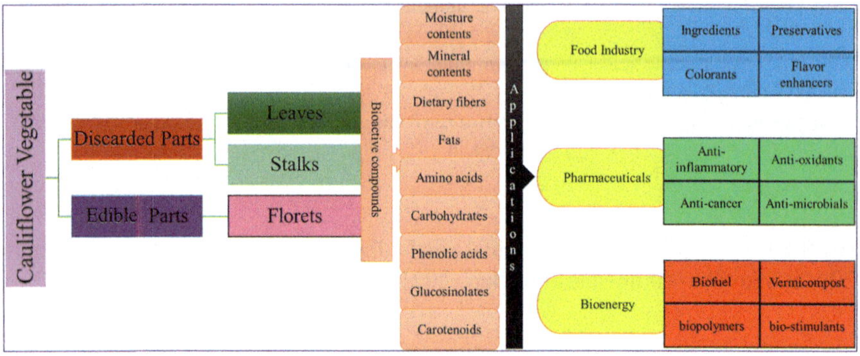

Fig. 3 A schematic representation of by-product valorization of cauliflower

Table 6 Biological activities of biocompounds found in the by-product of cauliflower

Bio compound	Biological activity	Reference
Protein	Improve the sensory profile of ready-to-eat expanded products	Stojceska et al. (2008)
Cellulose	composite, pharmaceutical, and cosmetic application	Eryilmaz (2024)
Hemicellulose	Preparation of hydrogels, microspheres, films, wound healing, and tissue engineering	Patil et al. (2022)
Dietary fiber	Antioxidant and anticarcinogenic	Podsędek (2007)
Vitamin C and E	Antioxidant	Podsędek (2007)
Non-starch polysaccharide	Source of dietary fiber	Femenia et al. (1998)
Carotenoids	Antioxidant	Femenia et al. (1998)
Phenolic compounds	Antioxidant and anticarcinogenic	Huynh et al. (2014)
Isothiocyanates	Anticarcinogenic effects	Arumugam and Razis (2018)
Glucosinolates (GLS)	Anticarcinogenic effects, tumor inhibitors	Arumugam and Razis (2018)
Phytosterols	Antioxidant	Tolve et al. (2020)
Sulforaphane	Antioxidant, antihypertensive, anticancer, and anti-inflammatory effects	Awan et al. (2017)
Indoles	Antioxidants lower the cholesterol level in the blood	Awan et al. (2017)
Organic acids	Microbial nematicides. Antioxidant and cytotoxic properties	Kim et al. (2018)
Lignin	Antimicrobial and cytotoxic properties	Majumdar et al. (2021)

3.1 Antioxidant Capacity

In recent years, nutrition's role in human health has gained global attention. Numerous epidemiological investigations have recommended that high consumption of plant-derived diets is linked to a reduced vulnerability to various chronic diseases, such as atherosclerosis and cancer (Zahra et al. 2023). These pharmacological effects have been credited to the presence of bioactive compounds with potent antioxidant properties. Antioxidants represent a broad and diverse assembly of biocompounds that share the common role of inhibiting, preventing, or retarding the oxidation process caused by an oxidizable substrate (Magama et al. 2022). As a result, these antioxidants reduce the risk of developing non-communicable diseases. For example, some phytochemicals found in the leaf of cauliflower, including phenolic compounds, flavonoids, vitamin C and chlorophylls, are reported with significant antioxidant properties and have also been proposed to lower the risk of various chronic diseases such as cancer (Nilsson et al. 2006). The action of antioxidants can be direct (catalase, metal chelators, and flavonoids) or indirect (glucosinolates and their by-products) by enhancing the natural defense system of the body (Magama et al. 2022). The combined action of multiple antioxidants reflects the overall antioxidant capacity of plants. The action of these antioxidants establishes the first line of defense (vitamin E and carotenoids) and second line of defense (vitamins, carotenoids, phenolics, and flavonoids contents) by scavenging free radicals and inhibiting the initiation of oxidative chain reactions (Ijaz et al. 2024b; Van Breusegem and Mittler 2023).

Biologically beneficial antioxidants found in brassica vegetables are glucosinolates, sulforaphane, indoles, carotenoids, organic acid, lignin, phenolic compounds, flavonoids, and vitamins (C and E) (Montone et al. 2018). The bioresidues of cauliflower are also found with many phenolic compounds, flavonoids, carotenoids, lignin, ferulic acid, gallic acid, *trans*-cinnamic acid peptide, and chlorophylls. These phytochemicals have antioxidant, antimicrobial, and cytotoxic activities (Majumdar et al. 2021). Recent studies reported that leaves reflected maximum antioxidant capacity among various parts of cauliflower, consistent with high amounts of phytochemicals. This reflection suggests that the leaves, with enriched protective compounds and chlorophylls, contribute significantly to their notable antioxidant potential (Kim and Ishii 2006). The bioactive rich profile of cauliflower by-products with potent antioxidants potentially exerts health-promoting properties, which may be helpful to cope with the damage caused by reactive oxygen species (ROS) in the human body (Karuppanapandian et al. 2011; Zhang et al. 2020). Conclusively, these studies suggested that cauliflower by-products are rich sources of antioxidants and highly appealing from both pharmaceutical and nutraceutical perspectives.

3.2 Anti-inflammatory Properties

Inflammation is a protective response exhibited by living tissues against harmful stimuli (pathogens, pollens, pollutants, damaged cells) and initiates the therapeutic process. The substances or biocompounds that reduce inflammation in the living body are known as anti-inflammatories. These anti-inflammatories can inhibit the process of inflammation without any change in homeostasis. Recent approaches focus on utilizing natural plant products, including vegetables, to overcome inflammation (Kumar Dhingra et al. 2015).

Cruciferous species (broccoli, cabbage, and cauliflower) and their derivatives have been studied to exhibit potent anti-inflammatory properties that can be utilized for therapeutic purposes (Nisar et al. 2023). The bioactive compounds present in cauliflower, such as vitamins, carotenoids, polyphenols, isothiocyanates, and glucosinolates, are demonstrated to have anti-inflammatory effects (Batool et al. 2019). These secondary metabolites have the potential to control the oxidative and inflammatory mediators, modulate the immune response, and maintain a healthy balance of gut microbiota (Batool et al. 2020). These phytoactive ingredients may mitigate environments associated with inflammation and improve overall health. Furthermore, pre-clinical and clinical assessments have highlighted the potential of these plant-derived compounds as natural anti-inflammatory agents (Sturm and Wagner 2017).

Isothiocyanates are reported to have anti-inflammatory abilities by influencing the inflammatory pathways. To mitigate inflammation, isothiocyanates modulate the signaling pathways, including mitogen-activated protein kinase (MAPK) pathways and nuclear factor kappa B (NF-κB). In this way, the isothiocyanates reduce the production of pro-inflammatory chemokines and cytokines, suppressing inflammation (Badshah et al. 2022). Additionally, the bioactivities of isothiocyanates facilitated through the activation of the nuclear factor-erythroid 2-related factor 2 (Nrf2) regulates the production of antioxidant and phase II enzymes (Ernst et al. 2011).

3.3 Anticancer Properties

Another study reported that the residue of brassica plant extracts, including cauliflower, has a remarkable cytotoxic effect against various human cancer cell lines (liver HepG2, colon HCT116, and breast MCF-7). These inhibitory properties can be accredited to the rich profile of biocompounds such as flavonoids, phenolic acids, vitamins, glucosinolates, isothiocyanates, and indoles, which have the potential to suppress the development of cancer cells and boost the immune system (Badshah et al. 2022; Ernst et al. 2011; Jamuna et al. 2017). These biochemicals also aid in the protection against chronic diseases such as atherosclerosis, chronic inflammation, and cancer by exhibiting antioxidative capabilities, stimulating the detoxifying enzymes, suppressing oxidative degradation, and blocking the cancer-causing agents (Ackland et al. 2005; Mandrich and Caputo 2020). For example, glucosinolates can stimulate the Nrf2 transcription factor, suppress tumor necrosis factor-α (TNFα) and interleukin-1β (IL-1β), and modulate inflammatory signaling pathways, thus triggering apoptosis. While sulforaphane stimulates detoxification enzymes, delays cellular proliferation, and establishes anti-inflammatory effects. Additionally, indoles regulate specific signaling pathways vital in the development and propagation of cancer. At the same time, phenolic acids display antioxidative and anti-inflammatory properties (Jamuna et al. 2017; Llorach et al. 2002).

4 Food and Non-food Applications of Extracts and Bioactive Compounds from By-products

4.1 Food Applications

Sections 2 and 3 clearly show that along with florets, the other non-used parts (leaves, stem, and peel) are also rich sources of bioactive molecules with mutually functional and biological activities (Fig. 3). The presence of secondary metabolites and other nutrients in by-products of cauliflower makes them a substantial supplementary constituent in food items (Table 7). Various studies have shown that the by-products of cauliflower can be used as functional ingredients, flavor enhancers, and coloring and preservative agents with antioxidant, antimicrobial, and preservative properties (Bhattacharjee and Singhal 2018; Faustino et al. 2019; Ijaz et al. 2023a; Iqbal et al. 2017; Moldes et al. 2017).

Table 7 Products made from the by-product of cauliflower with food additive properties

Product made	By-products	Properties	Reference
Dhokla, pancake, and idli	Cauliflower	Improve the nutritional content	Dorta and Lobo (2022)
Fortifying pizza	Cauliflower	Enhance the nutritional content	Nartea et al. (2023)
Quinoa salad	Cauliflower	Bioaccessibility of phytocompounds	Chauhan (2015)
Beef sausage	Flour	Improve sensory and nutritional composition, fat replacement	Abul (2012)
Meat product (patties)	Leaves	Improve the stability, firmness, and yield	Zhang et al. (2020), Femenia et al. (1998)
Malted wheat biscuits	Leaf powder	Improve sensory and nutritional composition	Wani and Sood (2014)
Noodles	Leaves	Enhance nutritional and organoleptic	Wani et al. (2013)
Rice bran (extrudates and phulkas)	Stalk	Improved the sensory properties of rice	Rafiuddin et al. (2021)
Ready-to-eat expanded snacks	Cauliflower	Enhanced sensory qualities	Stojceska et al. (2008)
Quiche "Lorraine"	Stalk and stem	Improve flavor, color, and fiber	Artés-Hernández et al. (2023)
Carrot paté	Stem and floret/curd	Improve flavor and fiber	Femenia et al.
Commercial chicken soup	Leaves and stems	Enhance flavor and antioxidants	Llorach et al. (2005)

4.1.1 Functional Ingredients

As individuals become increasingly conscious of their health, there has been a rise in the demand for functional food in recent years. Functional food is characterized as food that provides nutritional value and positively influences physical performance and mental state (Shonte et al. 2020). The American Dietetic Association has specified functional food as "whole, fortified, enriched, or enhanced" and is consumed as part of a diverse diet at appropriate levels regularly (Lau et al. 2021). Currently, the largest functional food markets are found in the Asia Pacific region and North America, accounting for 34% and 25% of the market share, respectively (Lau et al. 2021). The top three contributors to the total sales of functional foods and supplements are the USA, Europe, and Japan, collectively constituting around 82% of the market (Blandon et al. 2007). Functional food initially involved fortifying specific vitamins or minerals into food products. Functional food development has recently expanded to include fortifying one or more compounds that offer various health benefits (Siró et al. 2008). For example, the following phytocompounds, flavonoids, alkaloids, phenolics, and carotenoids mainly exist in functional foods, and these bioactive compounds can be useful for the treatment of various lethal diseases such as cardiovascular disease, neurological disorders, diabetes, and cancer (Mittal et al. 2024).

Various studies have reported that the by-products of cauliflower can be used as functional ingredients. For example, the by-products of cauliflower are used to make different functional foods such as muffins, noodles, pizza, pancakes, dhokla, and bakery foods are synthesized due to the presence of high fiber, mineral elements, glucosinolates, phytosterols, and carotenoids. This biowaste enhances the bioactive compounds in food and increases their shelf life (Sorrenti et al. 2023). The functional food applications of by-products are mainly due to their pasting properties, high viscosity, low moisture content (to avoid microbial attack), low fat and carbohydrates content, and also increased crude fiber, crude protein, and crude ash content (Femenia et al. 1997; Kapusta-Duch et al. 2019; Sedlar et al. 2021; Sharoba et al. 2013; Zhang et al. 2020). Some studies reported that cauliflower by-products are also helpful in meeting the recommended daily mineral intake (Abul 2012). All the above examples of prepared food products can be viewed as superfoods showing remarkable palatability, high nutritional value, easy bioavailability, and beneficial in controlling environmental pollution with economic benefits.

4.1.2 Flavor Enhancers

The first selection process of food choices is mainly their availability, nutrition, healthiness, affordability, and sensory qualities. Among these, flavor is one of the critical attributes of consumer acceptance. Numerous vegetables have different flavors, which are determined by the presence of compounds that have smell and trigeminal senses (Iqbal et al. 2018). These aromatic compounds are mainly amino acids, sugars, phenolics, and flavonoid contents (Akbar et al. 2024). The by-product of cauliflower can enhance the flavor of food products. Various studies reported that the odor and bitterness of cauliflower are mainly due to the sulfur-containing food phytochemicals such as glucosinolates, sinigrin, isothiocyanate, and other sulfur-containing bioactive compounds (Biondi et al. 2021).

Along with phytochemicals, amino acids such as thiamine and cysteine have also been reported to possess significant aroma precursors for food flavors (Ijaz et al. 2024a; Fadel et al. 2023). It is reported that adding cauliflower by-products into snacks increased the dietary fiber, protein content, bulk density, color, and flavors quality. It is reported that sensory qualities such as taste and flavors are best up to 10% of the addition of cauliflower by-products (Stojceska et al. 2008).

4.1.3 Preservatives

The demand for clean, labeled products for healthier diets is gaining attention to avoid chemical additives and search for new natural ingredients. To respond to these demands, food industries are introducing affordable natural additives with preservative properties into their food products. These natural ingredients have preservative properties due to bioactive phytochemicals (Novais et al. 2022). These bioactive phytochemicals are present in green plants, particularly in leafy vegetables, which have common diet ingredients and are related to health benefits. Vegetables are primarily rich in bioactive compounds (phenolic acids, carotenoids, flavonoids, anthocyanin, amino acids, vitamins, etc.), recognized with preservative properties. Incorporating these phytochemicals into food will enhance the quality and shelf life by inhibiting the microbial growth and reducing the oxidative stresses with altering the products' organoleptic properties (Pateiro et al. 2023). Recent studies indicated that glucosinolates and other antioxidants found in residual materials of cruciferous vegetables displayed antifungal and antibacterial properties, along with antioxidant, bioherbicidal, antimutagenic, and anticarcinogenic potential. The modes of antimicrobial activity involve neutralizing intracellular enzymes and inhibiting ATP production in pathogenic cells (Iqbal et al. 2019; Pateiro et al. 2023).

The by-products of cauliflower are known for their rich sources of bioactive compounds and, as mentioned above, also act as effective preservatives with potent antioxidant properties (Iqbal et al. 2019). In this regard, the enriched phytochemical properties make them a suitable candidate for use as an alternative to chemical additives. Various studies have reported that the by-product of cauliflower can be used to preserve various food products such as seafood, meat, bakery, and fermented dairy products to reduce oxidative processes and control microbial growth, as shown in Table 7 (Ahmed and Ali 2013). Consequently, incorporating cauliflower bioresidues into various food products has revealed a promising nutritional profile without changing their sensory appeal (Chaudhary et al. 2014; Messinese et al. 2024).

4.1.4 Coloring Agents

Food products' most studied and widely applied components are their flavors and nutritional values (Verbeke 2006). In addition to the flavor, the appearance of food products has also gained attention as a valuable sensory aspect. Moreover, colorants (food additives) are used to enhance the color of food items and make them more attractive (Coultate and Blackburn 2018). Colorants are one of the primary sensory aspects that attract the consumer and increase the food item's acceptability. Color is associated with the food's flavor and taste thresholds and influences food choice, preference, and acceptability (Sigurdson et al. 2017). Thus, cauliflower by-products with chromatic phytochemicals (anthocyanin, carotenoids, phenolic compounds, and carotenoids derivatives) may be used as natural colorants to improve the chroma, texture, and shelf life of beverages and food (Dorta and Lobo 2022; Kumar et al. 2022; Li et al. 2006; Lopez-Romero et al. 2017; Müller 1997).

4.2 Non-food Applications

Cauliflower by-products have been studied to hold a variety of non-food applications, encompassing potential applications in bioengineering, animal nutrition, antimicrobial activity, recovery of bioactive compounds, and as a reservoir of bioactive compounds for the innovation of novel goods. The abundant composition of cauliflower by-products, which includes bioactive compounds such as glucosinolates, carotenoids, phytosterols, and other components valuable to health, substantiates their potential applications in different fields. Within the field of bioengineering, isothiocyanate, a phytochemical abundantly present in cauliflower, has been scrutinized for its anticancer, antioxidant, antimicrobial, food preservative, and plausible delivery mechanisms (Gupta and Prakash 2022; Macharia et al. 2022).

The bioresidues of cauliflower are a rich reservoir of phytochemicals, which can be used as modifiers of gut bacterial communities. Maintaining the gut microbiota (GM) is crucial for a healthy life, and any alteration in gut microbial communities can lead to some chronic diseases. Consequently, dietary fiber fermentation and phytochemicals have been found to modulate the GM. This modulation is due to the antioxidant, anti-inflammatory, and prebiotic properties of phytochemicals present by-products (Dingeo et al. 2020). Additionally, research has investigated the health-promoting components of cauliflower by-products, revealing that they comprise functional elements such as flavonoids, squalene, vitamin E, glucoraphanin, indole-3-carbinol, carotenoids, phospholipids, sterols, and ferulic acid which contributes to various health advantages including anti-inflammatory, anticancer, anti-obesity, antimicrobial, antidiabetic, neuroprotective, and cardioprotective effects. Moreover, the antimicrobial efficacy of cauliflower by-products has been exhibited against foodborne bacteria, yielding promising outcomes in reducing specific bacterial populations (Javed et al. 2023; Meitha et al. 2020).

Natural products offer unlimited opportunities for discovering novel drug candidates due to the unparalleled richness of chemical diversity. Along with their health benefits, the by-product residues can be utilized in animal nutrition. The application of plants by-products as livestock feeds is an essential prerequisite for maximizing profitability. These by-products hold significant potential, particularly for ruminant animals, due to their ability to metabolize fiber (Joshi 2020). The cauliflower stem is subjected to characterize cellulosic fibers with significant composite properties.

Furthermore, cauliflower waste and other brassica plants, including their non-edible parts, have adsorbent properties and can be useful in removing heavy metals from contaminated water. For example, the dried powder of these vegetables can be used to remove Pb(II) and Cd(II) from synthetic solutions (Matei et al. 2021). Additionally, brassica vegetable biowaste has been used for non-health purposes such as bioenergy, biomass, bioethanol, and fuel briquette production (Jaiswal 2020). Cauliflower and other family members of the brassica plant have many uses beyond nutraceutical and pharmaceutical applications. They can produce vermicompost, organic acids, biopolymers, and bioenergy, removing heavy metals from polluted water. These findings offer new opportunities for incorporating cauliflower leftover in different sectors, contributing to establishing sustainable and valued-enhanced products.

5 Valorization of By-products for Non-health Purposes

Bioenergy production from renewable biological sources is becoming a sustainable solution to address energy consumption, decreasing fossil fuel assets, and environmental concerns (Zhang et al. 2010). Various studies have investigated the valorization of by-products for non-health purposes, particularly as a bioenergy source (Kumar et al. 2023a, 2023b; Ozogul et al. 2021; Shahbaz et al. 2021). Recent findings have shown that bioresidues of cauliflower can be utilized for bioenergy production, producing chief bioactive chemicals such as benzene, toluene, and phenols. Similarly, the biowaste of cauliflower (like kimchi cabbage waste) can help make organic acids and act as a microbial nematicide. Furthermore, it is reported that the putrescin-treated cauliflower plant under high temperatures and elevated CO_2 levels indicated the rise of biomass and biosynthesis of sugar contents (Tahir et al. 2022). This biomass production and sugar synthesis make the by-product of cauliflower an ideal candidate for use in bioenergy production.

The biomaterials of cauliflower by-products are also used as plant biostimulators or biofertilizers to boost plant growth by improving nutrient uptake, enhancing stress tolerance, and inhibiting pathogenic attacks. The plant's bio-stimulant abilities are due to bioactive compounds, including inulins, chlorogenic acid, sulfur-containing compounds, chiro-inositol, and sesquiterpene lactones. The application of by-products as plant bio-stimulants not only aligns with the principles of the circular economy but also approves sustainable agricultural practices. In another study, researchers reported that the by-product of cauliflower can be converted into biochar with specific agricultural properties through pyrolysis. The biochar prepared through pyrolysis is a rich source of nutrients with water and holding properties, improves microbial communities, and helps in nitrogen fixation (Pokharel et al. 2020). The by-products of cauliflower are also helpful in synthesizing vermicompost products. For the vermicomposting process, the biowaste is blended with cow dung in the presence of earthworm (*Eisenia fetida*). The vermicompost has become a rich source of nitrogen, phosphorus, and potassium (Mago et al. 2022). Similarly, the biowaste of cauliflower (like kimchi cabbage waste) can help produce organic acids (fumaric acid, lactic acid, and acetic acid) through saccharification and fermentation. These organic acids can be effective microbial nematicides (Kim et al. 2018).

These studies represent the significance of cauliflower by-products in the food sector and the sustainable energy domain, highlighting their flexibility and value within a circular economy. Therefore, the utilization of cauliflower by-products as feedstock for food, non-food, and biofuel production is evident that it provides an opportunity to diminish the necessity of fossil energy and lessen the environmental pollution caused by biowastes (Shinali et al. 2024).

6 Conclusion

This chapter reviews the unexplored potential of cauliflower by-products, examining their economic status, biochemical configuration, and inclusive diversity of biocompounds. Cauliflower is well known for its dietary benefits and possesses various health-promoting characteristics. The by-products of cauliflower, mainly consisting of leaves, stalks, and florets, create substantial biowaste. However, these discarded parts of cauliflower serve as plentiful reservoirs of bioactive phytochemicals, including glucosinates, phenolic compounds, flavonoids, and carotenoids displaying antimicrobial, antioxidant, anticancer, and anti-inflammatory properties that are progressively valuable in the fields of pharmaceuticals, nutraceuticals, and food production. Furthermore, this chapter evaluated the nutritional profile of cauliflower by-products, highlighting their dietary fiber, cellulose, hemicellulose, vitamin, amino acid, and mineral contents. Besides their human health-promoting and nutritional applications, this chapter also explores the usage of cauliflower by-products in bioenergy generation (biomass, biogas, bioethanol, and biochar), offering a sustainable substitute to non-renewable energy resources while reducing agricultural residue and encouraging a circular economic structure. In conclusion, this section extensively examines the bioactive ingredients in cauliflower by-products, highlighting their potential for both edible and non-edible purposes and explaining their agro-economic and environmental rewards. This brief overview is a valuable reference for researchers, policymakers, and health professionals aiming to harness the comprehensive range of benefits of cauliflower by-products.

References

Abbasi BA, Iqbal J, Mahmood T, Khalil AT, Ali B, Kanwal S, Shah SA, Ahmad R (2018) Role of dietary phytochemicals in modulation of miRNA expression: natural swords combating breast cancer. Asian Pac J Trop Med 11(9):501–509

Abul-Fadl MM (2012) Nutritional and chemical evaluation of white cauliflower by-products flour and the effect of its addition on beef sausage quality. J Appl Sci Res 8(2):693–704

Ackland ML, van de Waarsenburg S, Jones R (2005) Synergistic antiproliferative action of the flavonols quercetin and kaempferol in cultured human cancer cell lines. In Vivo 19(1):69–76

Ahmed FA, Ali RF (2013) Bioactive compounds and antioxidant activity of fresh and processed white cauliflower. Biomed Res Int 2013(1):367819

Akbar K, Yaseen T, Abbasi BA, Iqbal J, Ullah Z, Ijaz S, Mustafa AEZM, Elshikh MS, Ercisli S, Harsonowati W, Iqbal R (2024) Role and genetic diversity of arbuscular mycorrhizal fungi in improving the productivity and enhancing nutrients absorption in chickpea (Cicer arietinum L.) genotypes. Genet Resour Crop Evol, pp 1–15

Ali S, Khan MR, Iqbal J, Shah SA, Abbasi BA, Yaseen T, Batool R, Ali I, Hussain MD, Kazi M (2022) Chemical characterization and evaluation of the nephroprotective potential of *Parrotiopsis jacquemontiana* (Decne) Rehder and Periploca hydaspidis Falc crude extract in CCl4-induced Male Sprague-Dawley Rats. Saudi J Biol Sci 29(2):702–712

Ali S, Sayed AS, Saeed RF, Iqbal J, Ijaz S, Munazir M (2023) Ethnomedicinal plant use value in Lower Swat, Pakistan. Ethnobot Res Appl 25:1–22

Anjum NA, Gill SS, Ahmad I, Pacheco M, Duarte AC, Umar S, Khan NA, Pereira ME (2012) The plant family Brassicaceae: an introduction. The plant family Brassicaceae: contribution towards phytoremediation, pp 1–33

Artés-Hernández F, Martínez-Zamora L, Cano-Lamadrid M, Hashemi S, Castillejo N (2023) Genus Brassica by-products revalorization with green technologies to fortify innovative foods: a scoping review. Food Secur 12(3):561

Arumugam A, Razis AFA (2018) Apoptosis as a mechanism of the cancer chemopreventive activity of glucosinolates: a review. Asian Pacific J Cancer Prev APJCP 19(6):1439

Awan KA, Butt MS, Haq I, Ashfaq F, Suleria H (2017) Storage stability of garlic fortified chicken bites. J Food Chem Nanotechnol 3(2):80–85

Badshah H, Mumtaz AS, Hussain I, Ali B, Iqbal J, Ali S (2022) Wild morels in Pakistan: environmental and trading statues. In: Biodiversity, conservation and sustainability in Asia: Volume 2: Prospects and challenges in South and Middle Asia. Springer, Cham, pp 691–702

Baloch AB, Xia X, Sheikh SA (2015) Proximate and mineral compositions of dried cauliflower (*Brassica oleracea* L.) grown in Sindh, Pakistan. J Food Nutr Res 3(3):213–219

Bastings JJ, van Eijk HM, Olde Damink SW, Rensen SS (2019) D-amino acids in health and disease: a focus on cancer. Nutrients 11(9):2205

Batool R, Aziz E, Salahuddin H, Iqbal J, Tabassum S, Mahmood T (2019) Rumex dentatus could be a potent alternative to treatment of microbial infections and of breast cancer. J Tradit Chin Med 39(6):772–779

Batool R, Aziz E, Iqbal J, Salahuddin H, Tan BKH, Tabassum S, Mahmood T (2020) In vitro antioxidant and anti-cancer activities and phytochemical analysis of *Commelina benghalensis* L. root extracts. Asian Pac J Trop Biomed 10(9):417–425

Bhattacharjee P, Singhal RS (2018) Broccoli and cauliflower: production, quality, and processing. Handbook of vegetables and vegetable processing, pp 535–558

Biondi F, Balducci F, Capocasa F, Visciglio M, Mei E, Vagnoni M, Mezzetti B, Mazzoni L (2021) Environmental conditions and agronomical factors influencing the levels of phytochemicals in Brassica vegetables responsible for nutritional and sensorial properties. Appl Sci 11(4):1927

Blandon J, Cranfield J, Henson S (2007) Functional food and natural health product issues: the Canadian and international context. International food economy research group department of food, agricultural and resource economics, pp 1–55

Cartea ME, Francisco M, Soengas P, Velasco P (2010) Phenolic compounds in Brassica vegetables. Molecules 16(1):251–280
Castelão-Baptista JP, Barros A, Martins T, Rosa E, Sardão VA (2021) Three in one: the potential of Brassica by-products against economic waste, environmental hazard, and metabolic disruption in obesity. Nutrients 13(12):4194
Chaudhary A, Sharma U, Vig AP, Singh B, Arora S (2014) Free radical scavenging, antiproliferative activities and profiling of variations in the level of phytochemicals in different parts of broccoli (*Brassica oleracea* italica). Food Chem 148:373–380
Chauhan A (2015) Product development and sensory evaluation of value added food products made by incorporating dried cauliflower green leaves. Nutr Food Sci
Coultate T, Blackburn RS (2018) Food colorants: their past, present and future. Color Technol 134(3):165–186
Dingeo G, Brito A, Samouda H, Iddir M, La Frano MR, Bohn T (2020) Phytochemicals as modifiers of gut microbial communities. Food Funct 11(10):8444–8471
Dorta E, Lobo G (2022) Recovery of wasted vegetables and fruits for food additives. In: Fruits and vegetable wastes: valorization to bioproducts and platform chemicals. Singapore, Springer Nature Singapore, pp 33–60
Ernst IM, Wagner AE, Schuemann C, Storm N, Höppner W, Döring F, Stocker A, Rimbach G (2011) Allyl-, butyl-and phenylethyl-isothiocyanate activate Nrf2 in cultured fibroblasts. Pharmacol Res 63(3):233–240
Eryilmaz O (2024) Revalorization of cellulosic fiber extracted from the waste stem of *Brassica oleracea* var. botrytis L. (cauliflower) by characterizing for potential composite applications. Int J Biol Macromol 266:131086
Fadel HHM, Lotfy SN, Asker MMS, Mahmoud MG, Abd El-Aleem FS, Ahmed MYS, Saad R (2023) Effect of amino compounds on the volatile compounds and antioxidant activity of Maillard reaction products of enzymatic hydrolyzed cabbage leaves. Appl Food Res 3(2):100327
FAOSTAT, F (2019) Food and agriculture organization of the United Nations-Statistic Division. https://www.fao.org/faostat/en/#data
Faustino M, Veiga M, Sousa P, Costa EM, Silva S, Pintado M (2019) Agro-food byproducts as a new source of natural food additives. Molecules 24(6):1056
Femenia AIOF, Lefebvre AC, Thebaudin JY, Robertson JA, Bourgeois CM (1997) Physical and sensory properties of model foods supplemented with cauliflower fiber. J Food Sci 62(4):635–639
Femenia A, Robertson JA, Waldron KW, Selvendran RR (1998) Cauliflower (*Brassica oleracea* L), globe artichoke (*Cynara scolymus*) and chicory witloof (*Cichorium intybus*) processing by-products as sources of dietary fibre. J Sci Food Agric 77(4):511–518
Ferreira MS, Santos MC, Moro TM, Basto GJ, Andrade RM, Gonçalves ÉC (2015) Formulation and characterization of functional foods based on fruit and vegetable residue flour. J Food Sci Technol 52:822–830
Gul F, Khan I, Iqbal J, Abbasi BA, Shahbaz A, Capasso R, Amaro-Estrada I, Jardan YAB, Cossio-Bayugar R, Mahmood T (2022) Phytochemistry, biological activities and in silico molecular docking studies of *Oxalis pes-caprae* L. compounds against SARS-CoV-2. J King Saud Univ Sci 34(6):102136
Gupta C, Prakash D (2022) Phytochemicals as the source of natural immunomodulator and their role in cancer chemoprevention. In: Immunomodulators and human health. Singapore, Springer Nature Singapore, pp 209–241
Helkar PB, Sahoo AK, Patil NJ (2016) Review: food industry by-products used as a functional food ingredients. Int J Waste Resourc 6(3):1–6
Huynh NT, Smagghe G, Gonzales GB, Van Camp J, Raes K (2014) Enzyme-assisted extraction enhancing the phenolic release from cauliflower (*Brassica oleracea* L. var. botrytis) outer leaves. J Agric Food Chem 62(30):7468–7476

Ijaz S, Iqbal J, Abbasi BA, Ullah Z, Yaseen T, Kanwal S, Mahmood T, Sydykbayeva S, Ydyrys A, Almarhoon ZM, Sharifi-Rad J (2023a) Rosmarinic acid and its derivatives: current insights on anticancer potential and other biomedical applications. Biomed Pharmacother 162:114687

Ijaz S, Iqbal J, Abbasi BA, Tufail A, Ullah Z, Sharifi-Rad J, Yaseen T, Uddin S (2023b) Biofortification: lessons from the past and strategies for future food security. In: Legumes biofortification. Springer, Cham, pp 521–545

Ijaz S, Iqbal J, Abbasi BA, Kanwal S, Mahmoodi M, Raish M, Mahmood T (2024a) Investigation of bioactive constituents and evaluation of in vitro bioactivities of different Setaria glauca extracts. J King Saud Univ Sci 36(8):103321

Ijaz S, Iqbal J, Abbasi BA, Kanwal S, Tavafoghi M, Ahmed MZ, Mahmood T (2024b) Investigation of bioactive constituents and evaluation of different in vitro antimicrobial, antioxidant, and cytotoxicity potentials of different *Portulacaria afra* extracts. J King Saud Univ Sci 36(2):103033

Ijaz S, Iqbal J, Abbasi BA, Tufail A, Yaseen T, Uddin S, Usman K, Ullah R, Bibi H, Inam P, Sagindykova E (2024c) Current stage of preclinical and clinical development of guggulsterone in cancers: challenges and promises. Cell Biol Int 48(2):128–142

Ijaz S, Iqbal J, Abbasi BA, Tufail A, Ullah Z, Yaseen T, Ali I, Uddin S, Iqbal R (2024d) Heavy metal–polluted arable land and its consequences: a global scenario. In: Biochar-assisted remediation of contaminated soils under changing climate. Elsevier, pp 71–99

Ingallina C, Di Matteo G, Spano M, Acciaro E, Campiglia E, Mannina L, Sobolev AP (2023) Byproducts of globe artichoke and cauliflower production as a new source of bioactive compounds in the green economy perspective: an NMR study. Molecules 28(3):1363

Iqbal J, Abbasi BA, Mahmood T, Kanwal S, Ali B, Shah SA, Khalil AT (2017) Plant-derived anticancer agents: a green anticancer approach. Asian Pac J Trop Biomed 7(12):1129–1150

Iqbal J, Abbasi BA, Batool R, Mahmood T, Ali B, Khalil AT, Kanwal S, Shah SA, Ahmad R (2018) Potential phytocompounds for developing breast cancer therapeutics: nature's healing touch. Eur J Pharmacol 827:125–148

Iqbal J, Abbasi BA, Ahmad R, Batool R, Mahmood T, Ali B, Khalil AT, Kanwal S, Shah SA, Alam MM, Bashir S (2019) Potential phytochemicals in the fight against skin cancer: current landscape and future perspectives. Biomed Pharmacother 109:1381–1393

Izadpanah F, Frede K, Soltani F, Baldermann S (2024) Comparison of carotenoid, chlorophyll concentrations and their biosynthetic transcript levels in different coloured cauliflower. Hortic Plant J 10(3):743–754

Jaiswal AK (ed) (2020) Nutritional composition and antioxidant properties of fruits and vegetables. Academic Press

Jamuna KS, Suma MS, Ramesh CK, Mahmood R, Nanjundaswamy L (2017) Studies on in vitro antiproliferative activities in cruciferous vegetables. J Appl Hortic 19(3):230–234

Javed R, Ijaz S, Hameed H, Nazish M, Sharif MS, Afreen A, Alarjani KM, Elshikh MS, Mehboob S, Abdul Razak S, Waheed A (2023) Phytochemical-mediated biosynthesis of silver nanoparticles from Strobilanthes glutinosus: exploring biological applications. Micromachines 14(7):1372

Joshi VK (2020) Value addition to fruit and vegetable processing waste-an appraisal. Int J Food Ferment Technol 10(2):35–58

Kapusta-Duch J, Szeląg-Sikora A, Sikora J, Niemiec M, Gródek-Szostak Z, Kuboń M, Leszczyńska T, Borczak B (2019) Health-promoting properties of fresh and processed purple cauliflower. Sustain For 11(15):4008

Karuppanapandian T, Moon JC, Kim C, Manoharan K, Kim W (2011) Reactive oxygen species in plants: their generation, signal transduction, and scavenging mechanisms. Aust J Crop Sci 5(6):709–725

Khalid S, Ullah MI, Malik AU (2023) Food loss and waste, types, sources, pollution potential, and country-wise comparison. In: Waste problems and management in developing countries. Apple Academic Press, pp 291–328

Kim SJ, Ishii G (2006) Glucosinolate profiles in the seeds, leaves and roots of rocket salad (*Eruca sativa* Mill.) and anti-oxidative activities of intact plant powder and purified 4-methoxyglucobrassicin. Soil Sci Plant Nutr 52(3):394–400

Kim HM, Park JH, Choi IS, Wi SG, Ha SH, Chun HH, Hwang IM, Chang JY, Choi HJ, Kim JC, Park HW (2018) Effective approach to organic acid production from agricultural kimchi cabbage waste and its potential application. PLoS One 13:e0207801

Koul B, Yakoob M, Shah MP (2022) Agricultural waste management strategies for environmental sustainability. Environ Res 206:112285

Kumar Dhingra A, Chopra B, Dass R, Kumar Mittal S (2015) An update on anti-inflammatory compounds: a review. Anti-Inflamm Anti-Allergy Agents Med Chem (Formerly Current Medicinal Chemistry-Anti-Inflammatory and Anti-Allergy Agents) 14(2):81–97

Kumar K, Yadav AN, Kumar V, Vyas P, Dhaliwal HS (2017) Food waste: a potential bioresource for extraction of nutraceuticals and bioactive compounds. Bioresour Bioprocess 4:1–14

Kumar S, Sharma S, Kumar V, Sharma R, Minhas A, Boddu R (2022) Cruciferous vegetables: a mine of phytonutrients for functional and nutraceutical enrichment. In: Current advances for development of functional foods modulating inflammation and oxidative stress. Academic Press, pp 401–426

Kumar K, Debnath P, Singh S, Kumar N (2023a) An overview of plant phenolics and their involvement in abiotic stress tolerance. Stress 3(3):570–585

Kumar S, D'Silva TC, Kumar D, Isha A, Khan SA, Chandra R, Malik A, Vijay VK (2023b) Valorization of by-products produced during the extraction and purification of biofuels. Biofuel Extraction Techniques, pp 307–332

Lau KQ, Sabran MR, Shafie SR (2021) Utilization of vegetable and fruit by-products as functional ingredient and food. Front Nutr 8:661693

Lemaire A, Limbourg S (2019) How can food loss and waste management achieve sustainable development goals? J Clean Prod 234:1221–1234

Li L, Lu S, Cosman KM, Earle ED, Garvin DF, O'Neill J (2006) β-Carotene accumulation induced by the cauliflower or gene is not due to an increased capacity of biosynthesis. Phytochemistry 67(12):1177–1184

Llorach R, Tomas-Barberan FA, Vallejo F, Garcia-Viguera C, Ferreres F (2002) Valorization of by-products of the handling and commercialization of vegetables: possible use of Brassica spp by-products as source of antioxidant health-promoting phenolics. Natural products in the new millennium: prospects and industrial application, p 337–343

Llorach R, Espín JC, Tomás-Barberán FA, Ferreres F (2003) Valorization of cauliflower (Brassica oleracea L. var. botrytis) by-products as a source of antioxidant phenolics. J Agric Food Chem 51(8):2181–2187

Llorach R, Tomás-Barberán FA, Ferreres F (2005) Functionalisation of commercial chicken soup with enriched polyphenol extract from vegetable by-products. Eur Food Res Technol 220:31–36

Lopez-Romero JC, Ansorena R, Gonzalez-Aguilar GA, Gonzalez-Rios H, Ayala-Zavala JF, Siddiqui MW (2017) Applications of plant secondary metabolites in food systems. In: Plant secondary metabolites, vol 2. Apple Academic Press, pp 221–258

Lucarini M, Di Cocco ME, Raguso V, Milanetti F, Durazzo A, Lombardi-Boccia G, Santini A, Delfini M, Sciubba F (2020) NMR-based metabolomic comparison of Brassica oleracea (var. italica): organic and conventional farming. Food Secur 9(7):945

Macharia JM, Mwangi RW, Rozmann N, Zsolt K, Varjas T, Uchechukwu PO, Wagara IN, Raposa BL (2022) Medicinal plants with anti-colorectal cancer bioactive compounds: potential game-changers in colorectal cancer management. Biomed Pharmacother 153:113383

Magama P, Chiyanzu I, Mulopo J (2022) A systematic review of sustainable fruit and vegetable waste recycling alternatives and possibilities for anaerobic biorefinery. Bioresour Technol Reports 18:101031

Mago M, Gupta R, Yadav A, Garg VK (2022) Sustainable treatment and nutrient recovery from leafy waste through vermicomposting. Bioresour Technol 347:126390

Majumdar S, Seth S, Bhattacharyya DK, Bhowal J (2021) Evaluation of therapeutic properties of lignins extracted from cauliflower wastes for their potent valorization through sustainable approach. Waste Biomass Valoriz 12:3849–3873

Mandrich L, Caputo E (2020) Brassicaceae-derived anticancer agents: towards a green approach to beat cancer. Nutrients 12(3):868

Matei E, Râpă M, Predescu AM, Țurcanu AA, Vidu R, Predescu C, Bobirica C, Bobirica L, Orbeci C (2021) Valorization of agri-food wastes as sustainable eco-materials for wastewater treatment: current state and new perspectives. Materials 14(16):4581

Meitha K, Pramesti Y, Suhandono S (2020) Reactive oxygen species and antioxidants in postharvest vegetables and fruits. Int J Food Sci 2020(1):8817778

Messinese E, Pitirollo O, Grimaldi M, Milanese D, Sciancalepore C, Cavazza A (2024) By-products as sustainable source of bioactive compounds for potential application in the field of food and new materials for packaging development. Food Bioprocess Technol 17(3):606–627

Mithen RF, Dekker M, Verkerk R, Rabot S, Johnson IT (2000) The nutritional significance, biosynthesis and bioavailability of glucosinolates in human foods. J Sci Food Agric 80(7):967–984

Mittal RK, Mishra R, Sharma V, Purohit P (2024) Bioactive exploration in functional foods: unlocking nature's treasures. Curr Pharm Biotechnol 25(11):1419–1435

Moldes AB, Vecino X, Cruz JM (2017) Nutraceuticals and food additives. In: Current developments in biotechnology and bioengineering. Elsevier, pp 143–164

Montone CM, Capriotti AL, Cavaliere C, La Barbera G, Piovesana S, Chiozzi RZ, Laganà A (2018) Characterization of antioxidant and angiotensin-converting enzyme inhibitory peptides derived from cauliflower by-products by multidimensional liquid chromatography and bioinformatics. J Funct Foods 44:40–47

Müller H (1997) Determination of the carotenoid content in selected vegetables and fruit by HPLC and photodiode array detection. Zeitschrift für Lebensmitteluntersuchung und-Forschung A 204:88–94

Naczk M, Shahidi F (2006) Phenolics in cereals, fruits and vegetables: occurrence, extraction and analysis. J Pharm Biomed Anal 41(5):1523–1542

Nartea A, Fanesi B, Pacetti D, Lenti L, Fiorini D, Lucci P, Frega NG, Falcone PM (2023) Cauliflower by-products as functional ingredient in bakery foods: fortification of pizza with glucosinolates, carotenoids and phytosterols. Curr Res Food Sci 6:100437

Nilsson J, Olsson K, Engqvist G, Ekvall J, Olsson M, Nyman M, Åkesson B (2006) Variation in the content of glucosinolates, hydroxycinnamic acids, carotenoids, total antioxidant capacity and low-molecular-weight carbohydrates in Brassica vegetables. J Sci Food Agric 86(4):528–538

Nisar A, Jagtap S, Vyavahare S, Deshpande M, Harsulkar A, Ranjekar P, Prakash O (2023) Phytochemicals in the treatment of inflammation-associated diseases: the journey from preclinical trials to clinical practice. Front Pharmacol 14:1177050

Novais C, Molina AK, Abreu RM, Santo-Buelga C, Ferreira IC, Pereira C, Barros L (2022) Natural food colorants and preservatives: a review, a demand, and a challenge. J Agric Food Chem 70(9):2789–2805

Oliveira TC, Caleja C, Oliveira MBP, Pereira E, Barros L (2023) Reuse of fruits and vegetables biowaste for sustainable development of natural ingredients. Food Biosci 53:102711

Oriekhoe OI, Ilugbusi BS, Adisa O (2024) Ensuring global food safety: integrating blockchain technology into food supply chains. Eng Sci Technol J 5(3):811–820

Ozogul F, Cagalj M, Šimat V, Ozogul Y, Tkaczewska J, Hassoun A, Kaddour AA, Kuley E, Rathod NB, Phadke GG (2021) Recent developments in valorisation of bioactive ingredients in discard/seafood processing by-products. Trends Food Sci Technol 116:559–582

Pateiro M, Domínguez R, Munekata PE, Nieto G, Bangar SP, Dhama K, Lorenzo JM (2023) Bioactive compounds from leaf vegetables as preservatives. Food Secur 12(3):637

Patil PH, Pardeshi CV, Mahajan HS, Surana SJ (2022) Hemicellulose-based delivery systems: focus on pharmaceutical and biomedical applications. Hemicellulose biorefinery: a sustainable solution for value addition to bio-based products and bioenergy, pp 467–507

Petkowicz CL, Williams PA (2020) Pectins from food waste: characterization and functional properties of a pectin extracted from broccoli stalk. Food Hydrocoll 107:105930

Picchi V, Fibiani M, Scalzo RL (2020) Cauliflower. In: Nutritional composition and antioxidant properties of fruits and vegetables. Academic Press, pp 19–32

Podsędek A (2007) Natural antioxidants and antioxidant capacity of Brassica vegetables: a review. LWT-Food Sci Technol 40(1):1–11

Pokharel P, Ma Z, Chang SX (2020) Biochar increases soil microbial biomass with changes in extra-and intracellular enzyme activities: a global meta-analysis. Biochar 2:65–79

Press S (2015) Quinoa: the everyday superfood: 150 gluten-free recipes to delight every kind of eater. Callisto Media

Rafiuddin M, Suneetha WJ, Kumari BA (2021) Standardization and evaluation of cauliflower stalks powder incorporated products. Cutt Edge Res Agric Sci 7:9–16

Ramos-Andrés M, Aguilera-Torre B, García-Serna J (2021) Hydrothermal production of high-molecular weight hemicellulose-pectin, free sugars and residual cellulose pulp from discarded carrots. J Clean Prod 290:125179

Ribeiro TDC, Abreu JP, Freitas MCJ, Pumar M, Teodoro AJ (2015) Substitution of wheat flour with cauliflower flour in bakery products: effects on chemical, physical, antioxidant properties and sensory analyses. Int Food Res J 22(2):532

Šamec D, Urlić B, Salopek-Sondi B (2019) Kale (*Brassica oleracea* var. acephala) as a superfood: review of the scientific evidence behind the statement. Crit Rev Food Sci Nutr 59(15):2411–2422

Sanlier N, Guler SM (2018) The benefits of Brassica vegetables on human health. J Hum Health Res 1(1):1–13

Sedlar T, Čakarević J, Tomić J, Popović L (2021) Vegetable by-products as new sources of functional proteins. Plant Foods Hum Nutr 76:31–36

Shahbaz A, Abbasi BA, Iqbal J, Fatima I, Zahra SA, Kanwal S, Devkota HP, Capasso R, Ahmad A, Mahmood T (2021) Chemical composition of *Gastrocotyle hispida* (Forssk.) bunge and *Heliotropium crispum* Desf. and evaluation of their multiple in vitro biological potentials. Saudi J Biol Sci 28(11):6086–6096

Shahidi F, Ambigaipalan P (2015) Phenolics and polyphenolics in foods, beverages and spices: antioxidant activity and health effects – a review. J Funct Foods 18:820–897

Sharoba AM, Farrag MA, El-Salam A (2013) Utilization of some fruits and vegetables wastes as a source of dietary fibers in cake making. J Food Dairy Sci 4(9):433–453

Shinali TS, Zhang Y, Altaf M, Nsabiyeze A, Han Z, Shi S, Shang N (2024) The valorization of wastes and byproducts from cruciferous vegetables: a review on the potential utilization of cabbage, cauliflower, and broccoli byproducts. Food Secur 13(8):1163

Shonte TT, Duodu KG, de Kock HL (2020) Effect of drying methods on chemical composition and antioxidant activity of underutilized stinging nettle leaves. Heliyon 6(5)

Sigurdson GT, Tang P, Giusti MM (2017) Natural colorants: food colorants from natural sources. Annu Rev Food Sci Technol 8(1):261–280

Sikorska-Zimny K, Beneduce L (2021) The glucosinolates and their bioactive derivatives in Brassica: a review on classification, biosynthesis and content in plant tissues, fate during and after processing, effect on the human organism and interaction with the gut microbiota. Crit Rev Food Sci Nutr 61(15):2544–2571

Singh S, Devi MB (2015) Vegetables as a potential source of nutraceuticals and phytochemicals: a review. Int J Med Pharm Sci 5(2):1–14

Siró I, Kápolna E, Kápolna B, Lugasi A (2008) Functional food. Product development, marketing and consumer acceptance—a review. Appetite 51(3):456–467

Śmiechowska A, Bartoszek A, Namieśnik J (2010) Determination of glucosinolates and their decomposition products—indoles and isothiocyanates in cruciferous vegetables. Crit Rev Anal Chem 40(3):202–216

Sorrenti V, Burò I, Consoli V, Vanella L (2023) Recent advances in health benefits of bioactive compounds from food wastes and by-products: biochemical aspects. Int J Mol Sci 24(3):2019

Stojceska V, Ainsworth P, Plunkett A, İbanoğlu E, İbanoğlu Ş (2008) Cauliflower by-products as a new source of dietary fibre, antioxidants and proteins in cereal based ready-to-eat expanded snacks. J Food Eng 87(4):554–563

Sturm C, Wagner AE (2017) Brassica-derived plant bioactives as modulators of chemopreventive and inflammatory signaling pathways. Int J Mol Sci 18(9):1890

Tahir MH, Mubashir T, Schulze M, Irfan RM (2022) Thermochemical conversion of cabbage waste to bioenergy and bio-chemicals production. Int J Energy Res 46(14):20206–20215

Tlais AZA, Fiorino GM, Polo A, Filannino P, Di Cagno R (2020) High-value compounds in fruit, vegetable and cereal byproducts: an overview of potential sustainable reuse and exploitation. Molecules 25(13):2987

Tolve R, Cela N, Condelli N, Di Cairano M, Caruso MC, Galgano F (2020) Microencapsulation as a tool for the formulation of functional foods: the phytosterols' case study. Food Secur 9(4):470

Tsimogiannis D, Oreopoulou V (2019) Classification of phenolic compounds in plants. In: Polyphenols in plants. Academic Press, pp 263–284

Tukassar A, Shukat R, Butt MS, Nayik GA, Ramniwas S, Al Obaid S, Ali Alharbi S, Ansari MJ, Konstantinos Karabagias I, Sarwar N (2023) Compositional profiling and sensory analysis of cauliflower by-products-enriched muffins. Food Sci Nutr 11(10):6020–6031

Ubando AT, Felix CB, Chen WH (2020) Biorefineries in circular bioeconomy: a comprehensive review. Bioresour Technol 299:122585

Van Breusegem F, Mittler R (2023) Oxidative stress responses in plants. Elsevier

Verbeke W (2006) Functional foods: consumer willingness to compromise on taste for health? Food Qual Prefer 17(1–2):126–131

Wadhwa M, Bakshi MPS, Makkar HPS (2016) Wastes to worth: value added products from fruit and vegetable wastes. CABI Rev 2015:1–25

Wani TA, Sood M (2014) Effect of incorporation of cauliflower leaf powder on sensory and nutritional composition of malted wheat biscuits. Afr J Biotechnol 13(9)

Wani TA, Monica Sood MS, Amin QA, Nuzhat Wani NW, Harleen Kour HK (2013) Nutritional and organoleptic evaluation of noodles prepared by supplementation with cauliflower leaves

Yoneyama T, Suzuki A (2020) Light-independent nitrogen assimilation in plant leaves: nitrate incorporation into glutamine, glutamate, aspartate, and asparagine traced by 15N. Plan Theory 9(10):1303

Zahra N, Iqbal J, Arif M, Abbasi BA, Sher H, Nawaz AF, Yaseen T, Ydyrys A, Sharifi-Rad J, Calina D (2023) A comprehensive review on traditional uses, phytochemistry and pharmacological properties of *Paeonia emodi* Wall. ex Royle: current landscape and future perspectives. Chin Med 18(1):23

Zhang L, Tan Q, Lee R, Trethewy A, Lee YH, Tegeder M (2010) Altered xylem-phloem transfer of amino acids affects metabolism and leads to increased seed yield and oil content in Arabidopsis. Plant Cell 22(11):3603–3620

Zhang H, Liang Y, Li X, Kang H (2020) Antioxidant extract from cauliflower leaves effectively improve the stability of pork patties during refrigerated storage. J Food Process Preserv 44(7):e14510

Bioactive Phytochemicals from Watercress (*Nasturtium officinale R. Br.*) By-products

Süleyman Doğu, Turan Akdağ, Zeliha Üstün Argon, and Zinar Pınar Gümüş

Contents

Abstract In recent years, phytochemicals have been chemically and biologically investigated, and their effects on human health have been intensively researched. Aquatic plants are the main producers of the aquatic environment and ecology. *Nasturtium officinale R. Br.* is a green leafy plant rich in omega-3 fatty acids and is one of the most important ecological and economic plants. *Nasturtium officinale R. Br.* is a plant food with high pharmacological potential and a rich chemical composition. The plant's considerable group of bioactive complexes are vitamins, isothiocyanates, glucosinolates, polyphenols, and carotenoids. Nowadays, one of the

S. Doğu · T. Akdağ · Z. Ü. Argon (✉)
Necmettin Erbakan University, Meram Vocational School, Medicinal and Aromatic Plants Program, Medicinal and Cosmetic Plants Application and Research Center, Meram, Turkey
e-mail: sdogu@erbakan.edu.tr

Z. P. Gümüş
Central Research Testing and Analysis Laboratory Research and Application Center, Ege-Matal, Ege University, Izmir, Turkey
e-mail: zinar.pinar.gumus@ege.edu.tr

M. F. Ramadan (ed.), *Bioactive Phytochemicals in By-products from Leaf, Stem, Root and Tuber Vegetables*, https://doi.org/10.1007/978-3-031-80700-8_4

significant problems for various industrial sectors of the world is wastewater generation. *Nasturtium officinale R. Br.* is primarily used in salads, juices, soups manufacturing, and fresh consumption. Besides, it may be cooked and consumed as a vegetable. Nowadays, several products made with *Nasturtium officinale R. Br.,* such as noodles, pesto, juices, salt, and smoothies, are sold in markets in many countries. Also, different meat dishes and pasta are used by their producers.

Keywords *Nasturtium officinale* · Bioactive complex · Polyphenols · Human health

1 Introduction

Since prehistoric times, nature has been a rich source of many drugs and pharmaceutical pathways to treat many diseases, including cancer and metabolic diseases. In recent years, phytochemicals have been chemically and biologically investigated, and their effects on human health are being intensively researched.

They are found in the aquatic environment in different sizes and shapes and have beautiful visual appearances. As a macrohydrophyte plant, *Nasturtium officinale R. Br.* grows in constantly flowing, clean, cold freshwater. The freshwater plant *Nasturtium officinale R. Br.* is a semi-submerged, rhizomatous, and perennial hydrophyte belonging to the genus *Nasturtium* (Günlü and Kesici 2002). *Nasturtium officinale R. Br.* varies depending on the pH of the soil. It cannot grow well under basic conditions. It has soft, juicy, feathery leaves and cluster-shaped white flowers blooming between April and May as a cold-resistant plant. *Nasturtium officinale R. Br.* is a green leafy plant rich in omega-3 fatty acids and one of the most important ecological and economic plants. Studies have also conducted research on these plants, especially in health, environment, and biotechnology. Thus, these plants' activities and properties will be better identified in the present history (Doğan 2020).

Due to its valuable nutritional properties, *Nasturtium officinale R. Br.* is accepted by the EFSA (European Food Safety Authority) as a safe plant and recorded in the "Leafy vegetables, medicinal plants and edible flowers" section (Klimek-Szczykutowicz et al. 2018). *Nasturtium officinale R. Br.* leaves are used as salad greens or can be consumed as a regular processed vegetable (Gonçalves et al. 2009).

Nasturtium officinale R. Br. is a plant food with high pharmacological potential with its rich chemical composition. The considerable group of bioactive complexes in the plant are vitamins, isothiocyanates, glucosinolates, polyphenols, and carotenoids (Jeon et al. 2017). This plant, which has high nutritional quality, is quite useful. It has a high mineral and vitamin content. Its calorie capacity is low and includes a sufficient amount of fiber.

Moreover, *Nasturtium officinale R. Br.* is consumed to protect kidney and digestive health and clean the toxins from the body (Shahani et al. 2017). It has also been reported that it is suitable for pain, ulcers, jaundice, and fever. It is considered an excellent functional food for cancer prevention. The traditional medicine of some countries applied extensive uses of *Nasturtium officinale R. Br.* Also, Turkish folk medicine uses *Nasturtium officinale R. Br.* to relieve abdominal pain. Nowadays, manufacturers recognize the potential value of *Nasturtium officinale R. Br.* in contributing to healthy foods and cosmetic yields (Panahi et al. 2021).

2 Economic Values of *Nasturtium officinale R. Br.* By-product

As commonly described as a crop for consumption, *Nasturtium officinale R. Br.* is used in salads, juice, soup, or as a vegetable. Mature *Nasturtium officinale R. Br.* includes small amounts of fats and sugars, vitamin E, minerals and vitamins B, and folic acid. The proportions of iron (2 mg/100 g), protein (3%), vitamin C (62 mg/100 g), and carotenes are relatively high. The peppery pungency of the foliage and stems comes from glucosinolate (Cabicompendium 2022). The leaves are traditionally used for odontalgic, diuretic, depurative, stomach, and expectorant events (Chiej 1994).

A study reported that *Nasturtium officinale R. Br.* extract has flavonoid and phenolic contents of 96.2 mg gallic acid and 63.2 mg catechin, respectively. These results showed that *Nasturtium officinale R. Br.* extract has antioxidant properties from catching free radicals and metal chelating. The study also declared that *Nasturtium officinale R. Br.* extract could protect against free radical-related diseases (Bahramikia and Yazdanparast 2010). *Nasturtium officinale R. Br.* is also cooked and used raw for medicinal conditions such as diabetes, lung diseases, immune deficiency, asthma, and chronic inflammation (Zeb 2015; Ozen 2009). From knowledge, *Nasturtium officinale R. Br.* (especially isothiocyanates) could prevent the onset of lung, stomach, pancreatic, and breast cancer. This feature may lead to new avenues of research in the pharmaceutical industry, with the possibility of having pharmacognostic effects on cancer prevention (Ding et al. 1998). *Nasturtium officinale R. Br.* is considered to have disinfectant attributes. It is effective when included in acne-healing subjects. The extract of *Nasturtium officinale R. Br.* is used for cleansing external ulcers and in night creams for freckles. *Nasturtium officinale R. Br.* extract is utilized as a moisture regulator, perfuming agent, and skin activator (Michalun and DiNardo 2014). From this knowledge, it is proposed that *Nasturtium officinale R. Br.* may be used widely in pharmacology, cosmetics, medicine, the food industry, and other industries.

Nowadays, one of the significant problems for various industrial sectors of the world is wastewater generation. It is revealed that *Nasturtium officinale R. Br.* may be used in wastewater purification (effective in improving water quality and purifying water) (Torbati et al. 2014). *Nasturtium officinale R. Br.* has also been shown to accumulate a large amount of heavy metals such as Zn, Cu, and Cd (Duman and Ozturk 2010).

3 Chemical Composition and Bioactive Compounds of *Nasturtium officinale R. Br.* Extracts from By-products

The major bioactive contents of *Nasturtium officinale R. Br.* are found to be glucosinolates and phenolics. Basic phenolic substances were demonstrated to be isorhamnetin and chlorogenic acid (Kyriakou et al. 2022). A study investigated the chemical composition of isorhamnetin effects on health, and it was reported that isorhamnetin has anti-inflammatory, antitumor, antioxidant, and anti-osteoporosis effects (Gong et al. 2020).

Basic flavonoids of *Nasturtium officinale R. Br.* are found to be as kaempferol, quercetin, isorhamnetin, caftaric acid, chlorogenic acid, cafeoiltartaric acid, and quercetin-3-*O*-rutinoside (Yalçınkaya et al. 2019). From the literature, the phytochemical analysis of *Nasturtium officinale R. Br.* showed that it contains terpenoids/steroids, flavonoids, alkaloids, saponins, proteins, essential/volatile oils, tannins, folic acid, glycosides, vitamin K, E, A, and C, iron, iodine, calcium potassium, and sodium (Haro et al. 2018; Gonçalves et al. 2009; Pandey et al. 2018; Al-Snafi 2019). The phenolics are gallic acid, caffeic acid, quercetin-3-*O*-rutinoside, dicaffeoyltartaric acid, isorhamnetin, and chlorogenic acid. Isorhamnetin and quercetin-3-*O*-rutinoside display about 70% of the total phenolics composition of *Nasturtium officinale R. Br.* and are frequently observed to decrease the inflammatory and cardiovascular risks (Aires et al. 2013). Quercetin-3-*O*-rutinoside and caffeic acid were directly included in elevating the antioxidant activity and reducing free radicals. Clemente et al. (2019) demonstrated that *Nasturtium officinale R. Br.* may protect against oxidative stress.

The essential oil of *Nasturtium officinale R. Br.* may be obtained from the leaves and flowers but mainly from the roots (1.2%, 1.0%, and 1.5% (v/w), respectively). The major volatile compounds described in *Nasturtium officinale R. Br.* were α-terpinolene, limonene, caryophyllene oxide, and p-cymene-8-ol (Klimek-Szczykutowicz et al. 2018). In a study, Gerard and Magno (2011) declared that flavonoids, saponins, terpenoids, and other glycosides of *Nasturtium officinale R. Br.* have an intense antibacterial activity and may be used as a treatment for bacterial-related skin diseases. A recent study proposed that the juice of *Nasturtium officinale R. Br.*, due to its chemical composition rich in polyphenols such as caffeoylmalic and feluric acids, may play a role in the prevention of obesity and type II diabetes (Pinela et al. 2018).

In a study, the composition of the essential oils of stems, leaves, and flowers of *Nasturtium officinale R. Br.* was determined as follows: limonene, α-pinene, *trans*-β-ociemen, β-pinene, β-myrcene, octanal, α-terpinolene, α-copaen, *p*-cymene-8-ol, myristicin, β-caryophyllene, δ-cadinol, and β-sesquiphellandrene. In the study, they also concluded that *Nasturtium officinale R. Br.* species may be an important source for radical scavenging. The extracts and essential oil from the different parts of *Nasturtium officinale R. Br.* may be used as natural sources for different sectors (Amiri 2012).

4 Biological and Functional Properties of *Nasturtium officinale R. Br.* from By-products

High-fat rats were investigated to determine the effect of *Nasturtium officinale R. Br.* (hydroalcoholic) extract on lipid profile. In the research, 500 mg/kg bw/day administration of the extract of *Nasturtium officinale R. Br.* (via intragastric method) was performed, and lipid profile was measured as follows: lowered total cholesterol, triglyceride, and low-density lipoprotein cholesterol (34.2%, 30.1%, and 52.9%, respectively) and elevated high-density lipoprotein cholesterol. Moreover, decreased aspartate aminotransferase and alanine aminotransferase levels were observed in the high-fat diet groups (Bahramikia and Yazdanparast, 2008). In another study, similar findings were reported (Yazdanparast et al. 2008). From the studies, it could be proposed that *Nasturtium officinale R. Br.* has hypolipidemic effects on metabolism.

An experimental study demonstrated that methanol extract of *Nasturtium officinale R. Br.* caused an important decline in the glucose level after a week of treatment (Hoseini et al. 2009). In another study performed by Fenton-Navarro et al., the physicochemical properties (%) of *Nasturtium officinale R. Br* extracts were determined as follows: carbohydrates 9.26–25.44 g, dry material 0.5–1.61, crude fiber 12.06–15.42, ether extract 0.27–20.35, rough protein 33.51–47.91, cinder 13.67–23.64, and moisture percentage at 98.39–99.53 (Fenton-Navarro et al. 2018). Studies have shown that the hepatoprotective anticancer, antihyperlipidemic, and nephroprotective properties of *Nasturtium officinale R. Br.* were observed in vitro and in vivo experimental studies (Sadeghi et al. 2019). Additionally, some studies indicate that the *Nasturtium officinale R. Br* has significant antioxidant and anti-inflammatory activities under certain conditions (Sadeghi et al. 2014).

The *Nasturtium officinale R. Br.* (hydroalcoholic) extract's anticancer effect has been investigated against the growth and development of cancerous cells (Hela) and fibroblasts. The extract was applied at concentrations between 0.625 and 2 mg/ml, and, in the study examining cell death rates after 24, 48, and 72 hours of incubation, it was observed that the survival rate of Hela cells declined over time with elevating concentrations of *Nasturtium officinale R. Br.* extract (Moradiu et al. 2017). Similar findings were observed in previous studies (Rose et al. 2005). *Nasturtium officinale R. Br.* extracts were applied to dermatological topics (e.g., prevention of psoriasis). In chemical and thermal burn cases, *Nasturtium officinale R. Br.* oil was used on rabbits for wound healing. In the study, normal architecture rapidly healed and reduced the closure time of related injuries (Zinadah 2008). The protection effects of IAMG (indole 3-acetonitrile-4-methoxy-2-S-β-D-glucopyranoside) contained in *Nasturtium officinale R. Br.* were examined with regard to photodamage, which was caused by ultraviolet B on normal human skin. At the same time, UVB-induced elevation in MMP-1 and reduction in type I-procollagen by IAMG trial showed that IAMG increased dermal fibroblast migration. It was also reported that IAMG obtained from *Nasturtium officinale R. Br.* may strongly reduce UVB-induced photodamage (Kim and Bae 2019).

5 Food and Non-food Applications of Bioactive Compounds from *Nasturtium officinale R. Br.* By-products

As a semiaquatic plant, *Nasturtium officinale R. Br.* is well known for its attribution to human health. Because of its chemical composition and bioactive compounds, *Nasturtium officinale R. Br.* may be utilized in food and non-food applications. *Nasturtium officinale R. Br.* has not yet been fully used in the food industry. Moreover, it has a lot of applications in local gastronomy (mainly leaves are included in salad manufacturing).

Due to the accumulation capacity of *Nasturtium officinale R. Br.,* heavy metals such as Cu, Zn, and Cd are biologically filtered and can contribute to the protection of the environment (Blüthner 2020). *Nasturtium officinale R. Br.* was reflected as a powerful natural antioxidant agent with applications in medicine, the food industry, and gastronomy. Kaempferol and quercetin are described in *Nasturtium officinale R. Br.* and may be used in the aging processes, atherosclerosis, and prevention of DNA damage (Mazandarini et al. 2012). Zafar et al. (2017) assumed that *Nasturtium officinale R. Br.* extract has antimicrobial activity toward *Bacillus cereus, Klebsiella pneumonia*, and *Enterococcus faecalis*.

Furthermore, saponins, glycosides, terpenoids, and other flavonoids belonging to *Nasturtium officinale R. Br.* have a potent antibacterial activity on pseudomonas aeruginosa and *Staphylococcus aureus* and may be advised as an option for bacterial skin diseases (Gerard and Magno 2011). *Nasturtium officinale R. Br.* is primarily used in salads, juices, soups manufacturing, and fresh consumption. Besides, it may be cooked and consumed as a vegetable (Cruz et al. 2009). Nowadays, several products made with *Nasturtium officinale R. Br.,* such as noodles, pesto, juices, salt, and smoothies, are sold in markets in many countries.

References

Aires A, Carvalho R, Rosa EAS, Saavedra MJ (2013) Phytochemical characterization and antioxidant properties of babyleaf watercress produced under organic production system. CYTA-J Food 11(4):343–351

Al-Snafi AE (2019) A review on *Lawsonia inermis*: a potential medicinal plant. Int J Curr Pharm Res 11(5):1–13

Amiri H (2012) Volatile constituents and antioxidant activity of flowers, stems and leaves of Nasturtium officinale R. Br. Nat Prod Res 26(2):109–115

Bahramikia S, Yazdanparast R (2008) Effect of hydroalcoholic extracts of *Nasturtium officinale* leaves on lipid profile in high-fat diet rats. J Ethnopharmacol 115(1):116–121

Bahramikia S, Yazdanparast R (2010) Antioxidant efficacy of Nasturtium officinale extracts using various in vitro assay systems. J Acupunct Meridian Stud 3(4):283–290. https://doi.org/10.1016/S2005-2901(10)60049-0

Blüthner WD (2020) *Nasturtium officinale R.Br.*: Watercress. In: Novak J, Blüthner WD (eds) Medicinal, aromatic and stimulant plants, Handbook of plant breeding, vol 12. Springer, Cham

Cabicompendium (2022) 35646, CABI Compendium. CABI International, Nasturtium officinale (watercress). https://doi.org/10.1079/cabicompendium.35646

Chiej R (1994) Encyclopaedia of medicinal plants. MacDonald, Edinburgh

Clemente M, Miguel M, Gribner C, Moura PF, Angelica A et al (2019) Watercress, as a functional food, with protective effects on human health against oxidative stress: a review study. Int J Med Plants Nat Prod 5(3):12–16

Cruz RMS, Vieira MC, Silva CLM (2009) Effect of cold chain temperature abuses on the quality of frozen watercress (*Nasturtium officinale R. Br.*). J Food Eng 94(1):90–97

Ding J, Mimi CYU, Jean HH, Siev-hong L, Fung-lung C (1998) Total isothiocyanate content in cooked vegetables frequently consumed in Singapore. J Agric Food Chem Singapore 46(3):1055–1058

Doğan M (2020) Su teresinin (*Nasturtium officinale R. Br.*) beslenme-diyet potansiyeli ve antioksidan özellikleri: bir derleme. IAAOJ Health Sci 6(3):222–233

Duman F, Ozturk F (2010) Nickel accumulation and its effect on biomass, protein content and antioxidative enzymes in roots and leaves of watercress *(Nasturtium officinale R. Br.)*. J Environ Sci 22(4):526–532. https://doi.org/10.1016/S1001-0742(09)60137-6

Fenton-Navarro B, UrquizaMartínez MV, Fiscal Castro BB, Medrano Castillo OI, López-Rodríguez M, Padilla Arellanes S et al (2018) Antioxidant and hypoglycemic effects of watercress (*Nasturtium officinale*) extracts in diabetic rats. Afr J Tradit Complement Altern Med 1(2):68–79

Gerard LP, Magno CP (2011) Antibacterial activity of extracts of twelve common medicinal plants from the Philippines. J Med Plants Res 5(16):3975–3981

Gonçalves EM, Cruz RMS, Abreu M, Brandão TRS, Silva CLM (2009) Biochemical and colour changes of watercress (Nasturtium officinale R. Br.) during freezing and frozen storage. J Food Eng 93(1):32–39

Gong G, Guan YY, Zhang ZL, Rahman K, Wang SJ, Zhou S et al (2020) Isorhamnetin: a review of pharmacological effects. Biomed Pharmacother 128:110301

Günlü A, Kesici E (2002) Su Teresi'nin (*Nasturtium officinale* R. Br.) Besinsel Değeri ile İlgili Bazı Kimyasal Bileşenlerin Mevsimsel Değişimi. SDÜ Fen Bil Ens Der 6(2):69–79

Haro G, Iksen I, Rumanti RM, Marbun N, Sari RP, Gultom RPJ (2018) Evaluation of antioxidant activity and minerals value from watercress (*Nasturtium officinale R. Br.*). Rasayan J Chem 11(1):232–237

Hoseini HF, Gohari AR, Saeidnia S, Majd NS, Hadjiakhoondi A (2009) The effect of Nasturtium officinale on blood glucose in diabetic rats. Pharmacologyonline 3:866–871

Jeon J, Bong SJ, Park JS, Park YK, Arasu MV, Al-Dhabi et al (2017) De novo transcriptome analysis and glucosinolate profiling in watercress *(Nasturtium officinale R. Br.)*. BMC Genomics 18(1):401

Kim Y, Bae KS (2019) Protective effects of indole 3-acetonitrile-4-methoxy-2-S-β-D-glucopyranoside from *Nasturtium officinale R. Br.* against ultraviolet B-induced photodamage in normal human dermal fibroblasts. Natural product communications, 1–4

Klimek-Szczykutowicz M, Szopa A, Ekiert H (2018) Chemical composition, traditional and professional use in medicine, application in environmental protection, position in food and cosmetics industries, and biotechnological studies of Nasturtium officinale (watercress) – a review. Fitoterapia 129:283–292. https://doi.org/10.1016/j.fitote.2018.05.031

Kyriakou S, Michailidou K, Amery T, Stewart K, Winyard PG, Trafalis DT et al (2022) Polyphenolics, glucosinolates and isothiocyanates profiling of aerial parts of *Nasturtium officinale (Watercress)*. Front Plant Sci 13:998755. https://doi.org/10.3389/fpls.2022.998755

Mazandarini M, Momeji A, Moghaddam PZ (2012) Evaluation of phytochemical and antioxidant activities from different parts of *Nasturtium officinale R. Br* in Mazandaran. Iranian J Plant Physiol 3(2):659–664

Michalun MV, DiNardo JC (2014) Skin care and cosmetic. Ingredients Dictionary Ed. 4. Milady Publishing Company, p 317

Moradiu R, Ebrahimi S, Taravati A, Asrardel F, Khorasani HR, Aghajanpour M (2017) Cytotoxic effects of the hydroalcoholic extract of *Rorippa nasturtium aquaticum* on hela cell line. IBBJ 3(2):73–79

Ozen T (2009) Investigation of antioxidant properties of *Nasturtium officinale* (watercress) leaf extracts. Acta Pol Pharm 66(2):187–193

Panahi Kokhdan E, Khodabandehloo H, Ghahremani H, Doustimotlagh AH (2021) A narrative review on therapeutic potentials of watercress in human disorders. Evid Based Complement Alternat Med 2021:5516450. https://doi.org/10.1155/2021/5516450

Pandey Y, Bhatt SS, Debbarma N (2018) Watercress (Nasturtium officinale): a potential source of nutraceuticals. Int J Curr Microbiol App Sci 7(02):2685–2691

Pinela J, Prieto MA, Barros L, Carvalho AM, Oliveira MBPP et al (2018) Cold extraction of phenolic compounds from watercress by high hydrostatic pressure: process modelling and optimization. Sep Purif Technol 192:501–512

Rose P, Huang Q, Ong CN, Whiteman M (2005) Broccoli and watercress suppress matrix metalloproteinase9 activity and invasiveness of human MDA-MB-231 breast cancer cells. Toxicol Appl Pharmacol 209(2):105–113

Sadeghi H, Mostafazadeh M, Sadeghi H, Naderian M, Barmak MJ, Talebianpoor MS et al (2014) In vivo antiinflammatory properties of aerial parts of Nasturtium officinale. Pharm Biol 52(2):169–174

Sadeghi H, Azarmehr N, Razmkhah F, Sadeghi H, Danaei N, Omidifar N et al (2019) The hydroalcoholic extract of watercress attenuates protein oxidation, oxidative stress, and liver damage after bile duct ligation in rats. J Cell Biochem 120(9):14875–14884

Shahani S, Behzadfar F, Jahani D, Ghasemi M, Shaki F (2017) Antioxidant and antiinflammatory effects of *Nasturtium officinale* involved in attenuation of gentamicin-induced nephrotoxicity. Toxicol Mech Methods 27(2):107–114

Torbati S, Khataee AR, Movafeghi A (2014) Application of watercress (*Nasturtium officinale R. Br.*) for biotreatment of a textile dye: investigation of some physiological responses and effects of operational parameters. Chem Eng Res Des 92(10):1934–1941. https://doi.org/10.1016/j.cherd.2014.04.022

Yalçinkaya E, Özgüç S, Törer YO, Zeybek U (2019) The importance of the medicinal plant *Nasturtium officinale* l. in the anticancer activity research. Health Sci Quart 3(2):159–164. https://doi.org/10.26900/jsp.3.016

Yazdanparast R, Bahramikia S, Ardestani A (2008) *Nasturtium officinale* reduces oxidative stress and enhances antioxidant capacity in hypercholesterolaemic rats. Chem Biol Interact 172:176–184

Zafar R, Zahoor M, Shah AB, Majid F (2017) Determination of antioxidants and antibacterial activities, total phenolic, polyphenol and pigment contents in *Nasturtium officinale*. Pharmacology 1:11–18

Zeb A (2015) Phenolic profile and antioxidant potential of wild watercress (*Nasturtium officinale* L.). Springer Plus 4:714. https://doi.org/10.1186/s40064-015-1514-5

Zinadah OA (2008) Effects of watercress oil on the thermal and chemical burn injuries in rabbits. *JKAU*. Med Sci 15:3–17

Bioactive Phytochemicals from Spinach (*Spinacia oleracea* L.) By-products

Moncef Chouaibi

Contents

Abstract This chapter shows the physicochemical composition of spinach by-products, the phytochemical compounds, and the potential uses of spinach by-products in food industries. The results indicate that the spinach leaf is rich in

M. Chouaibi (✉)
Higher School of Food Industries of Tunisia, University of Carthage, Carthage, Tunisia
e-mail: moncef.chouaibi@esiat.ucar.tn

M. F. Ramadan (ed.), *Bioactive Phytochemicals in By-products from Leaf, Stem, Root and Tuber Vegetables*, https://doi.org/10.1007/978-3-031-80700-8_5

minerals and sugars. Interestingly, these findings suggest that spinach extract constitutes a high concentration of phytochemical compounds consisting mainly of chlorophyll and lutein pigments. There are many extracted methods of these compounds, such as conventional maceration, supercritical carbon dioxide, ultrasonic process, and drying techniques. Therefore, the results reveal that supercritical carbon dioxide is the most preferable technique for chlorophyll and lutein extractions. Besides, spinach extracts were incorporated into food products such as cheese, juices, beverages, cereal products, and pasta.

Keywords Spinach · Phytochemical compounds · Extraction methods · Potential uses

1 Introduction

Spinach (*Spinacia oleracea* L.) is an imperative cool season annual leaf vegetable, now cultivated throughout the temperate regions of the world, especially in North Africa regions (Tunisia, Algeria, and Egypt). It is most productive in cool seasons and climates. It is rich in minerals, especially iron, potassium, and calcium. However, calcium is probably unavailable, as it constitutes a complex with oxalic acid, leading to the creation of calcium oxalate. Interestingly, spinach leaves can be consumed fresh as a salad, purée, and juice, boiled or baked into various dishes.

Spinach (*Spinacia oleracea* L.) is a widely distributed vegetable throughout the world. The cultivated area of spinach in Tunisia attained 300 hectares, with a yield of 2.5 tons per hectare. This plant is a potted plant, annual or biannual, from the family of *Chenopodiaceae* or *Amaranthaceae* with classifications. Originating from Iran, it is grown in all regions and temperatures for its nutritional quality. It is known to be a good source of pigments (lutein and its isomer zeaxanthin) and chlorophyll, as mentioned by Derrien et al. (2017).

Interestingly, chlorophyll and lutein pigments are compounds important to life on earth, being essential for photosynthesis. They used the color of vegetables and several fruits. The spinach extracts thus obtained can be used on the ingredients market for various purposes, such as natural and biological coloring, as well as components that can be incorporated into the composition of functional foods. Indeed, these two pigments were beneficial to human health (antioxidant, anti-inflammatory, anti-mutagenic), as indicated by numerous studies (Özkan and Ersus Bilek 2015).

Indeed, just like a healthy diet, these compounds are responsible for beneficial effects on physiological functions by increasing well-being and reducing the risks of the various pathologies mentioned above. This phenomenon is explained by the presence of phytochemical compounds contained in plants. The latter, in fact, possess several groups of molecules with proven biological activity acting on the cells of our body. These molecules belong, for example, to the family of polyphenols, terpenoids, alkaloids, and sulfur compounds. Phytochemical compounds are known for their ability to stimulate the immune response and also their antioxidant, anti-free radical, and anti-inflammatory properties.

Thus, it is interesting to valorize the bioactive compounds contained in vegetables by using products. Vegetable waste downgraded or rejected by the market gardening industry includes a high amount of functional compounds that are beneficial to human health but remain little exploited. The valorization of these plant by-products is, therefore, likely to provide extracts rich in functional compounds, giving them beneficial properties for human and animal health. Spinach is a leafy vegetable rich in interesting bioproducts such as polyphenols, lutein, and chlorophyll pigments, possessing antioxidant, anti-inflammatory, and antigenotoxic activities essentially in the eyes for lutein and in the colon for chlorophyll.

There are many techniques applied to extract the bioactive compounds, such as a green extraction technique, using ethanol and water as extraction solvents, which were applied to extract these interesting molecules from spinach by-products. These factors gave a high yield of chlorophyll and lutein pigments. Nevertheless, this technique was not efficient for the isolation of the two molecules. Likewise, this technique consumed high quantities of chemical reagents and solvents. However, some works applied supercritical carbon dioxide using the optimization of physicochemical parameters such as flow rate, solid-liquid ratio, concentration time, extraction temperature, and pressure extraction. The yield of these pigments was increased compared to conventional techniques using natural solvents.

Given the hydrophobic nature of several bioactive molecules, the extraction of natural substances requires the use of chemical solvents, which remains the most used technique in the industry. Indeed, they dissolve solutes of interest easily, and their volatility greatly facilitates their elimination at the end of the process. Yet, these organic solvents are harmful and potentially carcinogenic, in addition to being involved in environmental pollution. The extracts obtained will have a high amount of functional compounds, which will allow them to ameliorate health as well as to integrate the functional ingredients market. The classic methods that will be applied for the extraction of phytochemical compounds were selected for their innovative characteristics and for meeting the ecological constraints best.

2 Objectives

The agro-industry produces a significant amount of spinach residues which can be valorized for their interested compounds, particularly carotenoids (lutein), chlorophyll pigments, and polyphenols, which can be destined for functional food, cosmetics, and pharmaceutical products. Petrochemical solvents can also have carcinogenic, mutagenic, and reprotoxic properties, i.e., affecting reproductive capacities through an action on the gonads or acting as an endocrine disruptor. The implementation of more sustainable chemistry in the pharmaceutical and agri-food world, therefore, aims to limit the harmful impact of toxic solvents still used intensively by manufacturers. Indeed, we have observed through the available literature that the production of lutein and chlorophyll on an industrial scale always involves the use of petrochemical solvents, allowing very efficient extraction of these compounds.

3 Chemical Composition of Spinach

Spinach is known to possess interesting biological activities such as antioxidant, anti-inflammatory, antiproliferative, anti-obesity, hypoglycemic, and finally, blood lipid-reducing properties (Djouadi and Derouiche 2021). In vitro and in vivo studies have been carried out in this direction on cellular, animal, and human models. Most often, these studies were carried out with aqueous or ethanolic extracts of spinach (antioxidant, anti-inflammatory effect), freeze-dried spinach extracts, and thylakoid extracts (anti-obesity, hypoglycemic effect, reduction of blood lipids), meals with fresh spinach, spinach extracts obtained with acetone, and glycolipid fractions (anti-cancer) (Roberts and Moreau 2016). Another study also indicated a modulation of the intestinal microbiota when taking spinach thylakoid extract (Montelius et al. 2013). These effects are explained by the composition of spinach, which contains 92% water, 2.9% proteins, 3.6% carbohydrates, and 0.4% fatty acids. Among these fatty acids, mono- and polyunsaturated fatty acids (linoleic, alpha-linolenic, and oleic acid) were mainly characterized. Some traces of saturated fatty acids (capric, myristic, and stearic acid) were also detected. Spinach also has a significant amount of fiber (2.2 g per 100 g of fresh spinach, or 9% of the recommended amount per day per serving). Finally, this plant has a high nutritional value due to the presence of high levels of minerals (Fe, Mg, and Mn) and vitamins (A, B9, K, 7VS). A 100 g portion of spinach allows you to reach or even exceed the daily recommendations for certain vitamins and minerals, as is the case for vitamin C (47% of daily requirements covered), vitamin K (604%), vitamin A (188%), folate (49%), Mg (20%), K (16%), and Fe (15%).

Spinach also contains phytochemicals such as carotenoids, polyphenols, and chlorophyll, as mentioned by Bunea et al. (2008). Phytochemical compounds are nutrients that are not essential for humans but are ubiquitously present in plants. They have health effects that have been proven by numerous authors (Roberts and Moreau 2016). Phenolic compounds are significantly present in spinach. Their proportions in spinach are 40–70% higher than in lettuce and salad, making it a very interesting source of bioactive compounds.

The chlorophyll content detected in spinach by-products is approximately 130 mg per 100 g of fresh spinach (Ferruzzi and Blakeslee 2007). The main carotenoid contained in spinach is lutein, the amount of which in this plant ranged from 3.9 and 9.5 mg per 100 g of fresh spinach compared to 6.4–15 mg per 100 g of fresh kale and 1.5–2.8 for broccoli (Burri 2013). The presence of lutein is masked by chlorophyll, which gives the leaf its green color. Lutein and chlorophyll are also involved in the health effects of spinach extracts through their antioxidant, anti-inflammatory, and anti-mutagenic effects.

4 Phytochemical Composition of Spinach By-products

4.1 Lutein

Fruit and vegetable residues generated by the market gardening food industry contain interesting compounds that can be incorporated into other products to generate extracts rich in bioactive compounds. Spinach transformation generates around 25% of waste, rich in bioactive compounds, notably lutein and chlorophyll (Derrien et al. 2017). A green extraction technique, using ethanol and water as extraction solvents, was applied to extract these compounds from spinach by-products. Different methods of optimization were chosen, such as response surface methodology, which was used to optimize the extraction parameters of lutein and chlorophyll using natural reagents. Various combinations of extraction parameters were evaluated to obtain an extraction with an extraction efficiency close to that obtained when using petrochemical solvents. The optimal extraction parameters were 93% ethanol concentration for 4.3 h at 43 °C and a solvent to raw materials ratio of 1/66 (Derrien et al. 2017). These independent variables or factors were validated with an experiment carried out with the optimal conditions determined by the model, and the findings indicated that the yield of lutein and chlorophyll was very good (96% for chlorophyll and 70% for lutein). The findings, therefore, stated the possibility of extracting lutein and chlorophyll from renewable materials and spinach by-products with the use of natural solvents as effectively as harmful petrochemical solvents.

Lutein extracts are often provided as a natural colorant (E161) or as ingredients for the composition of functional foods. Indeed, lutein has proven health benefits such as antioxidant, anti-inflammatory, and anti-mutagenic proteins and plays many key roles in the prevention of several diseases (Sharma et al. 2007). Lutein structure is made of conjugated double bonds, which are implicated in the reduction of light-related damages to the macula lutea, thus preventing age-related macular degeneration (AMD) (Carpentier et al. 2009). Lutein is also involved in the improvement of immune function, possibly inhibiting some cancers (Tanaka et al. 2012) and potentially impacting cognitive functions. Indeed, some authors have demonstrated a relationship between dementia in the elderly and lutein concentration in brain tissues (Johnson 2012). Currently, commercial lutein extracts are mainly produced from marigold flowers, which are rich in this molecule, with values of 21.6–97.6 mg/100 g (FW). Microalgae, a unicellular algal form containing large amounts of chlorophyll, are also a rich source of carotenoids.

4.2 Chlorophyll Pigments

Spinach by-products, like microalgae, cover high concentrations of chlorophyll amphiphilic compounds, giving spinach and many other plants their green color. Chlorophyll possesses several health benefits, as mentioned by many works (Ferruzzi and Blakeslee 2007; Sharma et al. 2007), and can be applied as an ingredient such as a green colorant. However, lutein and chlorophyll present similar polarity characteristics. Thus, their separation, one from the other to generate two distinct extracts or products, requires further purification steps, which are known to be time-consuming and costly at the industrial level.

Saponification is the standard analytical procedure to remove chlorophyll from green vegetables in order to prevent the interference problem between chlorophyll and carotenoids for chromatographic detection (Rodríguez-Bernaldo de Quirós and Costa 2006). Saponification can also be performed on an industrial scale in the extraction and purification procedures. For lutein production from marigolds, saponification improves the extraction yield by removing ionizable lipids and causing a softening and collapse of cell tissues. Saponification is also used for lutein production from microalgae. In this case, an initial alkaline treatment in an aqueous medium (4% aqueous KOH, 40 °C) is advantageously used to retain fatty acids and chlorophyll in the aqueous fraction. Chlorophyll-free carotenoids are then extracted with petrochemical solvents (Fernández-Sevilla et al. 2010). Hexane remains the most common solvent used for lutein extraction from marigolds or microalgae. However, its toxicity and recycling problems have led many to seek more eco-friendly extraction techniques. In this context, ethanol appears to be the safest and greenest solvent owing to its biodegradability, non-toxicity, and capacity for regeneration. The saponification conditions used in the separation process of carotenoid and chlorophyll result in the significant degradation of lutein.

5 Extraction Techniques

Different methods were applied to extract the lutein and chlorophyll compounds, such as conventional maceration, supercritical carbon dioxide, ultrasonic process, microwave-assisted extraction, and drying methods.

5.1 Conventional Maceration Method

This method aimed to obtain a safe and ecologically sustainable experiment to extract and separate the two pigments, lutein and chlorophyll, from spinach extracts. A saponification in an aqueous medium was applied to liberate the phytyl aliphatic chain from the porphyrin ring of chlorophyll. Polar chlorophyll derivatives were formed, remaining in the aqueous medium, whereas polar lutein remained in the plant residues and was extracted after that by maceration in 95% ethanol. Yet, this procedure may cause a significant degradation of lutein, resulting in low extraction yields. The saponification parameters were examined using an optimization procedure to assess the optimal conditions. These conditions for saponification were 16 h of aqueous maceration with 3.9% (W/V) sodium hydroxide at 60 °C, as mentioned by Derrien et al. This experiment can be easily scaled for high-volume commercial extraction of these interested compounds.

5.2 *Supercritical Carbon Dioxide*

Supercritical carbon dioxide technology is involved in the extraction of a large group of natural molecules (Fig. 1). The process optimization regarding the selection of the optimum extraction conditions requires a critical analysis. The predictive approach suggested that the solubility of lutein and chlorophyll was higher at 500 bar.

Industrial vegetable wastes contain high amounts of valuable molecules and constitute a valuable source of bioactive compounds providing health benefits. Spinach (*Spinacia oleracea*), a green leafy vegetable, is known for its valuable phytochemical content. However, its production yields about 25% of waste that can be valorized for the production of value-added food supplements. Optimization of sustainable extraction using supercritical CO_2 (SC-CO_2) was applied for the extraction of lutein and chlorophyll pigments, two healthy functional molecules, from spinach by-products. Box–Behnken experimental design was used to optimize the process variables: pressure (10, 30, and 50 MPa), co-solvent percentages (0, 5, and 10% (v/v), temperature (40, 50, 60 °C), and time (1, 3, 5 h). It was found that the

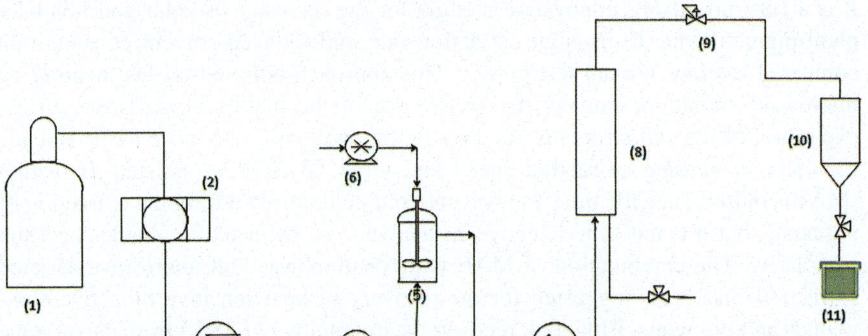

Fig. 1 Supercritical-CO_2 apparatus used for oil extraction from *Citrullus colocynthis* seeds: (1): Liquid CO_2 cylinder; (2): Microfilter; (3): Refrigerant; (4): Pump; (5): Mixer; (6): Pump; (7): Preheater; (8): Extractor; (9): Pressure regulator; (10): Cyclone separator; (11): B: beaker

optimum extraction parameters were at a temperature of 56 °C, extraction time of 3.6 h, pressure of 39 MPa, and 10% ethanol as cosolvent. Under these conditions, a yield of 72% lutein and 50% of chlorophyll was obtained (Derrian et al. 2017, 2019).

5.3 *Ultrasonic Process Extraction*

It uses ultrasonic waves to generate cavitation bubbles that improve extraction efficiency. Ultrasonic waves passing through a medium create alternating cycles of compression and decompression, which in turn cause the formed cavitation bubbles to compress and expand. As the bubbles grow larger, they can no longer be contained by the surface tension force and, therefore, collapse.

In this way, shear forces are generated, and they are capable of breaking the vegetable cell walls, facilitating the release of intracellular compounds. As reported, UAE is a technique suitable for the extraction of both polar and nonpolar compounds and also for those sensitive to heat as it is an athermic process, although known for its lower extraction efficiency compared to other nonconventional processes (e.g., SFE and PLE). Drosou et al. tested three different extraction methods (Soxhlet, MAE, and UAE) for the recovery of pigments from red grapes pomace, revealing the highest procyanidin (43,469 ppm) and anthocyanin contents (34,188 ppm) in UAE-assisted treatment combined with solvents.

5.4 *Microwave-Assisted Extraction*

It is a very promising, innovative method for the recovery of polar and nonpolar plant pigments due to its high extraction rate and reduced solvent consumption compared to conventional techniques. This approach is based on fast heating by microwave radiation, causing the breakdown of the plant cell walls due to the expansion of the cell structure. As a result, pigments migrate more easily outside the cells, increasing extraction speed and yield. Microwave-assisted extraction (MAE) could reduce the time and solvent volume required during the extraction of pigments, but it is not very selective, as analytes are extracted in a wide spectrum of polarity. The combination of MAE with vacuum (vacuum microwave-assisted extraction) has been suggested for the recovery of heat-sensitive bioactive compounds and pigments. Elik et al. recovered carotenoids (77.48%) from carrot juice processing waste using MAE at the optimal process conditions (165W of microwave power, 9.39 min of extraction time, and 8.06:1 g/g of oil to waste ratio).

5.5 Microwave Drying Technique

The drying kinetics of spinach was investigated in a microwave oven at various microwave output powers and sample amounts. Drying time decreased considerably with increased microwave output power and with a reduced sample amount of spinach as well as by using the microwave drying technique. This technique can be successfully utilized to dry spinach.

The effect of the microwave drying technique on moisture content, moisture ratio, drying rate, drying time, effective moisture diffusivity, and porosity of spinach (*Spinacia oleracea* L.) was investigated. By increasing the microwave output powers (180–900 W) and the sample amounts (25–100 g), the drying time decreased from 18 to 3.5 min and increased from 7.7 to 25 min, respectively. To determine the kinetic parameters, the drying data were fitted to various models based on the ratios of the differences between the initial and final moisture contents and equilibrium moisture content versus drying time. Among the models proposed, Page's model gave a better fit for all drying conditions applied. The activation energy was calculated using an exponential expression based on the Arrhenius equation. The relationship between the drying rate constant and effective moisture diffusivity was also estimated, and a linear relationship was found.

The color of an object is one of the most important quality factors and plays a significant role in the appearance, processing, and acceptability of food materials. At the point of sale, the first impact made by a consumer on a food is its visual appearance.

The effect of microwave output power and sample amount on the color change kinetics of Turkey spinach (*Spinacia oleracea* L.) was investigated using the microwave drying technique. The color parameters for the color change of the materials were quantified by the Hunter L (whiteness/darkness), a (redness/greenness), and b (yellowness/blueness) system. These values were also used for the calculation of the total color change (DE), chroma, hue angle, and Browning index. The microwave drying process changed the color parameters of L, a, and b, causing a color shift toward the darker region. The values of L and b decreased, while values of a and total color change (DE) increased during microwave drying. The mathematical modeling study of color change kinetic showed that L and b fitted to a first-order kinetic model, while a and total color change (DE) followed a zero-order kinetic model. However, chroma and Browning index (BI) followed a first-order kinetic model, while hue angle followed a zero-order kinetic model. To illustrate the relationship of the L, a, b, total color change (DE), chroma, hue angle, and Browning index depending on the ratio of the microwave output power to sample amount, the data adequately fitted to a quadratic (second-order polynomial) model. For the calculation of the activation energy for color change kinetic parameters, the exponential expression based on an Arrhenius equation was used.

The color change of spinach using the L, a, and b systems totally explained the real behavior of spinach samples undergoing microwave drying. The final values of L, a, b, total color change (DE), chroma, and hue angle were influenced by microwave drying. The values of the Browning index showed that microwave drying caused more brown compound(s)—the increase in value supported this result. The zero-order and first-order kinetic models were used to explain the color change kinetics, and it was observed that L, b, chroma, and Browning index were fitted to a first-order kinetic model. On the other hand, a total color change (DE) and hue angle followed a zero-order kinetic model.

6 Food Industry Uses

Spinach extract is a good source of protein, fiber, phytochemicals, and minerals, making it a suitable ingredient to be used in the formulation of foods with high nutritional or biological values.

6.1 Spinach Powder

This study aimed to optimize the optimal conditions to prepare spinach powder using a wet agglomeration process at selected conditions (at 60 °C air inlet temperature and 1.6 m/s air velocity) to improve the functional properties of the spinach juice powders produced by the spray drying technique and distilled water, maltodextrin, whey protein isolate, and gum Arabic solutions that were used as binder agents. The water amount and the water activity values of the obtained spinach juice agglomerates were in the range of 6.44%–11.68% and 0.254–0.412, respectively. However, a decrease in clarity (L*) values and an increase in yellowness/blueness (b*) values were detected in the powder formulations. Also, the mean particle diameters increased and decreased the wettability and solubility times. The predicted bulk and tapped density values for spinach juice agglomerates are in the range of 380–510 and 490–618 kg/m^3, respectively. The agglomeration process improved the values of cohesiveness, flowability, porosity, and hygroscopicity. Low friability and high solubility values were also obtained after agglomeration. The microstructure of powders changed depending on binder agents, and the glass transition temperature of the powders increased from 110.72 to 117.32–127.38 °C for the agglomerates. As a result, the agglomeration process of the spray-dried spinach juice powders was successfully carried out in the fluidized bed. The moisture content and water activity values were within safe limits, and their powder properties improved (Yuksel and Dirim 2020).

6.2 Juices and Beverages

The kefir-based smoothies with spinach extracts were considered ready-to-drink and innovative functional snacks. Microbiological, physicochemical, and pre-and post-gastrointestinal total antioxidant capacity analyses were assessed. The results revealed that the kefir-based smoothies with vegetables had higher nutritional compounds such ash, sugars, and dietary fiber values. The amount of fructose and glucose in smoothies with spinach extracts was low and was characterized by high sucrose and maltose. The microbiology data demonstrated that kefir-based vegetable smoothies had a low number of *Lactobacillaceae* viability (>log 7 cfu/g) for the required functional effect after two weeks of storage (Yilmaz-Ersan et al. 2024). Furthermore, the incorporation of spinach extract significantly increased the antioxidant activities and total phenolic content values. After in vitro gastric digestion analysis, smoothies with spinach extracts revealed higher total phenolic compound amounts compared to control. It was found that in vitro intestinal DPPH values were higher for the sample with spinach extracts (Yilmaz-Ersan et al. 2024). It was also found that the bioaccessibility indexes of plain kefir were determined to be the highest in both in vitro gastric and intestinal procedures. The current research provided novel insights into the in vitro digestion properties of kefir fortified with spinach by-products. However, other work is needed to identify the functional properties of the milk and plant matrices mixture using in vitro and in vivo tests (Yilmaz-Ersan et al. 2024).

6.3 In Bread Products

In this section, the incorporation of extracts into bread products was investigated. The physicochemical properties of spinach powder were studied and compared with wheat flour, and the spinach extract incorporation on the physicochemical characteristics, consistency dough rheology, and quality attributes of *chapatti* were evaluated. The obtained data indicated spinach extracts constituted 8.2% crude fiber, 19.2% protein, 1304 mg/100 g calcium, and 40.4 mg/100 g iron. Spinach extracts showed significantly augmented values for hygroscopicity, swelling power, and water solubility index values, that is, 6.4%, 7.1 g/g, and 4.2%, respectively, when compared with wheat flour. The incorporation of spinach extracts in wheat flour at a 20% substitution amount significantly decreased dough development properties, including water absorption, dough stability, and peak dough development time. Also, the color parameters of baked *chapatti* indicated a significant decrease in L^*, a^*, and chroma values with an increasing amount of spinach extract incorporation. However, sensory evaluation revealed that the incorporation of spinach extract at 7.5% had an optimum effect on the overall acceptability of the baked product. These findings further suggested that replacing wheat flour with low spinach extracts in baked products could be a viable dietary approach to enhance the optimum supply of micronutrients and combat micronutrient deficiencies among various population segments (Waseem et al. 2021).

6.4 Incorporation into Dairy Products

Due to the high utilization rate of dairy products, enrichment of these products will successfully decrease or prevent diseases related to nutrition deficiencies. Fruits and vegetables in different forms are excellent sources for the enrichment of dairy products because of their desired taste, color, aroma, fibers, and vitamin content. The effect of some fruits and vegetables on the rheological behavior, physicochemical attributes, color parameters, and sensorial and quality properties of dairy products, including cheese, ice cream, and yogurt was investigated. The physicochemical, color, texture, and sensorial properties of dairy products were affected by the addition of fruits or vegetables. Also, the addition of these products contributes to the higher content of vitamins, natural colorants, minerals, polyphenols, crude fiber, and carotenoids. In addition, some fruits and vegetables are considered potential dairy product stabilizing agents due to their desirable functional properties, such as water binding and holding, gelling, and thickening ability. In summary, the enrichment of cheese, ice cream, and yogurt with fruits and vegetables increases the market share of these products due to the high demand for goods for an improved diet, rich in compounds with antioxidant activity and biological properties.

It is important to note that nutritional characteristics, safety (microbiological and toxicological), and sensory properties should be taken into consideration when cheese added to fruits and vegetables products is manufactured. Spinach by-products are a good source of minerals (Fe, Mn, Zn, and Mg), vitamins (E, A, C, K, B1, B6, and B2), protein, fibers, and antioxidants, making it a suitable ingredient to be applied in the formulation of dairy products with high nutritional and biological properties (Slavin and Lloyd 2012). El-Sayed (2020) reported the effect of spinach nano-powder (0.5%–2%) supplementation on the quality of ultra-filtered soft cheeses. The author suggested that by increasing the amount of spinach extract with retentate, the amount of fibers, minerals, total phenolic content, and antioxidant activities of formulated products was improved (Salehi 2021). Therefore, the incorporation of spinach extracts into cheese represents a novel functional dairy product that can potentially provide the human body with better nutrients.

6.5 Food Colorant

The colorant can be extracted from the spinach by-products based on chlorophyll and lutein, as indicated by Özkan and Ersus Bilek (2015). Especially chlorophyll is a very unstable bioactive compound that needs to be stabilized by the addition of Zn^{2+}. Due to the formation of Zn–chlorophyll derivatives, the highest green color (a/

value) was obtained when 300 ppm $ZnCl_2$ was added to the spinach pulp at a pH of 7, followed by heat treatment at 110 °C for 15 min. Extraction is the most important step in isolating different types of bioactive compounds from plants. Various parameters, such as concentration of enzymes, temperature, and time of incubation, need to be optimized to improve the yield of bioactive compounds from plant tissues. Enzyme-assisted extraction of Zn–chlorophyll derivatives from spinach pulp under optimized conditions (8% enzyme concentration, 45 °C, and 30 min) showed a remarkable increase in the yield by 39%. Enzyme-assisted extraction of bioactive materials from plant tissue is a challenging area for food scientists, so interdisciplinary research with biochemists is essential. In addition, currently available enzyme preparations cannot completely hydrolyze plant cell walls, which limits extraction yields of compounds. Future investigations are needed on enzyme-assisted extractions to enhance the yields of bioactive materials and the efficiency of extraction methods.

6.6 Chips and Snakes

The formulation of spinach chips is as follows: first, prepare the tools and materials, second, clean the spinach and then separate the spinach leaves from the stalks, third, prepare the dough and the spinach chip flavor while heating the cooking oil, fourth, mix the spinach leaves with the dough and fry, fifth, drain the spinach chips, and lastly, put the spinach chips in the packaging. The results of the research also show good business prospects for spinach chips. The amount of money received per production is IDR910.000, and the income earned per production is IDR249.79. The value added was IDR9.725/kg. Therefore, it can be concluded that spinach chips can be a promising alternative source for the economy. If they continue to be developed, then spinach chips can become one of the typical souvenirs of Payakumbuh, like Sanjai. Therefore, spinach chips need to be introduced more widely so that more local people and tourists will recognize them.

6.7 Spinach Leaf Purée

The formulation of the spinach leaf purée indicated that the steaming of spinach leaves significantly increases ($p \leq 0.05$) ash, protein, and soluble iron by 55.36, 85.02, and 38.83%, respectively. On the other hand, this treatment led to a decrease in soluble sugars, crude fiber, and oxalates by 31.25, 12.41, and 75.65%, respectively. Similar results on the steam blanching of spinach leaves were obtained and discussed by Tedom et al. (2020). The increase in concentration of protein and iron, as well as the reduction of oxalate contents in the steamed spinach leaves, makes this treatment beneficial for infant food formulation. Steaming spinach leaves improves iron concentration, whereas soaking and toasting soybean grains do not affect their protein contents. The incorporation of steamed spinach leaf paste and soybean powder into fermented pumpkin flesh at different proportions has permitted the elaboration of two basic infant purees, PI and PII. Five others with the required viscosity for infant foods were derived from these basic purees. The formulated purees contained enough amounts of protein and iron and equally had an energy density.

6.8 *Antioxidant in Sunflower Oil*

The thermal stability of frying oil is an important topic in food science and technology. The low stability of frying oils can be enhanced with the addition of natural antioxidants. Food industries may use antioxidants from plant leaf extract, especially spinach leaves, which are very beneficial; the phenolic profile, lipid oxidation, and antioxidant activities of sunflower oil fortified with spinach extracts are enhanced. These results suggest that spinach extracts are beneficial in terms of lipid oxidation parameters and serve as a good source of important phenolic compounds (Zeb and Ullah 2019).

7 Conclusion

Spinach by-products are rich in minerals, vitamins, and phytochemical compounds such as lutein and chlorophyll pigments, as well as in total polyphenols. Spinach by-products can be valued for their content with other phytochemical molecules, such as polyphenols. There are different techniques for molecule extraction, such as maceration, saponification, microwave drying, and supercritical carbon dioxide. Indeed, it should be remembered that spinach contains a significant source of these phytomolecules, especially chlorophyll and lutein pigments. Besides, spinach by-products contain much fiber, and weight loss properties manifested by the appearance of satiety are also present in high quantities. The spinach by-products were incorporated into food products to enhance their stability and extend their shelf life.

References

Bunea A, Andjelkovic M, Socaciu C, Bobis O, Neacsu M, Verhé R, Van Camp J (2008) Total and individual carotenoids and phenolic acids content in fresh, refrigerated and processed spinach (*Spinacia oleracea* L.). Food Chem 108:649–656

Burri BJ (2013) Carotenoids: chemistry, sources and physiology. Encycl Hum Nutr 1:283–291

Carpentier S, Knaus M, Suh M (2009) Associations between lutein, zeaxanthin, and age-related macular degeneration: an overview. Crit Rev Food Sci Nutr 49:313–326

Derrien M, Badr A, Gosselin A, Desjardins Y, Angers P (2017) Optimization of a green process for the extraction of lutein and chlorophyll from spinach by-products using response surface methodology (RSM). LWT Food Sci Technol 79:170–177. https://doi.org/10.1016/j.lwt.2017.01.010

Djouadi A, Derouiche A (2021) Spinach mediated synthesis of zinc oxide nanoparticles: characterization, in vitro biological activities study and in vivo acute toxicity evaluation. Curr Res Green Sustain Chem 4:100214. https://doi.org/10.1016/j.crgsc.2021.100214

El-Sayed SM (2020) Use of spinach powder as functional ingredient in the manufacture of UF-Soft cheese. Heliyon 6(1):1–6

Fernández-Sevilla JM, Acién Fernández FG, Molina Grima E (2010) Biotechnological production of lutein and its applications. Appl Microbiol Biotechnol 86:27–40

Ferruzzi MG, Blakeslee J (2007) Digestion, absorption, and cancer preventative activity of dietary chlorophyll derivatives. Nutr Res 27:1–12

Johnson EJ (2012) A possible role for lutein and zeaxanthin in cognitive function in the elderly. Am J Clin Nutr 96:1161S–1165S

Montelius C, Osman N, Weström B, Ahrné S, Molin G, Albertsson P-Å, Erlanson-Albertsson C (2013) Feeding spinach thylakoids to rats modulates the gut microbiota, decreases food intake and affects the insulin response. J Nutr Sci 24:2–20

Özkan G, Ersus Bilek S (2015) Enzyme-assisted extraction of stabilized chlorophyll from spinach. Food Chem 176(0):152–157

Roberts JL, Moreau R (2016) Functional properties of spinach (Spinacia oleracea L.) phytochemicals and bioactives. Food Funct 7(8):3337–3353

Rodríguez-Bernaldo de Quirós A, Costa HS (2006) Analysis of carotenoids in vegetable and plasma samples: a review. J Food Compos Anal 19:97–111

Salehi F (2021) Quality, physicochemical, and textural properties of dairy products containing fruits and vegetables: a review. Food Sci Nutr 9:4666–4686

Sharma D, Kumar SS, Sainis KB (2007) Antiapoptotic and immunomodulatory effects of chlorophyllin. Mol Immunol 44:347–359

Slavin JL, Lloyd B (2012) Health benefits of fruits and vegetables. Adv Nutr 3(4):506–516

Tanaka T, Shnimizu M, Moriwaki H (2012) Cancer chemoprevention by carotenoids. Molecules 17:3202–3242

Tedom WD, Fombang ED, Ngaha WD, Ejoh RA (2020) Optimal conditions of steam blanching of spinach (*Spinacia oleracea*), a leafy vegetable consumed in Cameroon. Int J Nutr Sci Food Technol 6(4):3–9

Waseem M, Akthar S, Manzoor MF, Mirani AA, Ali Z, Ismail T, Ahmad N, Karrar E (2021) Nutritional characterization and food value addition properties of dehydrated spinach powder. Food Sci Nutr 9:1213–1221

Yilmaz-Ersan L, Ozcan T, Akpinar-Bayizit A, Sahin S (2024) Bioaccessibility and antioxidant capacity of kefir-based smoothies fortified with kale and spinach after in vitro gastrointestinal digestion. Food Sci Nutr 12:2153–2165

Yuksel H, Dirim SN (2020) Application of the agglomeration process on spinach juice powders obtained using spray drying method. Dry Technol 30:19–34. https://doi.org/10.1080/07373937.2020.1832515

Zeb A, Ullah F (2019) Effects of spinach leaf extracts on quality characterization and phenolic profile of sunflower oil. Eur J Lipid Sci Technol 121:1800325. https://doi.org/10.1002/ejlt.201800325

Part III
Phytochemicals from Stem Vegetables By-products

Bioactive Phytochemicals from Asparagus (*Asparagus officinalis* L.) By-products

Charu Gupta and Shikha Chauhan

Contents

C. Gupta (✉)
Amity Institute of Herbal Research & Studies, Amity University UP, Noida, India
e-mail: cgupta@amity.edu

S. Chauhan
Amity Institute of Pharmacy, Amity University UP, Noida, India

M. F. Ramadan (ed.), *Bioactive Phytochemicals in By-products from Leaf, Stem, Root and Tuber Vegetables*, https://doi.org/10.1007/978-3-031-80700-8_6

Abstract Asparagus spears (*Asparagus officinalis* L.) are regarded as a healthy food due to their high fiber content, low-calorie content, and phytochemical content, such as vitamins, flavonoids, fructans, saponins, and cinnamic acids. Various studies have recognized these compounds for their antioxidant, anti-inflammatory, and anti-cancer properties. However, because a large portion of the underground (>30 Tm/ha), which includes roots, and the aerial (~6 Tm/ha), which contains stems, leaves, and fruits, are wasted, asparagus cultivation and processing also generate a significant number of agri-food by-products. For asparagus growers, disposing of trash associated with the crop has become a major agronomic, financial, and environmental concern.

The same nutrients and phytochemicals found in asparagus spears may also be found in asparagus roots and other by-products, according to observations made in recent years. As a result, these ingredients have great potential for use as food ingredients with exceptional nutritional value and health-promoting qualities. Hence, these inedible components ought to be viewed as food by-products that are valuable to several industries and support sustainable agriculture rather than just being a waste product of farming. Phenolic–polysaccharide complexes are abundant in the cell walls of asparagus, particularly following the hardening phase that occurs after harvest.

This chapter thus explores the potential health benefits of bioactive phytochemicals from by-products of several species of *Asparagus,* including *A. officinalis* and *A. racemosus*, shedding light on utilizing agricultural waste to harness valuable compounds. Asparagus, a versatile vegetable, has gained attention not only for its culinary appeal but also for its rich phytochemical content.

Furthermore, the practical applications of these bioactive compounds are considered, focusing on integrating asparagus by-product extracts into functional foods, nutraceuticals, or pharmaceutical formulations. This adds economic value to the by-products and aligns with a broader sustainability narrative in the food industry.

Keywords Asparagus · By-products · Phytochemicals · Nutraceuticals · Food industry

Abbreviations

RS	Reducing sugars
TSS	Total soluble solids
Vc	Vitamin c

1 Introduction

Since ancient times, asparagus (*Asparagus officinalis* L.), a perennial herb in the Liliaceae family, has been used extensively for food and medicinal purposes. Its biological activities and health advantages have drawn increased interest from academia and the general public, earning it the moniker "the king of vegetables" in the global market. One of the most important traditional medicinal plants, it contains

Table 1 Nutritional composition of different parts of *Asparagus* (adapted from Guo et al. 2015)

Nutrient components	Leaves (wt%)	Mature stem (wt%)	Green stem (wt%)	Tender tip (wt%)
Moisture	67.14	91.01	93.58	88.77
Total sugar	5.9	5.7	19.2	7.3
Protein	21.8	15.9	31.3	35.8
Crude fat	7.4	1.0	2.2	2.3
Crude fiber	1.4	1.8	0.3	0.6
Reducing sugars	5.5	5.5	17.4	2.9

non-nutritive compounds with positive health effects. It contains bioactive metabolites such as fructooligosaccharides, polysaccharides, polyphenols, asparosides, shatavarins, sapogenins, racemosols, isoflavones, glycosides, anthocyanins, mucilage, and fatty acids, while saponin is one of the main active constituents of asparagus roots. Asparagus helps promote fertility, manage stress, and modulate hormones. Numerous phytochemical compounds found in asparagus have been shown in vitro and in vivo experiments to exhibit anticancer, antitumor, antioxidant, immunomodulatory, hypoglycemic, antihypertensive, and anti-epileptic effects. It also treats stomach ulcers, kidney disorders, and Alzheimer's disease. However, bridging the gap between bioactive components and human health is as critical as uncovering their corresponding mechanisms (Guo et al. 2020).

Food products' nutritional and functional qualities (such as milk, drinks, and baked goods) are improved when asparagus powder or extract is substituted in place of value addition. Because of the plant's damaging harvesting practices, habitat degradation, and deforestation, it is currently classified as endangered in its natural environment. Since it is a highly perishable commodity, it needs proper handling, preservation, and storage (Kohli et al. 2023).

There are about ten million asparagus eaters worldwide, and the growing trend of consumers focusing on eating a healthy diet will help to sustain the strong demand for this highly valued product, which ranks among the highest plant foods in terms of antioxidant content. The nutritional composition of different parts of *Asparagus* spp. is presented in Table 1.

2 *Asparagus officinalis*: Environmental Perspective

Asparagus farming generates a substantial quantity of by-products yearly, including roots and fronds. The latest data show that 68,000 tons of asparagus are produced in Spain annually; more than 45,000 tons originate from Andalusia (Observatorio de precios y mercados, CAPADS 2023). Considering that about 6 tons of fronds are gathered annually for each hectare under cultivation, the total volume of this by-product is equal to the production of 45,000 tons of asparagus. However, as 10% of the crop is grown each year and between 30 and 40 tons of roots accumulate in one hectare, the annual total volume of rhizome is approximately 25,000 tons. Many variables contribute to the significant environmental issue of removing these plant residues. The first is soil pollution, primarily caused by the waste materials left over when asparagus plants are replaced once they run out of supply. The common agricultural practice of incorporating the root and rhizome into the ground could be the main cause of the phenomenon known as "asparagus decay," which consists of a loss of crop quality and yield that appears after a few (5–7) years of crop, and it is maintained and accentuated over time (Elmer 2018; Zandstra et al. 2013). The "asparagus replant problem," which is exacerbated by the inability to replace old asparagus plants with new, young ones on the same crop fields, is another factor contributing to the issue (Morrison et al. 2011; Yergeau et al. 2006; Elena 2007). As a result, the plantation's ability to operate commercially is limited, and the hunt for new planting areas is ongoing.

However, burning fronds and roots—a method that farmers are doing increasingly—carries the risk of causing air pollution. After harvesting, leaving these residues on the fields has a detrimental effect on the subsequent asparagus crops because they release auto-toxic (allelopathic) substances into the soil. These substances build up over time and cause asparagus yields to decline gradually, eventually becoming an unprofitable crop within ten to fifteen years. The entire asparagus plantation must be lifted due to this occurrence called decay (abandoned). However, since the yields obtained are never higher than half of those of typical asparagus plants, it is not profitable to replant new plants right away after a particular plantation has been lifted. Asparagus must be replanted in these areas after an average of 4 or 5 years. The replanting problem is a phenomenon that makes it necessary to look for new land all the time to grow asparagus. The fact that removing those plant remnants from asparagus farming comes at a high financial expense is an additional issue for growers. The aerial portion of the plant must be removed if the plantation is under an integrated production system; its absorption into the land is not permitted. However, the roots continue to leave them when the plantation is raised since eliminating them would entail a cost the farmer cannot assume in most cases. In addition, in cases where it is possible to take the waste to the landfill, its transport entails great economic and environmental costs (Alcaide et al. 2023).

For all these reasons, it is crucial to look for waste management alternatives that increase the crop's sustainability within the framework of the circular economy. Within this framework, this study suggests generating bioactive compounds from asparagus by-products to increase their value.

3 Phytochemicals in Asparagus Species

Native medicinal plants called asparagus species have many biological benefits, such as being immune-stimulating, anti-inflammatory, antihepatotoxic, antibacterial, and reproductive agents. Asparagus roots have been used to treat various illnesses, including diabetes and dysentery, and boost milk production and appetite. Asparagus contains steroidal saponins, essential oils, vitamins A, B, B, C, and E, and minerals such as Mg, P, Ca, Fe, and folic acid. It also contains flavonoids, tannin, and various chemical compounds such as shatavarin IV, sarsasapogenin, kaempferol, quercetin, and rutin (Negi et al. 2010).

Various compounds have been isolated from different parts of *Asparagus* species, showing a range of biological activities. Isolated compounds include sarsasapogenin, shatavarin, racemosol, and asparagamine, with activities such as antibacterial, antifungal, and diuretic properties.

Asparagus roots are used as laxatives, tonics, and aphrodisiacs and are known to have antibacterial and antifungal properties. They have been used historically for their diuretic properties, cardiovascular benefits, and fertility promotion. Sarsasapogenin and shatavarin IV are major bioactive compounds in *Asparagus*, found in various plant parts, such as roots, leaves, and fruits. These compounds have significant medicinal properties and have been isolated and characterized from *Asparagus* species. Saponins, present in *Asparagus*, exhibit detergent, wetting, emulsifying, and foaming properties. Due to their amphiphilic nature, saponins are surface-active compounds with high molecular weight and exhibit solubility in water and alcohol (Negi et al. 2010).

Asparagus possesses diverse biological properties, making it a valuable medicinal plant. It exhibits properties such as antimicrobial, diuretic, and cardiovascular benefits, as well as antioxidant and anti-inflammatory effects. Asparagus is historically significant in traditional medicine and is used for various therapeutic purposes. It has been used in different cultures for its diuretic, tonic, aphrodisiac, and antibacterial properties, demonstrating its wide-ranging medicinal importance (Negi et al. 2010).

Saponins possess various biological properties, such as antioxidants, immunostimulants, antihepatotoxic, and antibacterial, and are valuable in diabetic retinopathy. Saponins act as plant defense systems, inhibiting mold growth and protecting plants from insects. Plants high in saponin have been shown to enhance ruminant development, feed efficiency, and overall health. The primary physiologically active components of the *Asparagus* genus are steroidal saponins. The literature review showed that the structures of bioactive substances differed considerably, ranging from steroidal sapogenins and saponins to carboxylic acids containing sulfur. Saponin-rich plants possess antimicrobial properties and can protect plants from various diseases. Asparagus root has been found to regulate cholesterol metabolism in hypercholesterolemic rats. This regulation can potentially contribute to improved antioxidant status. Studies have identified multiple steroidal constituents in different parts of *Asparagus curillus*, *Asparagus filicinus, Asparagus adscendens*, and *Asparagus officinalis*. These constituents play a role in the bioactivity of the plants. Ethanolic extracts from several *Asparagus* cultivars demonstrate antioxidant activity. This highlights the potential health benefits of *Asparagus*. Steroidal saponins from *Asparagus officinalis* exhibit cytotoxic activity. Asparagus roots and fruits have also shown antifungal activity. Studies have reported the carotenoid composition in the fruits of *Asparagus officinalis*, as well as the presence of oligofurostanosides and oligospiranosides in the roots of *Asparagus filicinus* and *Asparagus adscendens*. These phytochemicals contribute to the medicinal properties of

Asparagus. Extracts from *Asparagus cochinchinensis* have demonstrated bioactive and therapeutic effects, including inhibition of tumor necrosis factor-alpha-induced apoptosis. This suggests a potential role in cancer treatment. Methanolic extract of *Asparagus pubescens* root shows anti-nociceptive and anti-inflammatory effects. It also exhibits contraceptive and non-estrogenic effects (Negi et al. 2010).

3.1 Novel Compounds

The principal bioactive elements of *Asparagus* are a group of steroidal saponins. This plant also contains vitamins A, B_1, B_2, C, E, Mg, P, Ca, Fe, and folic acid. Essential oils, asparagine, arginine, tyrosine, flavonoids (kaempferol, quercetin, and rutin), resin, and tannin are other chemical components of asparagus.

With two *Asparagus* rhamnose and one glucose molecule, sarsasapogenin's glycoside shatavarin IV comprises two molecules. The principal bioactive chemical elements of *Asparagus* species, sarsasapogenin and shatavarin I–IV, are present in the roots, leaves, and fruits of *Asparagus* species. Synthesis of sarsasapogenin in the callus culture of *A. racemosus* was also reported (Kar and Sen 1985). A new isoflavone, 8-methoxy-5,6,4′-trihydroxyisoflavone-7-*O*-β-D-glucopyranoside was also reported from *A. racemosus* previously (Saxena and Chaurasia 2001). The isolation and characterization of a polycyclic alkaloid called asparagamine (Sekine et al. 1994) and a new 9,10-dihydrophenanthrene derivative named racemosol and kaempferol were also isolated from the ethanolic root extract of *A. racemosus* (Sekine et al. 1997). Curillins G and H, oligo-furostanosides, and sarsasapogenin, a spirostanoside, have been isolated from the roots and leaves of *A. curillus*, respectively.

The discovery of new compounds such as (9Z)-capsanthin-5,6-epoxide and asparosides A and B in different *Asparagus* species indicates ongoing research and the potential for new pharmaceutical developments. These compounds may have unique bioactivities. The comparative composition of white and green *Asparagus* is presented in Table 2.

Table 2 Comparative composition of white and green *Asparagus* (Zhang et al. 2002)

Sizes	Moisture (%)		Protein (%)		Ca (mg)		Vc (mg)		RS (g)		Total acid (%)		Fe (mg)		TSS (%)	
	W	G	W	G	W	G	W	G	W	G	W	G	W	G	W	G
Slender	93	92.2	1.8	2.9	13	26	29	52	2.4	1.7	0.080	0.190	0.9	1.4	4.5	5.3
Middle	92.9	92.3	1.8	2.6	14	24	28	51	2.2	1.8	0.086	0.192	1.1	1.5	4.2	5.4
Large	93	92.2	1.9	2.6	18	21	27	44	2.1	2	0.092	0.195	1.3	1.5	4.1	5.6

4 Bioactive Phytochemicals from Asparagus By-products

Because asparagus spears' lower portion includes a variety of compounds with great biological value, asparagus waste is an important area of concern. Therefore, it is crucial to extract the beneficial elements from asparagus by-products in order to increase their worth. In this work, bioactive peptides from asparagus waste were isolated, degraded, purified, and identified. Because Alcalase® is an inexpensive enzyme with the potential to be used in large-scale method implementation, it was specifically chosen as the enzyme to yield protein hydrolysate. Reversed-phase chromatography was used to fractionate the hydrolysate into ten fractions in order to simplify the peptide extract and improve identification. Two tests were conducted: one for antihypertensive effect (ACE) and one for antioxidant activity (ABTS-DPPH).

Bioinformatics was utilized to screen for bioactivity, and Peptidomimetics techniques were employed to determine which fractions had higher bioactivity scores. For ACE-inhibitor action, two peptides, PDWFLLL and ASQSIWLPGWL, were synthesized; their corresponding EC50 values were 1.76 μmol L^{-1} and 4.02 μmol L^{-1}. When antioxidant activity was measured using the DPPH test, MLLFPM had the lowest EC50 value at 4.14 μmol L^{-1}, followed by FIARNFLLGW and FAPVPFDF, with EC50 values of 6.76 μmol L^{-1} and 10.01 μmol L^{-1}, respectively. Furthermore, a validation study was conducted on the five peptides that were found. The findings show that peptides derived from by-products of asparagus have biological activity of interest and may be used as functional components (Montone et al. 2019).

Another recent investigation examined the phytochemical composition of asparagus fronds and roots. The aim was to utilize an ecologically friendly extraction technology that would be easy to use in the food business and would not interfere with food intake to generate bioactive extracts with unique technical and nutritional features. Upon analyzing asparagus by-products, it was found that the fronds had an average flavonoid content of 2.637 g/kg fresh weight, which is up to 5–6 times higher than that of the spears, and that the roots had up to 10 times more saponins (2.25 g/kg fresh weight), along with lower levels of phenolic acids (368 mg/kg fresh weight).

The results of this study suggest that the phytochemical contents are mostly determined by the location and stage of the vegetative cycle, with genetic variables having no discernible effect. Based on the extraction of two bioactive extracts—an antioxidant extract rich in flavonoids with an average yield of 10.7 g/kg fresh frond and a flavonoid richness of 17%, and a saponin extract with an average yield of 10.3 g/kg fresh root and a flavonoid richness of 51%—the proposal for the recovery and valorization of asparagus by-products is put forth. In the agri-food sector, these natural extracts have enormous techno-functional potential; several are already undergoing testing as additives for use in the making of soups, breads, and meat products (Alcaide et al. 2023).

Over 200 species of asparagus were identified in a 2019 research. The only species that is farmed for food purposes on a commercial scale is *Asparagus officinalis*.

The scientists conducted a comprehensive literature analysis on the chemical composition and phytochemicals of asparagus, with a focus on phenolic substances such as rutin and saponins such as diosgenin and protodioscin. The physiology of spear hardening and the use of hard-stem by-products as a source of dietary fiber and bioactive substances were also covered. The nutritional and phytochemical components found in asparagus spear and hard-stem by-products are abundant; the amount of these components varies based on several factors, including the kind of asparagus, spear portion, harvesting season, growing technique, etc. These were their principal discoveries and deductions.

Although studies have been done on drying, fermentation, and juice production, cold and MAP storage are the most often used processing techniques for preserving fresh asparagus spears. After harvesting, asparagus spears will unavoidably harden; nevertheless, several methods have been devised to delay this process. These methods include dark, anaerobic surroundings and low temperatures. Ammonium sulfate has also been used to inhibit PAL activity. When asparagus is produced and processed, a significant number of by-products (30–50%) are produced. These by-products may be used to create a variety of high-value goods, such as bioactive dietary fiber powders, bioactive extracts, and mainstream products (Chitrakar et al. 2019).

In a different study, ethanolic extracts from the leaves, pericarps, and rhizomes of *Asparagus albus* L. were tested for their phytochemical composition, antioxidant activity (DPPH· and FRAP assays), antimicrobial activity against human pathogenic isolates, and cytotoxic activity (human colon carcinoma HCT-116 cells). The largest quantities of flavonoids were found in the pericarp and leaf extract, respectively. But there were no flavonoids in the rhizome. However, the rhizome was packed with saponins. The flavonoid and saponin profiles were similar to those of the triguero Huetor-Tajar asparagus landrace that had been previously documented. Compared to the extracts from the rhizome and leaves, the pericarp ethanolic extract was demonstrated to have higher antioxidant activity.

Furthermore, the rhizome showed a more pronounced cytotoxic activity against HCT-116 cells in comparison to the leaf and pericarp. In contrast with most human pathogenic isolates, every extract showed different levels of antibacterial activity. With an inhibitory zone of 21 mm and 19 mm, respectively, the pericarp extract exhibited the best effectiveness against erythromycin-resistant *Streptococcus agalactiae* (ER) and multidrug-resistant *Pseudomonas aeruginosa* (MDR). Additionally, the leaves showed stronger inhibitory effects against all separated fungi and the greatest number of bacteria. According to Hamdi et al. (2017), the findings suggest that *A. albus* may be a novel crop of pharmacological interest due to its abundance of bioactive chemicals, which have significant positive effects on human health. Figure 1 illustrates the process of removing phytochemicals from by-products of asparagus.

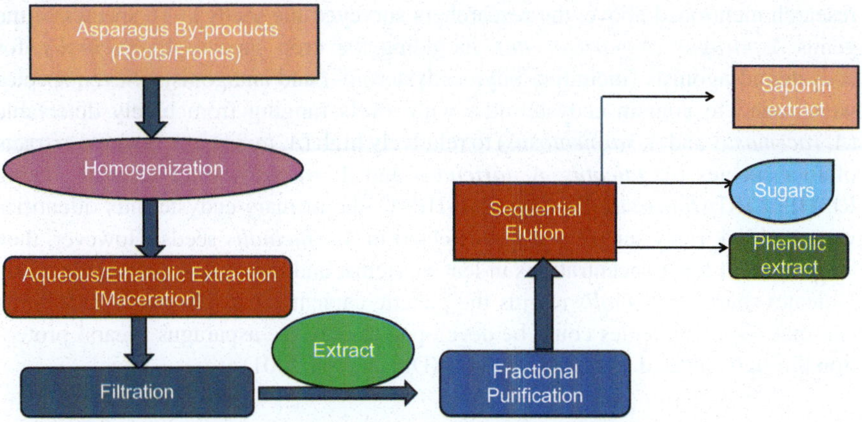

Fig. 1 Steps in extraction of phytochemicals from asparagus by-products (fronds/roots)

5 Applications of Asparagus By-products

5.1 *Bioactive Ingredient for the Development of Edible Films*

Another study showed that *Asparagus racemosus*, a bioactive ingredient in edible films, significantly improves lipid oxidative stability and storage quality of meat products, with higher sensory scores and lower microbial counts.

A novel calcium alginate edible film with preservation potential was developed using *Asparagus racemosus* as a bioactive ingredient to increase the lipid oxidative stability and storage quality of meat products. Chevon sausages were employed as a model system and were aerobically packaged in edible films combined with *A. racemosus,* viz. T0 (0.0%), T1 (1.0%), and T2 (2.0%), which were stored under refrigerated (4 ± 1 °C) conditions. Products packaged in T1 and T2 edible films exhibited significantly ($P < 0.05$) lower values for TBARS (mg malonaldehyde/kg), microbial counts (CFU/g), and FFA (% oleic acid) indicating the bioactive properties of the developed films. Higher sensory scores were noted for the goods packed in T1 and T2 films, with a significant difference ($P < 0.05$). *A. racemosus* added antioxidant and antimicrobial properties to the developed films, which improved the lipid oxidative stability and storage quality of the meat product in the model (Noor et al. 2018).

5.2 *As Predator Protectant*

Another study found that phytoecdysteroids in *Asparagus* species can potentially protect asparagus spears from predators and have antimicrobial properties, with the potential for future strategies to enhance protection through elevated levels. In the research mentioned above, the researchers surveyed the seeds of 16 species of the genus *Asparagus* (*Asparagaceae*), including the crop species *A. officinalis*, for ecdysteroid agonists (including phytoecdysteroids) and antagonists. Seven species were found to contain ecdysteroids, with levels ranging from barely detectable (*A. racemosus* and *A. sarmentosus*) to relatively high (*A. laricinus*). Positive extracts of four species (*A. falcatus, A. laricinus,* and *A. ramosis*) were separated using RP-HPLC/RIA/bioassay. By using NP-HPLC, the primary ecdysteroids' identities were verified. Ecdysteroids are not detected in *A. officinalis* seeds. However, they are present at low concentrations in leaves, stems, and roots (RIA detectable). This indicates that *A. officinalis* retains the genetic capacity to synthesize ecdysteroids and that future strategies could be developed to enhance asparagus spears' protection through elevated ecdysteroid levels (Dinan et al. 2001).

5.3 As Antioxidant and Anti-hypertensive Effects

Research has found that *Asparagus* by-products contain bioactive peptides with potential antioxidant and antihypertensive effects, making them suitable for use as functional ingredients.

Given that the lower part of asparagus spears contains many substances with high biological value, asparagus waste represents items of significant interest. Therefore, to provide more value for asparagus by-products, it is essential to extract their beneficial ingredients. Bioactive peptides from asparagus waste were isolated, broken down, purified, and identified in this study. Alcalase® was selected as the enzyme to create protein hydrolysate because of its low cost and potential for large-scale technique implementation. Reversed-phase chromatography separated the hydrolysate into ten fractions to simplify the peptide extract and improve identification. Antioxidant activity (ABTS-DPPH·) and antihypertensive activity (ACE) were tested twice each. Peptidomomics methods were used to identify the fractions with higher bioactivity scores, and bioinformatics was used to screen for bioactivity. Two peptides, PDWFLLL and ASQSIWLPGWL, were synthesized for ACE-inhibitor activity; the resulting EC50 values were 1.76 μmol L^{-1} and 4.02 μmol L^{-1}, respectively. By using the DPPH· assay to measure antioxidant activity, MLLFPM had the lowest EC50 value at 4.14 μmol L^{-1}, followed by EC50 values of 6.76 μmol L^{-1} and 10.01 μmol L^{-1} for FIARNFLLGW and FAPVPFDF, respectively. Additionally, a validation of the five discovered peptides was done. The results showed that peptides obtained from asparagus by-products are of interest in their biological activity and are suitable for use as functional ingredients (Montone et al. 2019).

5.4 As Cytotoxic and Lipase Inhibitory Activity

Recently, it has been proved in a study that *Asparagus acutifolius* exhibits potent antioxidant, cytotoxic, lipase inhibitory, and antimicrobial properties, with the leaf extract showing strong inhibitory activity against bacteria and fungi. HPLC-DAD-MS was used to evaluate the phytochemical composition of ethanolic extracts of *Asparagus acutifolius's* leaves, stems, pericarps, and rhizomes. At least eleven simple phenolics, one flavonone, two flavonols, and six steroidal saponins are present in *A. acutifolius* samples. While the primary components were rutin and caffeic acid, the stem extracts had the highest overall flavonoid and phenolic acid concentrations. The main flavonoids in the leaf, pericarp, and rhizome were ferulic and caffeic acids. The rhizome extracts of *A. acutifolius* had the highest concentrations of steroidal saponins discovered in all plant portions. The pericarp extracts had the strongest 2,2-azino-bis (3-ethylbenzothiazoline-6-sulfonic acid) (ABTS) scavenging activity, while the stem extracts demonstrated the highest antioxidant activity against DPPH· and ferric-reducing antioxidant power (FRAP). The rhizome and leaf extracts effectively killed HepG2 and HCT-116 cell lines.

Furthermore, moderate lipase inhibitory action was found in the pericarp and rhizome extracts. The antibacterial efficacy of the leaf and rhizome extracts against human pathogenic isolates was tested. The leaf extract exhibited a potent inhibitory activity against all the bacteria and fungi tested (Hamdi et al. 2021).

5.5 Hypolipidemic and Hepatoprotective Properties

According to a further investigation, the sections of asparagus (*Asparagus officinalis* L.) thrown out during industrial processing may have medicinal value when added to food. This study assessed the hepatoprotective and hypolipidemic properties of ethanolic (EEA) and aqueous (AEA) extracts from by-products of asparagus in mice given a high-fat diet (HFD). Mice on a continuous high-fat diet (HFD) developed evident hyperlipidemia and liver damage. However, when given to hyperlipidemic mice at a daily dose of 200 mg/kg for 8 weeks, both EEA and AEA significantly reduced the levels of body weight growth, serum total cholesterol, and serum low-density lipoprotein cholesterol. Additionally, the group receiving AEA treatment had higher serum levels of high-density lipoprotein cholesterol.

Furthermore, alanine and aspartate transaminase activities in serum were significantly reduced by both EEA and AEA. Lastly, in the liver cells of both EEA- and AEA-treated animals, superoxide dismutase activity and total antioxidant capacity were raised, whereas malondialdehyde level and lipid droplet distribution decreased. The findings of this study suggested that both EEA and AEA have strong hypolipidemic and hepatoprotective properties and could be used as supplements in healthcare foods and drugs or combination with other hypolipidemic drugs (Zhu et al. 2010).

5.6 As a Valuable Source of Fructans: Novel Feedstock for Sustainable Agriculture and Industry

Asparagus roots were found to be a valuable source of fructans, with up to 12.5% content, making them a useful feedstock for sustainable agriculture and industry. One of the finest places to get fructans is in the by-products of asparagus farming: the plant's roots. These polymers' prebiotic and immunostimulant properties make them intriguing food additives. To value these by-products as sources of fructans, this work aimed to define the fructan profile from the roots of many types of asparagus cultivated at different sites and picked at three vegetative stages. Following hot water extraction, fructans were segregated into three pools based on their molecular weight (MW). HPSEC examined their mean molecular weight, while HPAEC reviewed the extent of polymerization. Based on fresh weight, the fructan content ranged from 12.5% to 12.5%, contingent upon the variety and sampling date. Because the three pools' overall content and MW increased after the spear harvest period, their relative abundance also depended on the picking moment. HPAEC detected fructans up to 30 units, although the average MW of the three fractions was comparable across types with 4.8, 8.4, and 9 sugar units. These features put them in line with the widely used Orafti®-GR inulin, which is a commercially available food ingredient. Therefore, the concept of asparagus roots as cultivation waste must be changed to a new feedstock for sustainable agriculture and industry (Viera-Alcaide et al. 2022).

5.7 As a Functional and Nutritional Supplement

One of the most important traditional medicinal plants is asparagus (*Asparagus racemosus* L.), which contains non-nutritive yet health-promoting phytonutrients. While saponin is one of the primary active ingredients of asparagus roots, it also includes other bioactive metabolites such as fructooligosaccharides, polysaccharides, asparosides, shatavarins, sapogenins, aerosols, isoflavones, glycosides, mucilage, and fatty acids. Asparagus aids in hormone regulation, stress reduction, and the stimulation of reproduction. Alzheimer's illness, kidney problems, and stomach ulcers are also treated with it. Food products' nutritional and functional qualities (such as milk, drinks, and baked goods) are improved when asparagus powder or extract is substituted in place of value addition. Because of the plant's damaging harvesting practices, habitat degradation, and deforestation, it is currently classified as endangered in its natural environment. A highly perishable commodity requires careful handling, storage, and preservation. According to the study, irradiation and drying are currently the only methods of preservation utilized to lengthen the shelf life of asparagus roots. Asparagus roots have significant medicinal properties and can be added to food products for nutritional and functional enhancement, but proper handling and preservation are crucial due to their perishability (Kohli et al. 2023; Alcaide et al. 2020).

5.8 In Preventing Diabetic Complications Linked to Hyperglycemia and Hyperlipidemia

The bottom, inedible part of asparagus (*Asparagus officinalis* L.) spears, which make up around one-third to half of the entire length, is a by-product that is usually thrown away. Because of its remaining bioactive components, this by-product could be employed in dietary supplements with potential medical advantages. According to research, the aqueous extract of asparagus by-product (AEA) significantly lowers triglyceride and blood glucose levels in diabetic rats, indicating that it may be used to avoid diabetes problems that are associated with hyperglycemia and hyperlipidemia. This study employed a streptozotocin-induced diabetic rat model to evaluate the hypoglycemic effects of an aqueous extract of asparagus by-product (AEA). Continuous AEA injection for 21 days effectively reduced fasting blood glucose and cholesterol levels while significantly increasing hepatic glycogen levels and body weight in diabetic rats.

Following AEA medication, the area under the glucose curve in an oral glucose tolerance test showed a substantial decrease, as did the blood glucose levels measured 30, 60, and 120 minutes after glucose loading. According to Zhao et al. (2011), the study's findings showed that AEA has hypoglycemic and hypotriglyceridemic effects, which may help to reduce the consequences of diabetes that are linked to hyperglycemia and hyperlipidemia.

In a related study, asparagus woody stem juice showed putative hypoglycemic action. It may enhance diabetic rats' glucose tolerance, pointing to the plant's potential application in dietary supplements for medicinal purposes. The woody ends of spears of *Asparagus officinalis* L. contain a variety of bioactive chemicals. Nevertheless, they are often discarded as by-products. The researchers investigated the hypoglycemic impact of asparagus juice (AJ) made from recycled woody stems in a streptozotocin-induced diabetic rat model. Fasting blood glucose, glycosylated serum protein, and malondialdehyde levels in diabetic rats decreased considerably after 21 days of AJ treatment. Furthermore, AJ therapy can dramatically lower rat blood glucose levels at 30, 60, and 120 min after glucose loading, according to the oral glucose tolerance test (Zhang et al. 2014).

5.9 As a Valuable Source of Dietary Fiber

About half of the processed vegetable, asparagus (*Asparagus officinalis* L.), is made up of by-products from processing that may be used as a source of dietary fiber. Only the more soluble components (soluble sugars, uronic acids, and proteins) demonstrated substantial changes in the composition and functional qualities of fiber-rich powders, as determined by the treatment of these by-products. Variables such as the drying technique, solvent, and treatment intensity were investigated. High TDF concentrations (62–77%) were present in all of the powders rich in fiber, and as treatment intensity increased, the IF/SF ratio dropped, improving the physiological quality of the fiber. Numerous functional properties were impacted by the preparation procedure, such as water-holding capacity (WHC), glucose dialysis retardation index (GDRI), oil-holding capacity (OHC), and solubility (SOL).

Due to the influence of heat processing, WHC and GDRI were greater in fibers that were heavily removed. OHC and GDRI were much higher (5–8 mL oil/g powder and 25–45%, respectively), whereas WHC showed values similar to those found in other agricultural by-product reports (11–20 mL water/g powder). According to Fuentes-Alventosa et al. (2009), these qualities make fiber-rich powders made from asparagus by-products a useful source of dietary fiber that may be used in the creation of meals that are enhanced with fiber. Therefore, asparagus by-products can be a good source of dietary fiber; the content and functional qualities of fiber-rich powders are affected by the extraction methods used.

The phytochemical profiles of asparagus's edible part and its by-products from processing plants are similar. However, techniques used to reduce the size of such by-products before their utilization usually results in lower-quality final items due to heat generated during the procedures. The impact of low-temperature ball milling on particular physiochemical, functional, and microstructural characteristics of the milled product was evaluated in distinct research. The asparagus leaves were a by-product. Ball milling decreased the particle size from 62.3 to 7.5 μm in roughly 4 hours. Better hydration properties were displayed by the milled product, especially its increased water solubility (from 25.74 to 32.47%). The green hue of the milled product was enhanced, and its chlorophyll concentration increased.

Higher levels of antioxidant activity (as shown by a 45.37% drop in the EC50 value of DPPH radicals) and total phenolics and flavonoids (26.13% and 150.43%, respectively) were present. A shift from insoluble to soluble dietary fiber was also seen; the soluble fiber content increased from 10.33 to 19.45 g/100 g, and the associated SDF: IDF ratio decreased to 0.53 from 0.22. The microstructural results demonstrated that both size reduction and particle agglomeration occurred during milling. A greener, more bioactive dietary fiber powder is produced by low-temperature ball milling of asparagus leaf by-products, as evidenced by improvements in hydration, phenolics, flavonoids, and soluble fiber content (Chitrakar et al. 2020).

5.10 As a Functional Ingredient to Prevent Diseases Related to Oxidative Damage

Previous studies on the bioactive components of asparagus have demonstrated that a large number of the phytochemicals present in the edible region of the spears are also plentiful in the by-products. Asparagus contains phenols and saponins, which comprise flavonoids and hydroxycinnamic acids, as their main bioactive components. The researchers in this study have developed a technique for generating high-value compounds from asparagus by-products by hydrothermally treating the materials. After the hydrothermal treatment, two fractions were obtained: a fibrous residue that, when dried, produced the bioactive fiber from asparagus and an aqueous functional extract that included the majority of the soluble bioactive compounds from the by-product asparagus.

Particulate-specific chemicals (phenolics and/or saponins) can be extracted from partly purified aqueous extracts by the use of an adsorbent polymeric resin in a column purification stage. The biological activity of each of these presentations has been investigated. Based on preliminary results, the novel chemicals derived from asparagus by-products exhibit biological activity that renders them appealing for use as functional additives. Because of their strong antioxidant content, it is conceivable to propose that regular consumption of them might help prevent several illnesses linked to oxidative damage. According to Fuentes-Alventosa et al. (2013), their hydrothermal treatment method efficiently recovers beneficial chemicals from asparagus by-products, which may help prevent illnesses linked to oxidative damage.

Thus, this study indicated that the phytochemicals found in asparagus by-products, such as saponins and phenols (flavonoids and hydroxycinnamic acids), may be employed as functional components and may even help prevent illnesses associated with oxidative damage.

5.11 Suppression of Tumor Growth

The majority of asparagus (*Asparagus officinalis* L.) spears (around 30–40%) that have an inedible bottom half are usually thrown away as waste. However, as this by-product is purportedly rich in bioactive phytochemicals, it might be beneficial as a dietary supplement or in natural medicines. The results of this study showed that saponins extracted from old asparagus stem (SSA) might have the ability to suppress tumor development and metastasis. SSA decreased breast, colon, and pancreatic cancers' cell viability in a concentration-dependent manner; the half-maximum inhibitory dosages ranged from 809.42 to 1829.9 μg mL^{-1}. With an effective inhibitory dosage of 400 μg mL^{-1}, SSA was more efficient in preventing cell migration and invasion than its cytotoxic effects.

A mechanistic investigation showed that SSA markedly increased RhoA activity while decreasing Cdc42, Rac1, and RhoA activities in cancer cells. By altering the

Rho GTPase signaling system, SSA suppresses tumor cell motility, indicating a potential application for SSA as a natural medicine and dietary supplement for the prevention and treatment of cancer. The study's findings showed that the saponins from aged asparagus stems might have the ability to suppress tumor development and metastasis, which may lead to their application as a natural medicine and supplement in healthcare foods for the prevention and treatment of cancer (Wang et al. 2013).

6 Enhancing the Bioactive Components and Qualities of Asparagus Through the Use of Innovative and Efficient Processing Methods

Asparagus has long been known as a vegetable with significant bioactivities (antibacterial, neuroprotective, anti-tyrosinase, and asthma-fighting qualities) and a wealth of phytochemicals (polyphenols, saponins, alkaloids, and asparagusic acid). The functional, physicochemical, and microstructural characteristics of asparagus are significantly influenced by the processing techniques used, according to several research studies. High-pressure processing, green extraction techniques, representative shelf-life extension techniques, heating treatments (thermal, combined, and hydrothermal), and ultrasonic treatments (purification, extraction, and preservation) are some of the inventive applications of processing technology. The yields of bioactive chemicals, bioactivities, and product quality are all enhanced by these physical technologies. To fully use asparagus, standard technologies with cutting-edge effects and technologies (cold plasma, pulsed electric fields, membrane processing, and ohmic heating) should be considered.

Ultrasound, heating, and high-pressure processing are examples of novel processing methods that improve the production of bioactive compounds and product quality in asparagus and have the potential to be developed further (Yu and Fan 2021).

7 Conclusion

Massive volumes of biomass (leaves, stems, fruits, roots, and rhizomes) are produced during the production of asparagus. However, this biomass has little commercial value and poses an environmental risk. From the perspective of the economy, these by-products need to be provided additional value to be considered a chance rather than waste. They contain a wealth of phytochemicals, including flavonoids, fructans, and saponins, which are easily isolated and processed using eco-friendly methods. Many different industrial sectors are quite interested in those bioactive extracts. Utilizing this biomass will raise the standard of living in rural regions and increase the revenue of asparagus producers. Rich in phytochemicals, co-products from asparagus farming can be collected and purified using eco-friendly, sustainable methods, boosting growers' profits and improving living standards in rural areas.

Acknowledgments The authors thankfully acknowledge the research project funded by the Department of Science and Technology (DST-SERB), New Delhi, India, under the ASEAN-India S&T Development Fund (AISTDF) [Grant No. CRD/2020/000203] and grateful to the funding agency for financial support. The authors thank Dr. Ashok K. Chauhan, Founder President, Dr. Atul Chauhan, Chancellor, and Prof. Dr. Balvinder Shukla, Vice Chancellor, Amity University, Noida, UP, India, for the motivation and research facilities.

References

Alcaide IV, Hamdi A, Guilleín-Bejarano R, Jiménez-Araujo A, Rodríguez-Arcos R (2023) Sustainable valorization of co-products from asparagus cultivation by obtaining bioactive compounds. Front Plant Sci 14:1199436. https://doi.org/10.3389/fpls.2023.1199436

Alcaide I, Hamdi A, Rodríguez-Arcos R, Bejarano R, Araujo A (2020) Asparagus cultivation co-products: from waste to chance. Food Sci Nutr 6(1):1–4. https://doi.org/10.24966/FSN-1076/100057

Chitrakar B, Zhang M, Adhikari B (2019) Asparagus (Asparagus officinalis): processing effect on nutritional and phytochemical composition of spear and hard-stem byproducts. Trends Food Sci Technol 93:1–11. https://doi.org/10.1016/j.tifs.2019.08.020

Chitrakar B, Zhang M, Zhang X, Devahastin S (2020) Bioactive dietary Fiber powder from asparagus leaf by-product: effect of low-temperature ball milling on physico-chemical, functional and microstructural characteristics. Powder Technol 366:275–282. https://doi.org/10.1016/j.powtec.2020.02.068

Dinan L, Savchenko T, Whiting P (2001) Phytoecdysteroids in the genus Asparagus (Asparagaceae). Phytochemistry 56(6):569–576. https://doi.org/10.1016/S0031-9422(00)00438-6

Elena K (2007) Asparagus diseases. Eur J Plant Sci Biotechnol 1:76–83

Fuentes-Alventosa J, Jaramillo-Carmona S, Rodríguez-Gutiérrez G, Guillén-Bejarano R, Jiménez-Araujo A, Fernández-Bolaños J, Rodríguez-Arcos R (2013) Preparation of bioactive extracts from asparagus by-product. Food Bioprod Process 91:74–82. https://doi.org/10.1016/J.FBP.2012.12.004

Fuentes-Alventosa J, Rodríguez-Gutiérrez G, Jaramillo-Carmona S, Espejo-Calvo J, Rodríguez-Arcos R, Fernández-Bolaños J, Guillén-Bejarano R, Jiménez-Araujo A (2009) Effect of extraction method on chemical composition and functional characteristics of high dietary fibre powders obtained from asparagus by-products. Food Chem 113:665–671. https://doi.org/10.1016/J.FOODCHEM.2008.07.075

Guo Q, Cui SW, Kang J, Ding H, Wang Q, Wang C (2015) Non-starch polysaccharides from American ginseng: physicochemical investigation and structural characterization. Food Hydrocoll 44:320–327. https://doi.org/10.1016/j.foodhyd.2014.09.031

Guo Q, Wang N, Liu H, Li Z, Lu L, Wang C (2020) The bioactive compounds and biological functions of *Asparagus officinalis* L. – a review. J Funct Foods 65:103727. https://doi.org/10.1016/j.jff.2019.103727

Hamdi A, Jaramillo-Carmona S, Beji R, Tej R, Zaoui S, Rodríguez-Arcos R, Jiménez-Araujo A, Kasri M, Lachâal M, Bouraoui N, Guillén-Bejarano R (2017) The phytochemical and bioactivity profiles of wild *Asparagus albus* L. plant. Food Res Int 99(1):720–729. https://doi.org/10.1016/j.foodres.2017.06.027

Hamdi A, Jaramillo-Carmona S, Rodríguez-Arcos R, Jiménez-Araujo A, Lachâal M, Karray-Bouraoui N, Guillén-Bejarano R (2021) Phytochemical characterization and bioactivity of

Asparagus acutifolius: a focus on antioxidant, cytotoxic, lipase inhibitory and antimicrobial activities. Molecules 26(11):3328. https://doi.org/10.3390/molecules26113328

Kar DK, Sen S (1985) Sarsasapogenin in callus culture of *Asparagus racemosus*. Curr Sci 54:585

Kohli D, Champawat PS, Mudgal VD (2023) Asparagus (Asparagus racemosus L.) roots: nutritional profile, medicinal profile, preservation, and value addition. J Sci Food Agric 103(5):2239–2250. https://doi.org/10.1002/jsfa.12358

Montone CM, Chiozzi RZ, Marchetti N, Cerrato A, Antonelli M, Capriotti AL, Cavaliere C, Piovesana S, Laganà A (2019) Peptidomic approach for the identification of peptides with potential antioxidant and anti-hyperthensive effects derived from Asparagus by-products. Molecules 24(19). https://doi.org/10.3390/molecules24193627

Morrison WR, Tuell JK, Hausbeck MK, Szendrei Z (2011) Constraints on asparagus production: the association of *Ophiomyia simplex* (Diptera: agromyzidae) and Fusarium spp. Crop Sci 51:1414–1423. https://doi.org/10.2135/cropsci2011.01.0032

Negi JS, Singh P, Joshi GP, Rawat MS, Bisht VK (2010) Chemical constituents of Asparagus. Pharmacogn Rev 4(8):215–220. https://doi.org/10.4103/0973-7847.70921

Noor S, Bhat Z, Kumar S, Mudiyanselage R (2018) Preservative effect of *Asparagus racemosus*: a novel additive for bioactive edible films for improved lipid oxidative stability and storage quality of meat products. Meat Sci 139:207–212. https://doi.org/10.1016/j.meatsci.2018.02.001

Observatorio de precios y mercados, CAPADS (2023) Espárrago 2021–2022 (Junta de Andalucía, Sevilla, Spain: Junta de Andalucía). Available from http://www.cap.junta-andalucia.es/agriculturaypesca/observatorio

Saxena VK, Chaurasia S (2001) A new isoflavone from the roots of *Asparagus racemosus*. Fitoterapia 72:307–309

Elmer W (2018) Asparagus decline and replant problem: a look back and a look forward at strategies for mitigating losses. Acta Hortic 1223:195–204. https://doi.org/10.17660/ActaHortic.2018.1223.27

Sekine T, Fukasawa N, Kashiwagi Y, Ruangrungsi N, Murakoshi I (1994) Structure of asparagamine A: a novel polycyclic alkaloid from Asparagus racemosus. Chem Pharm Bull 42:1360–1362

Sekine T, Fukasawa N, Murakoshi I, Ruangrungsi N (1997) A 9,10-dihydrophenanthrene from *Asparagus racemosus*. Phytochemistry 44:763–764

Viera-Alcaide I, Hamdi A, Guillén-Bejarano R, Rodríguez-Arcos R, Espejo-Calvo J, Jiménez-Araujo A (2022) Asparagus roots: from an agricultural by-product to a valuable source of fructans. Food Secur 11(5):652. https://doi.org/10.3390/foods11050652

Wang J, Liu Y, Zhao J, Zhang W, Pang X (2013) Saponins extracted from by-product of Asparagus officinalis L. suppress tumour cell migration and invasion through targeting Rho GTPase signalling pathway. J Sci Food Agric 93(6):1492–1498. https://doi.org/10.1002/jsfa.5922

Yergeau E, Vujanovic V, St-Arnaud M (2006) Changes in communities of fusarium and arbuscular mycorrhizal fungi as related to different asparagus cultural factors. Microb Ecol 52:104–113. https://doi.org/10.1007/s00248-006-9047-7

Yu Q, Fan L (2021) Improving the bioactive ingredients and functions of asparagus from efficient to emerging processing technologies: a review. Food Chem 358:129903. https://doi.org/10.1016/j.foodchem.2021.129903

Zandstra BH, Morse S, Tocco RV, Morrice JJ (2013) Response of asparagus to repeated application of residual herbicides. Hortechnology 23:109–113. https://doi.org/10.21273/HORTTECH.23.1.109

Zhang W, Wu W, Wang Q, Chen Y, Yue G (2014) The Juice of Asparagus by-product exerts hypoglycemic activity in streptozotocin-induced diabetic rats. J Food Biochem 38:509–517. https://doi.org/10.1111/JFBC.12084

Zhao J, Zhang W, Zhu X, Zhao D, Wang K, Wang R, Qu W (2011) The aqueous extract of *Asparagus officinalis* L. by-product exerts hypoglycaemic activity in streptozotocin-induced diabetic rats. J Sci Food Agric 91(11):2095–2099. https://doi.org/10.1002/jsfa.4429

Zhang SH, Xia YQ, Zhu Q (2002) Study on nutrient composition analysis and processing quality improvement of asparagus. Sci Technol Food Ind 6

Zhu X, Zhang W, Zhao J, Wang J, Qu W (2010) Hypolipidaemic and hepatoprotective effects of ethanolic and aqueous extracts from Asparagus officinalis L. by-products in mice fed a high-fat diet. J Sci Food Agric 90(7):1129–1135. https://doi.org/10.1002/jsfa.3923

Bioactive Phytochemicals in Kohlrabi (*Brassica oleracea* var. *gongylodes*) By-products

Vanessa Situmbeko, Martha Ikechukwu, Natascha Cheikhyoussef, Ahmed A. Hussein, and Ahmad Cheikhyoussef

Contents

Abstract Kohlrabi (*Brassica oleracea* var. *gongylodes*) is a vegetable of the cruciferous family (Brassicaceae). Based on the color of kohlrabi, it is classified as pur-

V. Situmbeko
Department of Biochemistry, Microbiology & Biotechnology, Faculty of Agriculture, Engineering and Natural Sciences, University of Namibia, Windhoek, Namibia

M. Ikechukwu
Sam Nujoma Marine and Coastal Resources Research Centre, Sam Nujoma Campus University of Namibia, Henties Bay, Namibia

N. Cheikhyoussef
Ministry of Higher Education, Technology, and Innovation (MHETI), Windhoek, Namibia

A. A. Hussein
Chemistry Department, Cape Peninsula University of Technology, Cape Town, South Africa

A. Cheikhyoussef (✉)
Science and Technology Division, Multidisciplinary Research Services, Centre for Research Services, University of Namibia, Windhoek, Namibia
e-mail: acheikhyoussef@unam.na

M. F. Ramadan (ed.), *Bioactive Phytochemicals in By-products from Leaf, Stem, Root and Tuber Vegetables*, https://doi.org/10.1007/978-3-031-80700-8_7

ple or pale green kohlrabi. Kohlrabi by-products include mainly peels, sprouts, leaves, and seeds, with several phytochemicals being reported from these by-products, such as anthocyanin, flavonoids, glucosinolates, indoles, phenolic acids, and phytosterols. Phytochemicals from kohlrabi by-products possess several biological activities, including anti-hyperalgesic, anti-inflammatory, anti-hyperglycemic, antibacterial, and antioxidant activities. Other health benefits from kohlrabi by-products include blood pressure lowering, cholesterol reduction, cancer-fighting, chemopreventive properties, and weight loss adjuvant. Furthermore, kohlrabi by-products have demonstrated promising applications in silver and gold nanoparticle synthesis, fermented vegetables, functional food formulation and packaging with potent antioxidant, antibacterial, and UV-blocking properties and extended shelf life and preservation properties. This chapter presents the latest research and development on the kohlrabi by-products regarding bioactive phytochemicals, biological activities, and their potential industrial applications.

Keywords Anthocyanin · Antioxidant · Kohlrabi by-products · Glucosinolates · Functional food · Health benefits

Abbreviations

ABTS	2,2′-azino-bis(3-ethylbenzothiazoline-6-sulfonic acid)
ASTM	American Society for Testing and Materials
DPPH	1,1-diphenyl-2-picrylhydrazyl
GSLs	Glucosinolates
GK	Green kohlrabi
HC	Hydrogen-to-carbon
LED	Light-emitting diode
LPS	Lipopolysaccharide
NO	Nitric oxide
$ONOO^-$	Peroxynitrite
RK	Red kohlrabi
TPC	Total phenolic content
TSH	Thyroid-stimulating hormone
UFAs	Unsaturated fatty acids
UV	Ultraviolet

1 Introduction and Economic Values of Each Discussed By-product

Brassica oleracea var. *gongylodes* is known as kohlrabi, a biennial vegetable well-known for its mildly sweet flavor and several health benefits upon regular consumption and inclusion in the human diet (Smychkovich and Hashemi 2022). Kohlrabi has several worldwide cultivation landscapes in Europe, Asia, and North America. Kohlrabi by-products include mainly peels, sprouts, leaves, and seeds. The two varieties of "Soroksar White" and "Szentes Forcing" have been approved for the cultivation of kohlrabi, according to Kalloo (1993). However, several kohlrabi varieties with various colors of white, purple, and green are being cultivated (Choi et al. 2010). The practice of clonal multiplication for kohlrabi has been successfully achieved by using the cuttings of root buds formed after the decapitation of the original plant (Kalloo 1993). Kohlrabi by-products include peels, sprouts, leaves, and seeds with varieties of biological activities and health benefits associated with their consumption, including anticancer activity (Fahey et al. 2001; Kalhor Qom et al. 2020) and weight management (Liu et al. 2003), among others. This chapter aims to collate and present the latest literature on the Kohlrabi by-products' bioactive compounds and their functionalities and potential valorization beyond classical uses in food and flock medicine.

2 Chemical Composition and Bioactive Compounds of Extracts from By-products

Kohlrabi by-products are rich with several secondary metabolites, including anthocyanin, flavonoids, glucosinolates (glucoerucin, glucoiberin, and sinigrin), isothiocyanates, indoles, phenolic acids, amino acids, sulforaphane, and phytosterols (Choi et al. 2010; Gerendás et al. 2008; Paśko and Galanty 2023; You et al. 2008). This rich profile qualifies kohlrabi by-products as a functional food (Zagrodzki et al. 2020; Paśko et al. 2021). The rich and significant amount of anthocyanin available in the peel of kohlrabi gives it a unique color, especially purple or red. Ağagündüz et al. (2022) reported high content of vitamin C (62 mg/100 g), β-carotene (22 μg/100 g), and glucosinolate (829 mg/100 g) in kohlrabi bulbs, as well as smaller amount of folate, vitamin A, E, and K. Pasko et al. (2014) reported the presence of serotonin, melatonin, and certain indole derivatives in kohlrabi seeds, sprouts, and bulbs, respectively. Volatile organic compounds (VOCs) in four kohlrabi cultivars have been reported by Wieczorek and Jeleń (2019) and indicated the high level of isothiocyanates and nitrile concentration, as well as the sulfides, were the most abundant group of volatile components in all raw kohlrabi cultivars. The phytochemical composition of purple kohlrabi sprouts was changed after being exposed to UVB irradiation, which can impact the overall alteration in metabolite composition (Yeo et al. 2022), whereby an increased level of secondary metabolites (anthocyanins and phenolic acids) and the ABTS-free radical scavenging capability have been observed by Yeo et al. (2022). Another study on kohlrabi leaves VOCs and phenolic profile was reported recently by Ben Sassi et al. (2022) in Tunisia, showing their antioxidant and antibacterial properties. The methanolic extract of kohlrabi leaves contains hydroxycinnamic acids such as chlorogenic, ferulic, *p*-coumaric and sinapic acids, hydroxybenzoic acids of syringic and gallic acids, and four flavonoids including catechol, epigallocatechin, and epicatechin 3-*O*-gallate (Ben Sassi et al. 2022). They further indicated that kohlrabi leaf methanol extract has an antiradical activity as determined by DPPH (EC_{50} of 0.32 mg/mL) and ABTS (EC_{50} 0.45 mg/mL). Kohlrabi leaves offer a rich source of nutrient-dense leafy greens to consumers when compared to kale and other Kohlrabi by-products thanks to the high level of macro minerals (Ca, K, and Mg) of 427, 393, and 44 mg/100 g of the fresh weight, respectively (Smychkovich and Hashemi 2022).

3 Biological and Functional Properties of Extracts and Bioactive Compounds from By-products

Kohlrabi is characterized by its low-fat content while being rich in dietary fiber and a spectrum of essential nutrients, including vitamins C, B6, Mg, Ca, P, K, and vital minerals. Additionally, it is a valuable source of bioactive compounds (Table 1) known for their health-promoting attributes. These include phenylpropanoids (flavonoids and phenolic acids), glucosinolates, carotenoids, anthocyanins, and

Table 1 Bioactive compounds/extracts in kohlrabi by-products and their functional properties and health benefits

Compound	Functional properties	References
β-carotene	Prevent blindness, skin aging, heart disease, and cancer	Choi et al. (2010), Eggersdorfer and Wyss (2018), Maoka (2020)
Flesh and peel ethanol extracts	Antioxidant and anticancer activity	Yang et al. (2015)
Fibers (including pectin and cellulose)	Anti-obesity, appetite, and weight regulation	Choi et al. (2010), Hebbar et al. (2024), Akhlaghi (2024)
Glucosinolates	Antioxidant and anticarcinogenic properties	Hayes et al. (2008), Cartea and Velasco (2008)
Lutein	Antioxidant, anticancer, neuroprotective, and cardiovascular protective properties	Calvo (2005), Divisi et al. (2006), Takeda et al. (2017), Zeni et al. (2019), Akkewar et al. (2024)
Methanol (MeOH) extracts from RK and GK bulbs	Inhibited LPS-induced NO production and suppressed iNOS and COX-2 protein production	Jung et al. (2014)
Phenylpropanoids	Inhibits nitric oxide production in RAW 264.7 macrophage cells, anti-inflammatory agents	Korkina et al. (2011), Lee et al. (2014)
Sterol	Anti-diabetes, anticancer, cholesterol management, and cardiovascular prevention	Barkas et al. (2023), Lee et al. (2010), Munawar et al. (2023)
Sulforaphane and erucin	Anticancer activity	Higdon et al. (2007), You et al. (2008)
Zeaxanthin	Hepatoprotective, retina protection	Divisi et al. (2006), Bernstein et al. (2016), Akkewar et al. (2024)

phytosterols, which have been associated with both antibacterial and antioxidant properties (Paśko et al. 2021; Ağagündüz et al. 2022).

According to Sathasivam et al. (2023), phenylpropanoids are a varied group of organic compounds that combine the amino acids: phenylalanine and tyrosine. These compounds are synthesized within plants and possess considerable significance due to their reported beneficial effects on human health and applications in the pharmaceutical sector. Additionally, phenylpropanoids have been associated with mitigating health issues such as cancer, diabetes, and heart disease. Moreover, glucosinolates (GSLs), categorized into aliphatic, aromatic, and indolyl types based on their chemical composition, represent vital secondary metabolites characterized by their nitrogen and sulfur content (Cartea and Velasco 2008; Choi et al. 2010). GSLs have been noted for their many antioxidant properties and potential anticarcinogenic effects in human subjects. Carotenoids, a group of natural pigments prevalent in plant and animal tissues, contribute to various organisms' vibrant orange, yellow, and red hues (Ngamwonglumlert et al. 2020). Besides enhancing photosynthetic processes in plants, carotenoids serve as photoprotective agents by neutralizing reactive oxygen species. In addition, they possess antioxidant properties, offer ultraviolet (UV) protection, and facilitate vitamin synthesis, thereby holding promise for improving human health and mitigating conditions such as vision impairment, skin aging, heart disease, and cancer (de Mejia et al. 2020; Carrillo et al. 2022).

Park et al. (2012) investigated the chemical composition of the skin and flesh of pale green and purple kohlrabi varieties. Their analysis revealed the presence of 8 glucosinolates, 12 anthocyanins, 2 carotenoids, and 7 phenylpropanoids. The distribution of glucosinolates differed across various parts and cultivars of kohlrabi, with the flesh of purple kohlrabi exhibiting glucoerucin levels four times higher than those found in the skin (Park et al. 2012). Notably, cyanidin-3-(feruloyl)-diglucoside-5-glucoside emerged as the predominant anthocyanin among the 12 identified. Carotenoid concentrations were significantly higher in the skins compared to the flesh in both kohlrabi types.

Furthermore, purple kohlrabi demonstrated elevated levels of most phenylpropanoids compared to pale green varieties, with *trans*-cinnamic acid content notably 12.7 times higher in the flesh of purple kohlrabi. Hence, considerable variations were observed in the quantities of glucosinolates, anthocyanins, carotenoids, and phenylpropanoids, emphasizing the significant differences between the two kohlrabi variants (Park et al. 2012). Antidiabetic and anti-inflammatory effects of green and red kohlrabi cultivars were reported by Jung et al. (2014) and revealed that red kohlrabi (RK) had double total phenolic content (TPC) than green kohlrabi (GK) and showed significant DPPH, ABTS, and peroxynitrite (ONOO−) scavenging assays indicating the potential antioxidant effects.

A comparable investigation undertaken by Sathasivam et al. (2023) utilizing light-emitting diode (LED) technology unveiled the presence of 13 phenylpropanoid compounds, 8 glucosinolates (GSLs), and 5 distinct carotenoids. Notably, blue LED light produced the highest concentrations of GSL and phenylpropanoid. In contrast, carotenoid concentration was highest under white LED light.

In regions with northern climates, early transplanting of Kohlrabi presents a potential avenue for enhancing crop yield and quality of bulbs and leaves. Smychkovich and Hashemi (2022) undertook a two-year field study to assess the effects of varying transplanting dates on kohlrabi's yield and nutrient content, with a specific focus on quantifying the nutritional value of its leaves. The findings indicated a significant correlation between delayed transplanting and increased kohlrabi yield, with a potential gain of up to 307 kg per hectare for each day of delay. Notably, a rise in soil temperature by 2.8 °C observed between April 23 and May 14 resulted in heightened accumulation of essential elements such as Ca, Cu, Mn, and Fe within the kohlrabi bulbs, along with increased levels of Ca and Cu in the leaves (Smychkovich and Hashemi 2022). The study highlighted a positive correlation between nutrient concentrations in the leaves and bulbs, suggesting that simultaneous harvesting of both components could optimize quality. While leaf yield did not exhibit significant variation across transplanting dates, the number of leaves and total leaf area increased with delayed transplanting. Interestingly, no correlation was observed between leaf yield, leaf area, and bulb yield, indicating that reductions in yield and nutrient concentrations were not attributed to photosynthetic efficiency (Smychkovich and Hashemi 2022).

While kohlrabi bulbs have long been recognized as essential to a healthy diet, sprouts are a relatively recent addition to functional foods (Paśko et al. 2021). It is commonly known that eating Brassica vegetables can lower a person's chance of developing gastrointestinal cancers. In vivo studies have also shown that kohlrabi possesses anti-inflammatory, anti-hyperglycemic, and antioxidant properties (Amssayef and Eddouks 2021; Paśko et al. 2021; Rasal et al. 2005; Sharma et al. 2015). Sprouts are an example of functional food with an appealing, ready-to-use form and no unnecessary ingredients. These findings have resulted in a significant increase in global sprout consumption (López-Chillón et al. 2019). According to a study conducted by Paśko et al. (2021), kohlrabi seeds and bulbs contain 9.1 and 37.0% of total fatty acids, respectively, with palmitic and stearic acids dominating in seeds and stearic, palmitic, and capric acids in bulbs. Their analysis comprised the compounds beneficial to human health and those with possibly anti-nutritive properties, especially progoitrin and erucic acid (Paśko et al. 2021).

Kohlrabi seeds contained 90.9% unsaturated fatty acids (UFAs), primarily oleic and erucic acids. Oleic acid was the primary UFA in kohlrabi bulbs (26.7%), with erucic acid accounting for 13.6%. Erucic acid is one of the most common UFAs in most kohlrabi samples and is known for its potential toxicity (Paśko et al. 2018). They concluded that growth conditions can be changed to lower the amount of potentially hazardous chemicals, as demonstrated with progoitrin, and their first results imply that erucic acid might not be as harmful as originally believed. Another study by Paśko et al. (2022) explored the impact of kohlrabi sprouts diet in two models of rat hypothyroidism as animal model for sulfadimethoxine-induced thyroid disorders whereby kohlrabi sprouts diet decreased the TNF-α and WBC levels and adversely interfered with liver function in rats, because of the high dietary intake of glucosinolates in kohlrabi sprouts.

Furthermore, they reported a protective effect against thyroid-stimulating hormone (TSH) and triiodothyronine (T3) hormone in groups with thyroid dysfunction and reducing negative changes in the morphological image of the thyroid gland (Paśko et al. 2022). An earlier study in the last century by Langer (1960) reported the goitrogenic potential of kohlrabi after it was administered to guinea pigs for 24 h and reported lower accumulation of radioiodine in the thyroid gland. An earlier study in 1932 conducted on rabbits fed with 250 g kohlrabi for 5 weeks yielded a significant result in the thyroid gland, whereby twice the normal size and moderate hyperemia were confirmed at the end of the experiment (Marine et al. 1933). For further reading on the potential goitrogenic properties of kohlrabi, an up-to-date systematic review was reported by Galanty et al. (2024) on *Brassica* vegetable's effect on thyroid function.

The cytotoxic impact of kohlrabi sprouts and bulb extract on human breast MCF-7 cells was evaluated with IC50 389.5 μg/mL and classified as a moderate effect by Jamuna et al. (2017). They attributed the mechanism of cell death to necrosis (45.5 and 63.9%) rather than apoptosis (32 and 32.5%) in HepG2 cells. Meanwhile, in SW480 cells, sprouts induced necrosis as the predominant mechanism of cell death, while the bulbs stimulated apoptosis in the cells (Rizk and Zarzour 2013).

4 Food and Non-food Applications of Extracts and Bioactive Compounds from By-products

4.1 *Food Applications*

Kohlrabi can be eaten fresh in salads, cooked with other vegetables and meat, or even pickled for long-term use. Kohlrabi leaves are flavorful and can be compared with kale in appearance and texture, with high nutritional value and additional commodity to growers and consumers (Smychkovich and Hashemi 2022). Jam and fermented products of kohlrabi were reported (Hien et al. 2022; Karaoglan and Aktas 2023). Kohlrabi fermentation increased different organic acids, amino acids, and other metabolites, including L-glutamic acid, L-aspartic acid, and γ-aminobutyric acid (Nie et al. 2022).

4.2 *Non-Food Applications*

Kohlrabi by-products have demonstrated promise in non-food applications. An example of this is a recent study conducted by Khan et al. (2024) in which kohlrabi skin-derived anthocyanin (KA) and Zn-doped carbon dots (Zn-CDs) were utilized to create carrageenan-based pH-responsive food packaging films, where Zn-CDs offered antioxidant, antibacterial, and UV-blocking properties without compromising mechanical characteristics, while KA introduced pH-dependent color-changing attributes. Additionally, active packaging with Car/KA@Zn-CD films extended the shelf life and preserved the quality of packaged shrimp, showcasing the potential for sustainable and innovative biomass-derived fillers in intelligent packaging.

Silver and gold nanoparticles from purple kohlrabi peel have been synthesized (Sorescu et al. 2021) with promising applications in fermented vegetables, functional food formulation and food packaging due to their potent antioxidant, antibacterial, UV-blocking properties, and preservation properties (Ansar et al. 2020; Kuppusamy et al. 2015; Singh et al. 2018; Perveen et al. 2021). Sorescu et al. (2021) formulated silver nanoparticles (AgNPs) and gold nanoparticles (AuNPs) from the aqueous extract of purple kohlrabi peel and the corresponding metallic salts at different temperatures and reported DPPH for the AgNPs (79.33%) at room temperature and (71.29%) for the AuNPs at 50 °C.

5 Valorization of By-products for Non-health Purposes, i.e., Source of Bioenergy

5.1 Biofuels Application

Kohlrabi by-products, such as leaves and stems, can be repurposed as raw materials to produce products with downstream applications. They can be valorized and utilized with other materials and used to make products with applications in renewable energy, as illustrated in Fig. 1.

Fig. 1 Applications of kohlrabi by-products valorization in bioenergy

Biofuels are fuels made by fermentation and hydrolysis of plant material with yeast to produce ethanol and carbon dioxide. Biofuels are classified as first-, second-, and third-generation based on the raw material from which they are generated and the technology used to manufacture them (De Almeida and Colombo 2023). Ethanol and biodiesel are first-generation biofuels produced by fermenting sugars and starches from edible crops such as sugarcane and corn (Dharani et al. 2024). However, the sustainable production of this type of first-generation biofuels is a challenge because their feedstocks are crops consumed as food, creating competition for resources. The second-generation biofuels are fuels produced using lignin and cellulose-rich feedstocks from non-food crops. Produced renewable diesel and drop-in aviation fuels are examples of second-generation biofuels that are considered more sustainable energy sources as they utilize material readily available in abundance as feedstocks (De Almeida and Colombo 2023). Third-generation biofuels are synthesized using microorganisms such as microalgae and cyanobacteria biomass (Bhat et al. 2022; Cavelius et al. 2023). The fourth-generation biofuels are generated by genetically modifying microorganisms that are used in the third generation of biofuel to achieve a preferable hydrogen-to-carbon (HC) yield (Alalwan et al. 2019).

The abundance of lignocellulosic material with limited use presents an opportunity to explore the potential of alternative materials that can be used as feedstock to enhance biofuel production yield and efficiency. The lignocellulosic nature of the kohlrabi by-products qualifies them as suitable feedstock to produce second-generation biofuels. Although there is no documentation on using kohlrabi by-products for biofuel synthesis, they are ideal candidates for investigations. Many crops that belong to *Brassica* have been valorized to produce biofuels. For instance, the valorization of *Brassica oleracea* L. has been reported to achieve a high biodiesel yield whose properties meet the American Society for Testing and Materials (ASTM) standards (Cheema et al. 2021). Other brassicas such as rapeseed (*Brassica napus*), black mustard (*Brassica nigra*), and Indian mustard (*Brassica juncea*) have also been used as low-cost substrates for biodiesel production (Gidik et al. 2023; Yadav et al. 2022; Basili and Rossi 2018). Non-edible oil seeds from *Brassica oleracea* L. have been used as raw material which produced a high yield of biodiesel and the latter has been confirmed to meet ASTM standards (Cheema et al. 2021). Both biodiesel and bioethanol have also been synthesized from *Brassica carinata* oil and optimized using KOH as a catalyst, resulting in a maximum yield of ester (Basili and Rossi 2018).

5.2 Biogas Production

Kohlrabi by-products have potential applications in the production of biogas. The by-products of kohlrabi, however, are not commonly used as a feedstock in biodigester due to their glucosinolate content. Glucosinolates are known to affect microbial composition negatively and, in turn, methanogenesis. The sulfur compounds resulting from the hydrolysis of glucosinolates interfere with their metabolic processes, which are essential for efficient biogas generation (Peu et al. 2013). Besides, glucosinolates from *Brassica* are known to pollute biogas with hydrogen sulfide. Co-digestion of several substrates has been shown to reduce the negative impacts of kohlrabi by-products on biogas production and quality.

Moreover, co-digestion has been demonstrated to boost biogas yields because multiple anaerobic bacteria operate synergistically, targeting different materials and steps for optimal biogas production (Das et al. 2023). Although no studies have investigated the use of kohlrabi specifically as a feedstock for biodigester, other *Brassica* crops have been investigated. For instance, *Brassica juncea* and *Brassica napus* are viable co-substrates in anaerobic digestion with no interference, resulting in biogas with an acceptable hydrogen sulfide level (Peu et al. 2013). Since these *Brassica* crops have similar phytochemical properties to kohlrabi, co-digestion of kohlrabi with other substrates might have similar effects on biogas production and quality. Given the similarities shared among the *Brassica*, the possibility that by-products of kohlrabi can be harnessed as feedstock for biofuel and biogas production cannot be ruled out. The applications of by-products of kohlrabi in bioenergy synthesis can help to reduce waste and promote sustainable energy production.

References

Ağagündüz D, Şahin TÖ, Yılmaz B, Ekenci KD, Duyar Özer Ş, Capasso R (2022) Cruciferous vegetables and their bioactive metabolites: from prevention to novel therapies of colorectal cancer. Evid Based Complement Alternat Med 2022:1–20

Akhlaghi M (2024) The role of dietary fibers in regulating appetite, an overview of mechanisms and weight consequences. Crit Rev Food Sci Nutr 64(10):3139–3150

Akkewar AS, Mishra KA, Kamble MG, Kumar S, Dey J, Sethi KK (2024) A mechanistic review on growing multiple therapeutic applications of lutein and its global market research. Phytother Res 38., Advance online publication. https://doi.org/10.1002/ptr.8197

Alalwan HA, Alminshid AH, Aljaafari HA (2019) Promising evolution of biofuel generations. Subject review. Renew Energy Focus 28:127–139

Amssayef A, Eddouks M (2021) Antihyperglycemic effect of the Moroccan Collard Green (*Brassica oleracea* var. *viridis*) in Streptozotocin-induced diabetic rats. Endocr Metab Immune Disorders Drug Targets 21(6):1043–1052

Ansar S, Tabassum H, Aladwan NSM, Naiman Ali M, Almaarik B, AlMahrouqi S, Abudawood M, Banu N, Alsubki R (2020) Eco friendly silver nanoparticles synthesis by *Brassica oleracea* and its antibacterial, anticancer and antioxidant properties. Sci Rep 10(1):18564

Barkas F, Bathrellou E, Nomikos T, Panagiotakos D, Liberopoulos E, Kontogianni MD (2023) Plant sterols and plant stanols in cholesterol management and Cardiovascular prevention. Nutrients 15(13):2845

Basili M, Rossi MA (2018) Brassica carinata-derived biodiesel production: economics, sustainability and policies. The Italian case. J Clean Prod 191:40–47

Ben Sassi A, Ascrizzi R, Chiboub W, Cheikh Mhamed A, ElAyeb A, Skhiri F, Tounsi Saidani M, Mastouri M, Flamini G (2022) Volatiles, phenolic compounds, antioxidant, and antibacterial properties of kohlrabi leaves. Nat Prod Res 36(12):3143–3148

Bernstein PS, Li B, Vachali PP, Gorusupudi A, Shyam R, Henriksen BS, Nolan JM (2016) Lutein, zeaxanthin, and meso-zeaxanthin: the basic and clinical science underlying carotenoid-based nutritional interventions against ocular disease. Prog Retin Eye Res 50:34–66

Bhat RA, Singh DV, Tonelli FMP, Hakeem KR (2022) Plant and algae biomass. Springer

Calvo MM (2005) Lutein: a valuable ingredient of fruit and vegetables. Crit Rev Food Sci Nutr 45(7–8):671–696

Carrillo C, Nieto G, Martínez-Zamora L, Ros G, Kamiloglu S, Munekata PES, Pateiro M, Lorenzo JM, Fernández-López J, Viuda-Martos M, Pérez-Álvarez JÁ, Barba FJ (2022) Novel approaches for the recovery of natural pigments with potential health effects. J Agric Food Chem 70(23):6864–6883

Cartea ME, Velasco P (2008) Glucosinolates in Brassica foods: bioavailability in food and significance for human health. Phytochem Rev 7(2):213–229

Cavelius P, Engelhart-Straub S, Mehlmer N, Lercher J, Awad D, Brück T (2023) The potential of biofuels from first to fourth generation. PLoS Biol 21(3):e3002063

Cheema SI, Ahmad M, Ullah R, Mothana RA, Noman OM, Zafar M et al (2021) Implication, visualization, and characterization through scanning electron microscopy as a tool to identify nonedible oil seeds. Microsc Res Tech 84(3):379–393

Choi S-H, Ryu D-K, Park S-H, Ahn K-G, Lim Y-P, An G-H (2010) Composition analysis between kohlrabi (*Brassica oleracea* var. *gongylodes*) and radish (*Raphanus sativus*). Hortic Sci Technol 28(3):469–475

Das A, Das S, Das N, Pandey P, Ingti B, Panchenko V et al (2023) Advancements and innovations in harnessing microbial processes for enhanced biogas production from waste materials. Agriculture 13(9):1689

De Almeida MA, Colombo R (2023) Production chain of first-generation sugarcane bioethanol: characterization and value-added application of wastes. Bioenergy Res 16(2):924–939

Dharani L, Umapriya R, Arunkumar N, Gokila M, Sakthi Shankar R (2024) Generations of biofuel. In: Emerging sustainable technologies for biofuel production. Cham, Springer Nature Switzerland, pp 15–42

Divisi D, Di Tommaso S, Salvemini S, Garramone M, Crisci R (2006) Diet and cancer. Acta Biomedica-Ateneo Parmense 77(2):118

Eggersdorfer M, Wyss A (2018) Carotenoids in human nutrition and health. Arch Biochem Biophys 652:18–26

Fahey JW, Zalcmann AT, Talalay P (2001) The chemical diversity and distribution of glucosinolates and isothiocyanates among plants. Phytochemistry 56(1):5–51

Galanty A, Grudzińska M, Paździora W, Służały P, Paśko P (2024) Do Brassica vegetables affect thyroid function? – A comprehensive systematic review. Int J Mol Sci 25(7):3988

Gerendás J, Breuning S, Stahl T, Mersch-Sundermann V, Mühling KH (2008) Isothiocyanate concentration in Kohlrabi (*Brassica oleracea* L. var. *gongylodes*) plants as influenced by sulfur and nitrogen supply. J Agric Food Chem 56(18):8334–8342

Gidik B, Volkan GÜL, Önemli F, Girgel Ü (2023) Fatty acid composition and biodiesel quality of *Brassica nigra, Brassica napus* and *Sinapis arvensis* seeds. Bahçe 52(1):1–6

Hayes JD, Kelleher MO, Eggleston IM (2008) The cancer chemopreventive actions of phytochemicals derived from glucosinolates. European journal of nutrition 47:73–88

Hebbar S, Umakanth S, Thimmappa L, Galbao J (2024) Effect of high dietary fiber intake on insulin resistance, body composition and weight, among overweight or obese middle-aged women: study protocol for a double-blinded randomized controlled trail. F1000Research 13:396

Hien TT, Truc TT, Muoi NV (2022) Effect of salt concentration and pH value on the lactic fermentation process of kohlrabi (*Brassica oleracea* L.). Res Innov Food Sci Technol 10(4):399–412

Higdon JV, Delage B, Williams DE, Dashwood RH (2007) Cruciferous vegetables and human cancer risk: epidemiologic evidence and mechanistic basis. Pharmacol Res 55(3):224–236

Jamuna KS, Suma MS, Ramesh CK, Mahmood R, Nanjundaswamy L (2017) Studies on in vitro antiproliferative activities in cruciferous vegetables. J Appl Hortic 19(3):230–234

Jung HA, Karki S, Ehom NY, Yoon MH, Kim EJ, Choi JS (2014) Anti-Diabetic and anti-inflammatory effects of green and red kohlrabi cultivars (*Brassica oleracea* var. *gongylodes*). Prev Nutr Food Sci 19(4):281–290

Kalhor Qom N, Sheykhhasan M, Kowsari A (2020) Evaluation of kohlrabi (*Brassica oleracea* var. *gongylodes*) extract effect on mesenchymal stem cells viability and apoptosis. Res Mol Med 8(2):93–102

Kalloo G (1993) Kohlrabi. In: Genetic improvement of vegetable crops. Elsevier, pp 191–194
Karaoglan HA, Aktas AB (2023) Optimization of Spontaneous Fermentation Conditions of Kohlrabi by Response Surface Methodology. Fermentation 9(6):550
Khan A, Riahi Z, Tae Kim J, Rhim J-W (2024) Carrageenan-based multifunctional packaging films containing Zn-carbon dots/anthocyanin derived from kohlrabi peel for monitoring quality and extending the shelf life of shrimps. Food Chem 432:137215
Korkina L, Kostyuk V, De Luca C, Pastore S (2011) Plant phenylpropanoids as emerging anti-inflammatory agents. Mini Rev Med Chem 11(10):823–835
Kuppusamy P, Ichwan SJ, Parine NR, Yusoff MM, Maniam GP, Govindan N (2015) Intracellular biosynthesis of Au and Ag nanoparticles using ethanolic extract of *Brassica oleracea* L. and studies on their physicochemical and biological properties. J Environ Sci 29:151–157
Langer P (1960) Antithyroidal effect of some fresh plants on guinea pigs. Nature 185(4707):174–175
Lee JW, Lee DY, Cho JG, Baek NI, Lee YH (2010) Isolation and identification of sterol compounds from the red kohlrabi (*Brassica oleracea* var. *gongylodes*) sprouts. J Appl Biol Chem 53(4):207–211
Lee JW, Lee DY, Baek DR, Jeong RH, Lee DS, Kim YC et al (2014) Phenylpropanoids from red kohlrabi sprouts inhibits nitric oxide production in RAW 264.7 macrophage cells. Food Sci Biotechnol 23:965–969
Liu S, Willett WC, Manson JE, Hu FB, Rosner B, Colditz G (2003) Relation between changes in intakes of dietary fiber and grain products and changes in weight and development of obesity among middle-aged women. Am J Clin Nutr 78(5):920–927
López-Chillón MT, Carazo-Díaz C, Prieto-Merino D, Zafrilla P, Moreno DA, Villaño D (2019) Effects of long-term consumption of broccoli sprouts on inflammatory markers in overweight subjects. Clin Nutr 38(2):745–752
Maoka T (2020) Carotenoids as natural functional pigments. J Nat Med 74(1):1–16
Marine D, Baumann EJ, Webster B, Cipra A (1933) The occurrence of antigoitrous substances in plants. J Exp Med 57(1):121
de Mejia EG, Zhang Q, Penta K, Eroglu A, Lila MA (2020) The colors of health: chemistry, bioactivity, and market demand for colorful foods and natural food sources of colorants. Annu Rev Food Sci Technol 11:145–182
Munawar M, Khan MS, Saeed M, Younas U, Farag MR, Di Cerbo A et al (2023) Phytosterol: nutritional significance, health benefits, and its uses in poultry and livestock nutrition. Anim Biotechnol 34(7):3206–3215
Ngamwonglumlert L, Devahastin S, Chiewchan N, Raghavan V (2020) Plant carotenoids evolution during cultivation, postharvest storage, and food processing: a review. Compr Rev Food Sci Food Saf 19(4):1561–1604
Nie X, Chen H, Xiang L, Zhang Y, Liu D, Zhao Z (2022) GC-TOF-MS-based non-targeted metabolomic analysis of differential metabolites in Chinese ultra-long-term industrially fermented kohlrabi and their associated metabolic pathways. Meta 12(10):991
Park WT, Kim JK, Park S, Lee S-W, Li X, Kim YB, Uddin MR, Park NI, Kim S-J, Park SU (2012) Metabolic profiling of glucosinolates, anthocyanins, carotenoids, and other secondary metabolites in kohlrabi (*Brassica oleracea* var. *gongylodes*). J Agric Food Chem 60(33):8111–8116
Paśko P, Galanty A (2023) Bioactive compounds and biological activities of kohlrabi (*Brassica oleracea* Convar. Acephala var. *Gongylodes* L.). In: Murthy HN, Paek KY, Park SY (eds) Bioactive compounds in the storage organs of plants. Reference series in phytochemistry. Springer, Cham
Pasko P, Sulkowska-Ziaja K, Muszynska B, Zagrodzki P (2014) Serotonin, melatonin, and certain indole derivatives profiles in rutabaga and kohlrabi seeds, sprouts, bulbs, and roots. LWT-Food Sci Technol 59(2):740–745
Paśko P, Tyszka-Czochara M, Galanty A, Gdula-Argasińska J, Żmudzki P, Bartoń H, Zagrodzki P, Gorinstein S (2018) Comparative study of predominant phytochemical compounds and proapoptotic potential of broccoli sprouts and florets. Plant Foods Hum Nutr 73(2):95–100

Paśko P, Galanty A, Tyszka-Czochara M, Żmudzki P, Zagrodzki P, Gdula-Argasińska J, Prochownik E, Gorinstein S (2021) Health promoting vs. anti-nutritive aspects of kohlrabi sprouts, a promising candidate for novel functional food. Plant Foods Hum Nutr 76(1):76–82

Paśko P, Okoń K, Prochownik E, Krośniak M, Francik R, Kryczyk-Kozioł J et al (2022) The impact of kohlrabi sprouts on various thyroid parameters in iodine deficiency-and sulfadimethoxine-induced hypothyroid rats. Nutrients 14(14):2802

Perveen R, Shujaat S, Naz M, Qureshi MZ, Nawaz S, Shahzad K, Ikram M (2021) Green synthesis of antimicrobial silver nanoparticles with Brassicaceae seeds. Mater Res Expr 8(5):055007

Peu P, Picard S, Girault R, Labreuche J, Béline F, Dabert P (2013) Catch crops for agricultural biogas production, case study for *Brassicaceae* sp. 15th international conference, Ramiran 2013, Jun 2013, Versailles, France. pp 4. ⟨hal-02599079⟩

Rasal V, Shetty B, Sinnathambi A, Yeshmaina S, Ashok P (2005) Antihyperglycaemic and antioxidant activity of *Brassica oleracea* in streptozotocin diabetic rats. Int J Pharmacol 4(2):1–7

Rizk S, Zarzour V (2013) A comparative study of the antiproliferative effect of kohlrabi and green cabbage on colorectal cancer cell lines in vitro. FASEB J 27:639.13

Sathasivam R, Park SU, Kim JK, Park YJ, Kim MC, Nguyen BV, Lee SY (2023) Metabolic profiling of primary and secondary metabolites in kohlrabi (*Brassica oleracea* var. *gongylodes*) sprouts exposed to different light-emitting diodes. Plan Theory 12(6):1296

Sharma I, Aaradhya M, Kodikonda M, Naik PR (2015) Antihyperglycemic, antihyperlipidemic and antioxidant activity of phenolic rich extract of *Brassica oleraceae* var *gongylodes* on streptozotocin induced Wistar rats. Springerplus 4:212

Singh A, Sharma B, Deswal R (2018) Green silver nanoparticles from novel Brassicaceae cultivars with enhanced antimicrobial potential than earlier reported Brassicaceae members. J Trace Elem Med Biol 47:1–11

Smychkovich A, Hashemi M (2022) Yield and nutrient concentrations of kohlrabi bulbs and leaves as affected by spring transplanting dates. Agronomy 12(4):770

Sorescu AA, Nuta A, Ion RM, Grigore M, Nistor CL, Suica-Bunghez IR (2021) Purple kohlrabi peel, a natural material for eco-friendly silver and gold nanoparticles. Environ Eng Manag J 20(4):605–611

Takeda H, Tominari T, Hirata M, Watanabe K, Matsumoto C, Grundler FMW, Inada M, Miyaura C (2017) Lutein enhances bone mass by stimulating bone formation and suppressing bone resorption in growing mice. Biol Pharm Bull 40(5):716–721

Wieczorek MN, Jeleń HH (2019) Volatile compounds of selected raw and cooked Brassica vegetables. Molecules 24(3):391

Yadav AK, Kuila A, Garlapati VK (2022) Biodiesel production from *Brassica juncea* using oleaginous yeast. Appl Biochem Biotechnol 194(9):4066–4080

Yang MJ, Cha SS, Lee JJ (2015) Effects of purple kohlrabi (*Brassica oleracea* var. *gongylodes*) flesh and peel ethanol extracts on the antioxidant activity and antiproliferation of human cancer cells. Korean J Commun Living Sci 26(2):405–414

Yeo HJ, Lim SY, Park CH, Kim CY, Sathasivam R, Kim JK, Park SU (2022) Metabolic analyses and evaluation of antioxidant activity in purple kohlrabi sprouts after exposed to UVB radiation. Antioxidants 11(8):1443

You Y, Wu Y, Mao J, Zou L, Liu S (2008) Screening of Chinese brassica species for anti-cancer sulforaphane and erucin. Afr J Biotechnol 7(2):147–152

Zagrodzki P, Paśko P, Galanty A, Tyszka-Czochara M, Wietecha-Posłuszny R, Rubió PS, Bartoń H, Prochownik E, Muszyńska B, Sułkowska-Ziaja K, Bierła K, Łobiński R, Szpunar J, Gorinstein S (2020) Does selenium fortification of kale and kohlrabi sprouts change significantly their biochemical and cytotoxic properties? J Trace Elem Med Biol 59:126466

Zeni ALB, Camargo A, Dalmagro AP (2019) Lutein prevents corticosterone-induced depressive-like behavior in mice with the involvement of antioxidant and neuroprotective activities. Pharmacol Biochem Behav 179:63–72

Bioactive Phytochemicals from Celery (*Apium graveolens* L.) By-products

Moncef Chouaibi

Contents

Abstract This chapter describes the most bioactive compounds extracted from celery (*Apium graveolens* L.) leaves. Therefore, there are many bioactive compounds such as polyphenols, volatile organic compounds, dietary fibers, and essential oils. Besides, celery has a characteristic taste and smell due to the content of volatile compounds, i.e., terpenes, aldehydes, and phthalides. Moreover, celery leaves possess many clinical activities such as antioxidant, anti-inflammatory, and antibacterial. Therefore, the results indicated that celery, thanks to its content of numerous

M. Chouaibi (✉)
Higher School of Food Industries of Tunisia, University of Carthage, Carthage, Tunisia
e-mail: moncef.chouaibi@esiat.ucar.tn

M. F. Ramadan (ed.), *Bioactive Phytochemicals in By-products from Leaf, Stem, Root and Tuber Vegetables*, https://doi.org/10.1007/978-3-031-80700-8_8

biologically active substances, is important in health care and can have positive effects on endocrine disorders.

Keywords Extract · Bioactive compounds · Antioxidant activities · Extraction methods

1 Introduction

The most common plant in its family, Apiaceae, is celery (*Apium graveolens* L.). In regions with moderate climates, this annual or biennial plant is widely grown for food or for its seeds, which are frequently used as spices (Malhotra 2006). Detection and quantification in food matrices are further complicated by the fact that all portions of celery (root, stem, leaves, and seeds) are edible and can be eaten raw or prepared. Because of its unique scent character, it is also widely used for its essential oils (Malhotra 2006). Phthalides, specifically 3-butylphthalide, sedanolide, and senkyunolide A (sedanenolide), have been identified as the cause of celery's distinctive smell (Malhotra 2006).

The phytochemical profile of celery and the separation and identification of bioactive components have been the subject of much research in recent years. Coumarins, furanocoumarins, and their corresponding glucosides, phenolics, flavonoids, and phthalides, are found in the seeds and leaves (Motti et al. 2024). Additionally, it has long been used to treat dengue fever, inflammation, and joint or muscle discomfort, as well as to encourage menstrual flow and improve urine excretion. Antimicrobial, antifungal, antiparasitic, anti-inflammatory, anticancer, anti-ulcer, antioxidant, antidiabetic, anti-infertility, antiplatelet, anti-spasmolytic, hepatoprotective, cardioprotective, neuroprotective, cytoprotective, hypolipidemic, and analgesic activities are among the pharmacological effects of *A. graveolens* that have been demonstrated in *in vitro* or *in vivo* studies (Malhotra 2006; Motti et al. 2024).

2 Extraction Methods

2.1 Drying Process

Three methods were used to dry fresh celery and celery samples: convection method (CD), vacuum method (VD), and sublimation method (FD) until equilibrium moisture was reached. The convection method (CD) was carried out at drying medium temperatures of 50, 60, and 70 °C and airflow velocities of 0.5 m/s. Each drying batch contained 40 g of root celery, and three technological repetitions were made. The raw material was placed on screens in a single layer, and the drying medium flowed perpendicular to the material layer. Weight loss was measured at intervals of 5 min for the first hour, 10 min for the second, 20 min for the third, and then every half hour for the 21–23 drying period. The relative humidity of the air in the room containing the drying system was between fifty and sixty percent. The vacuum method was the second technique. In a vacuum laboratory drier, drying was done for 22 h at 60 °C and a vacuum of 10 kPa. The samples were frozen at 20 °C at a rate of 1 °C/min after sublimation drying. The dry material was heated using a contact method to a hotplate temperature of 22 °C. The procedure took place under pressure for 24 h.

2.2 *Vacuum Impregnation (VI)*

Vacuum impregnation was carried out using the Paslawska et al. (2018) method. The 250 g samples were put in a stainless steel tank with holes in it. In stage I, the vessel was placed inside the chamber and kept at a lower pressure of 6 kPa for 2 min. Stage II involved adding 1000 mL of the impregnating solution and leaving the material in the impregnation chamber for 4 min. After restoring air pressure, the samples were held for 20 min in the third stage. Using filter paper, the impregnating fluid was removed. The generated juices were used to create four infiltration solutions (Krecisz et al. 2023).

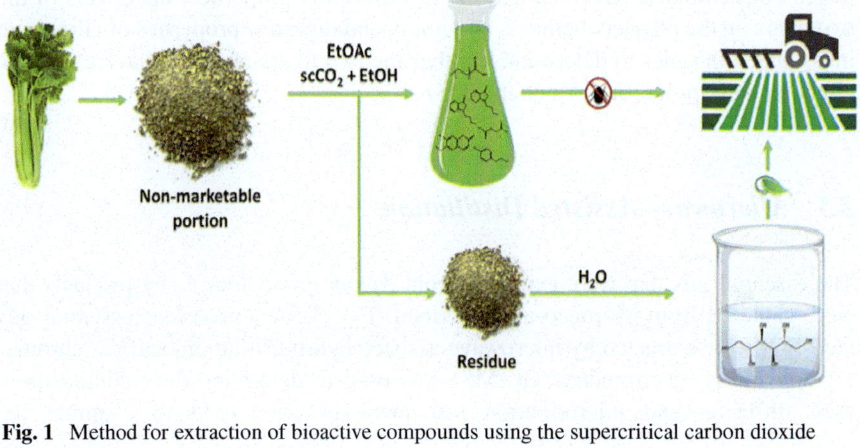

Fig. 1 Method for extraction of bioactive compounds using the supercritical carbon dioxide

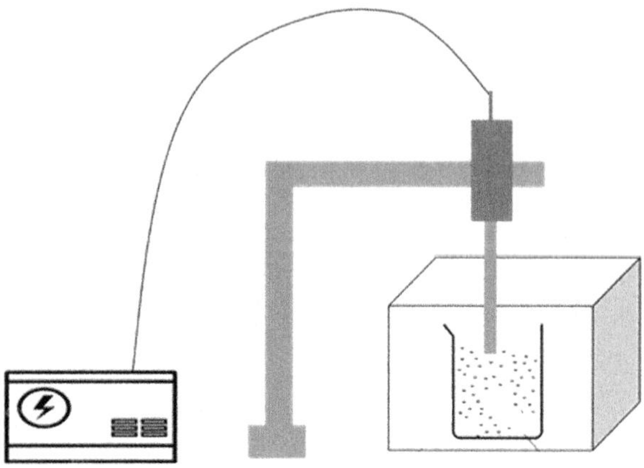

Fig. 2 Experimental implementation of ultrasonic pretreatment of celery leaves

2.3 Supercritical Carbon Dioxide

Using ethanol as a cosolvent, the most popular extraction technique for obtaining the bioactive chemicals was the supercritical carbon dioxide process (Fig. 1). This process's variables included the extraction temperature, duration, ethanol content, solid-to-liquid ratio, and CO_2 flow rate.

2.4 Ultrasonic Process

The two dietary fibers (SDF and IDF) from celery were extracted from the leaves using an ultrasonic process (Fig. 2). This study examines the interaction between two dietary fibers (SDF and IDF) from celery and glutinous rice starch (GRS). Based on rheological, thermal, and in vitro digestion properties, the effects of the two fibers on the physicochemical, structural, and digestive properties of GRS were investigated in order to determine whether the two fibers of celery have a positive effect on these properties (Yu et al. 2023).

2.5 Microwave-Assisted Distillation

The essential oils that were extracted from *Apium graveolens* L. by-products that were gathered from Morocco are described (Fig. 3). *A. graveolens* essential oils (AG-EO) were extracted by microwave-assisted hydrodistillation, and gas chromatography/mass spectrometry (GC/MS) was used to determine their composition. Disc diffusion and microdilution test methods were used to examine the

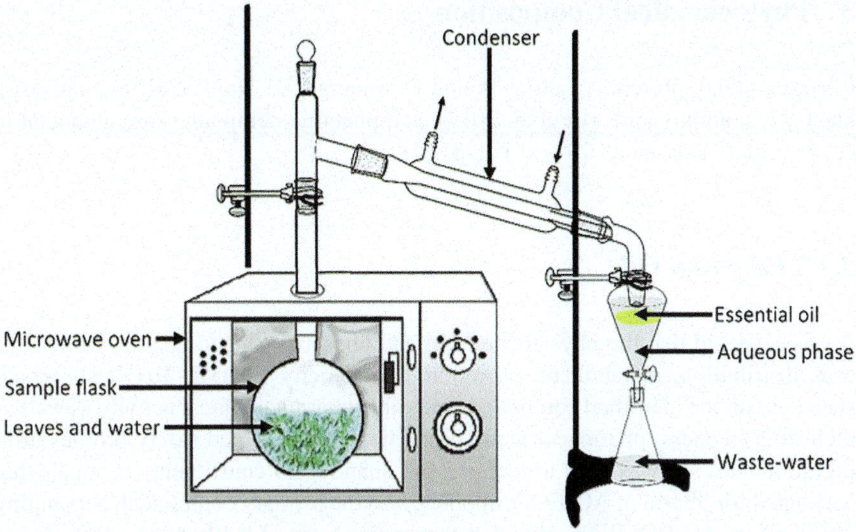

Fig. 3 Microwave-assisted hydrodistillation for essential oil extraction

antimicrobial capabilities, while spectrophotometric techniques were used to evaluate the antioxidant activities. Furthermore, anti-inflammatory properties *in vivo* were investigated. The antioxidative and antibacterial properties of the constituent compounds were clarified by using the molecular docking methodology, a computer method used to anticipate the binding of small molecules to particular proteins.

It was expected that the physicochemical and pharmacokinetic features of absorption, distribution, metabolism, excretion, and toxicity (ADME-Tox) tests would shed light on the identified compounds' possible pharmacological activity as well as their drug-likeness, pharmacokinetic properties, and expected safety profiles after ingestion. GC/MS was used to identify and quantify 15 constituent chemicals that accounted for 99.9% of AG-EO. Limonene was the primary component, accounting for 64.5% of AG-EO. With estimated scavenging rates of 8.49 μg/mL and 5.09 μg/mL for ABTS, AG-EO showed a strong DPPH-free radical scavenging activity. The total antioxidant activity (TAC) and reducing power (RP) of AG-EO were 245.9 mg AAE/mL and 3.42 μg/mL, respectively. Against every examined bacterium, AG-EO demonstrated strong antibacterial activity.

Furthermore, the anti-inflammatory activity proved AG-EO's extraordinary effectiveness. When combined, the essential oil extracted from celery waste has the potential to be widely used as a sustainable preservative in the food and agricultural sectors, successfully preventing aflatoxin contamination and fungal development in items that are kept in storage. Additionally, it shows promise as a viable option for pharmaceutical drug development (Nouioura et al. 2024).

3 Phytochemical Composition

Fiber, carotene, protein, vitamins A and C, minerals (P, Na, Ca, K, Fe, and Mg), karst acids, amino acids (Li et al. 2017), and phenolic compounds are abundant in celery (Kooti and Daraei 2017; Li et al. 2017).

3.1 Essential Oil

It was expected that the physicochemical and pharmacokinetic features of absorption, distribution, metabolism, excretion, and toxicity (ADME-Tox) tests would shed light on the identified compounds' possible pharmacological activity as well as their drug-likeness, pharmacokinetic properties, and expected safety profiles after ingestion. GC/MS was used to identify and quantify 15 constituent chemicals that accounted for 99.9% of AG-EO. Limonene was the primary component, accounting for 64.5% of AG-EO. With estimated scavenging rates of 8.49 μg/mL and 5.09 μg/mL for ABTS, AG-EO showed a high DPPH-free radical scavenging activity.

With reducing power (RP) and total antioxidant activity (TAC) of 3.42 μg/mL and 245.9 mg AAE/mL, respectively, AG-EO demonstrated strong antimicrobial activity against all tested microorganisms, and its anti-inflammatory activity further demonstrated its remarkable efficacy. When combined, the celery seed essential oil shows promise for widespread use as an environmentally friendly preservative in the food and agriculture industries, effectively inhibiting fungal growth and aflatoxin contamination in stored commodities, and is a potential candidate for pharmaceutical drug development. Of the entire mass of AG-EO, 14 compounds made up 99.99%. The main component of AG-EO, limonene, accounted for 64.5% of its mass. Sedanolide (15.3%), hexenyl salicylate (7.27%), naphthalene (6.30%), a-himachalene (1.88%), and β-pinene (1.53%) were among the other substances that were present in lesser amounts. The majority of the chemical makeup (72.4%) was made up of monoterpene hydrocarbons.

Limonene (76.9%), β-selinene (9.7%), sedanenolide (3.4%), 3-butylphthalide (3.6%), and α-selinene (1.4%) were the main chemical ingredients, and these results are in line with those previously reported (Zorga et al. 2020). In a similar vein, Senegal *A. graveolens* taken from Senegal have high levels of β-selinene and limonene. EOs' chemical makeup varies and is affected by a wide range of environmental conditions. These differences in *A. graveolens* oils depend on the following variables: the part of the plant used, the stage of the vegetative cycle, the age of the plants, the harvest time, and genetic characteristics (Barra 2009).

Monoterpene hydrocarbons, oxygenated monoterpenoids, and sesquiterpene hydrocarbons derived from a variety of medicinal plants are among the phytoconstituents found in abundance in essential oils. These substances are known to have significant pharmacological and biological potential. Their proven capabilities include insecticidal efficacy (Kokotkiewicz and Maria Luczkiewicz 2016), antiviral traits (Das et al. 2019), antioxidant effect (Saini et al. 2021), antibacterial potential (Bat-Ozmatara 2020), and cytotoxic impact (Modaresi et al. 2012).

3.2 Polyphenols

Caffeic acid, *p*-coumaric acid, and ferulic acid were the main phenolic acids found in the extracts, whereas apigenin, luteolin, and kaempferol were the flavonoids found. The Folin-Ciocalteu method was used to measure the total phenol concentration, while DPPH· and ABTS radicals were used to analyze the TAOC. The predominant flavonoid in this sample was apigenin, and *p*-coumaric acid was the most prevalent phenolic acid. High amounts of phenolic compounds and antioxidant capacity were reported. Total phenolic acids or total phenolics were determined, and overall flavonoid levels and antioxidant activity were positively correlated. Nagella et al. (2012) investigated the chemical composition and antioxidant properties of essential oils that were extracted from celery leaves. The celery leaves' essential oils were extracted, and they were then distilled and subjected to gas chromatography/mass spectrometry analysis. Essential oils made up 73.7% of the leaf content overall. The essential oils that were extracted from celery leaves and found to inhibit the DPPH· radical were then examined.

The characteristics of celery leaves and seeds were further examined. Components include caffeic acid, *p*-coumaric acid, ferulic acid, apigenin, luteolin, tannin, saponin, and kaempferol, which give celery its potent antioxidant traits. Different plant components in varying amounts can have varied therapeutic benefits. Due to its unique components and concentrations, celery can have a range of therapeutic effects. It is recommended that future research focuses on further medicinal and industrial uses of celery (Kooti and Daraei 2017).

3.3 Bioactive Compounds

Samples including F (fresh celery) and S, O, K, and OK (after drying) contained a total of 41 components, including 15 phenolic acids, 8 flavanols, 3 iridoids, 12 flavonols, and 3 coumarins. Twenty-two phenolic substances from the phenolic acid, flavonoid, iridoid, and coumarin families were found in non-impregnated celery samples. The chemicals identified in the celery samples under study aligned with the reported findings for the celery extract (Li et al. 2017). Two phenolic acids, feruloyl-caffeoyl quinic acid and syringoylquinic acid, were among the chemicals found in celery for the first time. Compounds from the flavonoid category were not identified in non-impregnated celery samples (Krecisz et al. 2023).

3.4 Dietary Fiber

Flavonoids and dietary fiber are potential medications that have been used to treat inflammatory bowel disease (IBD). The relationship between dietary fiber and flavonoids and intestinal health is unclear, however. Using non-group feeding cages, the therapeutic effects of powdered celery, kale, and red chicory were examined in colitis-affected mice (Fig. 4). In order to identify the component or components responsible for the protective effect, the effectiveness of whole celery, celery soluble dietary fiber (CSDF), celery insoluble dietary fiber (CIDF), and celery flavonoids (CF) in an IBD mouse model was evaluated. Mice were given 3% dextran sulfate sodium salt (DSS) to cause colitis. Many molecular biological techniques were used to assess the degree of colitis in mice and the makeup of their gut microbiota (Yu et al. 2023) (Fig. 4).

Fibers can combine with starch to decrease its digestibility and raise the RS percentage. Therefore, a variety of techniques, such as physical, chemical, and biological techniques, are employed to postpone the digestion of starch in order to create functional food with a low glycemic index and satisfy human health requirements. Physical modification, such as adding hydrocolloids and polysaccharides, is one of the many modification techniques that have been employed extensively due to its affordability, ease of use, and environmental protection (Yu et al. 2023).

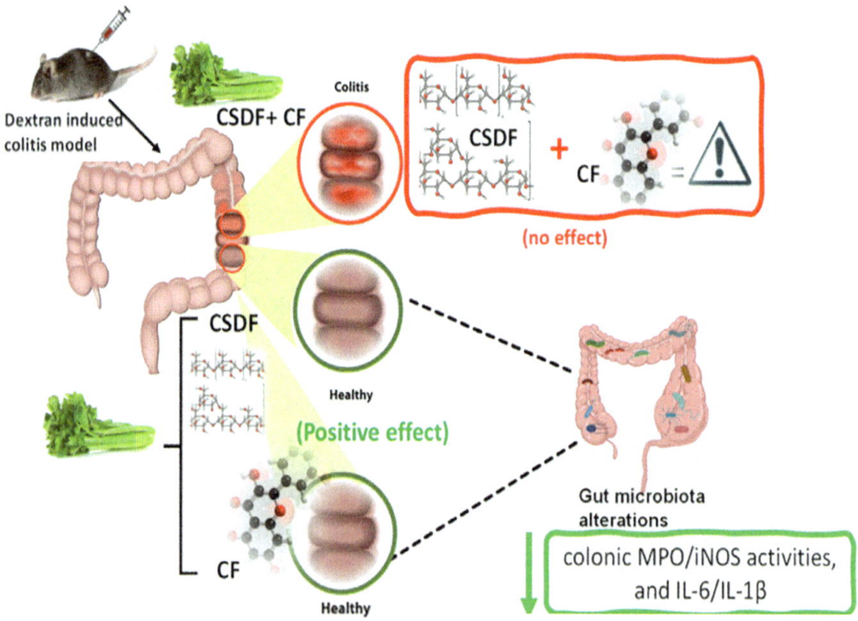

Fig. 4 Role of bioactive compounds extracted from celery leaves in the human colon

3.5 Volatile Organic Constituents (VOC)

Seventy of the seventy-three volatile organic compounds (VOCs) released into headspace in celery samples were identified. Nonane (8.17%), β-pinene (14.4%), myrcene (2.08%), β-cymene (4.10%), limonene (37.3%), β-(Z)-ocimene (3.96%), butylphthalide (10.9%), 3-isobutylidene phthalide (5.51%), and an unidentified phthalide derivative (8.06%) were the main chemicals (1.0%) found in fresh celery. There were only 45 volatile organic compounds (VOCs) in raw, fresh celery. As a result, 28 were exposed to the plant material through different fluids. According to Krecisz et al. (2023), among the substances artificially added to the celery VOCs are sulfuric compounds (such as methyl propyl disulfide or methyl tetrasulfide) and a wide range of C6-derived alcohols and aldehydes (such as (2E)-hexen-1-ol or (2E)-hexanal).

It has been reported that celery contains potentially harmful polyacetylenes, such as falcarinol and falcarindiol. According to research using ultraviolet spectroscopy, a substance that is identical to or comparable to 8-methoxypsoralen is present. Another molecule containing a furocoumarin glucoside, isoquercitrin, has been found using infrared spectroscopy. Apiumoside, a coumarin glucoside, has also been discovered. Acetaldehyde, propionic aldehyde, and isovalerianic aldehyde are additional organic components (Nagella et al. 2012).

Celery seed essential oils (EOs) are classified as nutraceuticals due to their anti-inflammatory, anticancer, antibacterial, and free radical scavenging properties, among other nutritional and medicinal advantages. "Xiaoguozi" (CJX), "Doterra" (ADO), "Florihana" (FFL), "Aroma zone" (FAR), and "Mixiangfang" (IMX) were assessed for their scent attributes and aroma compounds (AOCs). The findings indicated that there are significant differences in the aroma attribute and AOC composition of the five EOs. High concentrations of limonene-D, β-pinene, α-maaliene, senkyunolide, and 9-octadecenoic acid, (E) in IMX, α-pinene oxide, α-thujene, isosativene, senkyunolide, *n*-hexadecanoic acid, and β-myrcene were found in EOs (Xu et al. 2023).

References

Barra A (2009) Factors affecting chemical variability of essential oils: a review of recent developments. Nat Prod Commun 4:1147–1154. https://doi.org/10.1177/1934578X0900400827

Bat-Ozmatara M (2020) The antioxidant activity of Apium graveolens. Int J Food Eng Res 6(1):17–33

Das S, Singh VK, Dwivedy AK, Chaudhari AK, Upadhyay N, Singh A, Deepika, Dubey NK (2019) Antimicrobial activity, antiaflatoxigenic potential and in situ efficacy of novel formulation comprising of *Apium graveolens* essential oil and its major component. Pestic Biochem Physiol 160:102–111. https://doi.org/10.1016/j.pestbp.2019.07.013

Kokotkiewicz A, Maria Luczkiewicz M (2016) Celery (Apium Graveolens var. Dulce (Mill.) Pers.) Oils. In: Essential oils in food preservation, flavor and safety. Elsevier, pp 325–338

Kooti W, Daraei N (2017) A review of the antioxidant activity of celery (*Apium graveolens* L). J Evid Based Complem Alternat Med 22(4):1029–1034. https://doi.org/10.1177/2156587217717415
Krecisz M, Kolniak-Ostek J, Lyczko J, Stepien B (2023) Evaluation of bioactive compounds, volatile compounds, drying process kinetics and selected physical properties of vacuum impregnation celery dried by different methods. Food Chem 413:135490. https://doi.org/10.1016/j.foodchem.2023.135490
Li MY, Hou XL, Wang F, Tan GF, Xu ZS, Xiong AS (2017) Advances in the research of celery, an important Apiaceae vegetable crop. Crit Rev Biotechnol 38:172–183. https://doi.org/10.1080/07388551.2017.1312275
Malhotra SK (2006) Handbook of herbs and spices, vol 3. Woodhead Publishing, Cambridge, pp 317–336
Modaresi M, Ghalamkari G, Jalalizand A (2012) The effect of celery (*Apium graveolens*) extract on the reproductive hormones in male mice. APCBEE Proc 4:99–104. https://doi.org/10.1016/j.apcbee.2012.11.017
Motti P, Bastiaens L, Geelen D, Mangelinckx S (2024) An exploratory study of extractions of celery (*Apium graveolens* L.) waste material as source of bioactive compounds for agricultural applications. Ind Crop Prod 218:118848. https://doi.org/10.1016/j.indcrop.2024.118848
Nagella P, Ahmad A, Kim SJ, Chung IM (2012) Chemical composition, antioxidant activity and larvicidal effects of essential oil from leaves of *Apium graveolens*. Immunopharmacol Immunotoxicol 34(2):205–209
Nouioura G, El Fadili M, Ghneim HK, Zbadi L, Maache S, Zouirech O, Danouche M, Aboul-Soud MAM, Giesy JP, Lyoussi B et al (2024) Exploring the essence of celery seeds (*Apium graveolens* L.): innovations in microwave-assisted hydrodistillation for essential oil extraction using in vitro, in vivo and in silico studies. Arab J Chem 17:105726. https://doi.org/10.1016/j.arabjc.2024.105726
Paslawska M, Nawirska-Olszanska A, Stepien B, Klim A (2018) The influence of vacuum impregnation on nutritional properties of fluidized bed dried Kale (*Brassica oleracea* L. *Var Acephala*) leaves. Molecules 23:2764. https://doi.org/10.3390/molecules23112764
Saini RK, Song M-H, Ji-Woo Y, Shang X, Keum Y-S (2021) Phytosterol profiling of Apiaceae family seeds spices using GC-MS. Food Secur 10(10):2378. https://doi.org/10.3390/foods10102378
Xu CC, Li QL, Wang N, Liu DK, Guo CX (2023) Identifying and discriminating aroma attribute and bioactive components of five commercial essential oils of celery (*Apium graveolens* L.) seeds using E-nose, HS-GC-IMS, and GC-MS. LWT-Food Sci Technol 184:115094. https://doi.org/10.1016/j.lwt.2023.115094
Yu Y, Hao Z, Wang B, Deng C, Hu J, Bian Y, Wang T, Zheng M, Yu Z, Zhou Y (2023) Effects of two celery fibers on the structural properties and digestibility of glutinous rice starch: a comparative study. Int J Biol Macromol 264:130776. https://doi.org/10.1016/j.ijbiomac.2024.130776
Zorga J, Kunicka-Styczynska A, Gruska R, Smigielski K (2020) Ultrasound-assisted hydrodistillation of essential oil from celery seeds (*Apium Graveolens* L.) and its biological and aroma profiles. Molecules 25(22):5322. https://doi.org/10.3390/molecules25225322

Part IV
Phytochemicals from Root and Tuber Vegetables By-products

Bioactive Phytochemicals from Beetroot (*Beta vulgaris*) By-Products

Shumaila Ijaz, Javed Iqbal, Banzeer Ahsan Abbasi, Zakir Ullah, Siraj Uddin, Tabassum Yaseen, Farishta Zarshan, Madiha Imtiaz, Habib Ali, Muhammad Jamil, Sobia Kanwal, and Tariq Mahmood

Contents

Shumaila Ijaz and Javed Iqbal contributed equally with all other contributors.

S. Ijaz
School of Biomedical Engineering, Shenzhen University Medical School, Shenzhen University, Shenzhen, China

J. Iqbal (✉) · T. Yaseen · F. Zarshan · M. Imtiaz
Department of Botany, Bacha Khan University, Charsadda, Khyber Pakhtunkhwa, Pakistan
e-mail: tabassumyaseen@bkuc.edu.pk

B. A. Abbasi
Department of Botany, Rawalpindi Women University, Satellite Town, Rawalpindi, Pakistan

Z. Ullah · S. Uddin · T. Mahmood
Department of Plant Sciences, Faculty of Biological Sciences, Quaid-i-Azam University, Islamabad, Pakistan
e-mail: zakirullah@bs.qau.edu.pk; tmahmood@qau.edu.pk

H. Ali
Department of Agricultural Engineering, Khwaja Fareed University of Engineering and Information Technology, Rahim Yar Khan, Pakistan

M. Jamil
Department of Biotechnology and Genetic Engineering, Kohat University of Science and Technology, Kohat, Pakistan
e-mail: dr.jamil@kust.edu.pk

S. Kanwal
Department of Biology and Environmental Sciences, Allama Iqbal Open University, Islamabad, Pakistan
e-mail: sobia.kanwal@aiou.edu.pk

M. F. Ramadan (ed.), *Bioactive Phytochemicals in By-products from Leaf, Stem, Root and Tuber Vegetables*, https://doi.org/10.1007/978-3-031-80700-8_9

Abstract Beetroot, scientifically known as *Beta vulgaris*, is a widely recognized and extensively cultivated fruit due to its abundant nutritional content, including proteins, fiber, vitamins (B complex and C), minerals, and various bioactive phenolic substances belonging to the *Amaranthaceae* family. The primary constituents of these bioactive compounds are betalains, accompanied by other components such as coumarins, carotenoids, sesquiterpenoids, triterpenes, and flavonoids (such as astragalin, tiliroside, rhamnocitrin, kaempferol, and rhamnetin), which possess antioxidant properties. These beneficial compounds in beetroot and its derived products offer many health benefits and potential for preventing and managing various ailments and diseases. Moreover, the diverse phytochemical composition of beetroot makes it a promising source for developing nutraceutical chemicals that can be incorporated into functional foods. From a pharmacological and phytochemical perspective, beetroot exhibits antioxidant, antibacterial, anti-fungal, anti-cancer, hypocholesterolemic, and anti-inflammatory properties.

However, significant quantities of by-products such as leaves, stems, and peels are generated during the beetroot processing, often underutilized or discarded as waste. These beetroot by-products possess a wealth of bioactive phytochemicals with potential applications in food, pharmaceuticals, cosmetics, and bioenergy production. This chapter explores the chemical composition, bioactive compounds, biological activities, and potential value of beetroot by-products. The introduction emphasizes the economic significance and common uses of beetroots, underscoring the importance of valorizing the by-products. Chemical profiling studies have revealed that beetroot by-products exhibit high betalains, phenolic compounds, vitamins, and dietary fiber levels. Regarding food applications, extracts derived from beetroot by-products can be utilized as natural colorants, preservatives, and functional ingredients. Furthermore, these by-products have non-food applications in nutraceutical, cosmeceutical, and renewable materials.

Keywords Betalains · Phenolic compounds · Flavonoids · Vitamins · Dietary fiber · Bioactive by-products

Abbreviations

BP	Blood pressure
CB	Cooked beetroot
Cd	Cadmium
CVDs	Cardiovascular diseases
HDL-C	High-density lipoprotein cholesterol
LAB	Lactic acid bacteria
LDL-C	Low-density lipoprotein cholesterol
MAP	Mean arterial pressure
NAFLD	Non-alcoholic fatty liver disease
NO	Nitric oxide
Nrf2	Nuclear erythroid factor 2
PBQ	Phenyl-p-benzoquinone
RBJ	Raw beetroot juice
USDA	United States Department of Agriculture

1 Introduction

Beetroot, a member of the *Amaranthaceae* family, is a cultivated plant that can be grown annually or biennially. It is a variety of *Beta vulgaris* subsp. vulgaris conditiva and its edible taproots have their origins in the Middle East. Over time, beetroot has become popular worldwide, cultivated in regions across Asia and from America to Europe (Wruss et al. 2015). This vegetable is flavorful and highly nutritious, known for its health-promoting properties. It is believed to have anti-inflammatory and anti-oxidative effects (Georgiev et al. 2010) and exhibits anti-carcinogenic and anti-diabetic potential. Additionally, beetroot has been found to have hepatoprotective, hypotensive, and wound-healing properties (Clifford et al. 2015). Due to its numerous health benefits, beetroot is now utilized to create different delicious meals (Chhikara et al. 2019a, b; Domínguez et al. 2018). Notably, most current research on beetroot supplements has focused on the crucial role that inorganic NO_3 plays in its biomedical effect, particularly in studies that address its hypotensive and ergogenic qualities. Beetroot contains various minerals, including potassium, sodium, phosphorus, manganese, zinc, iron, calcium, and copper (Singh and Hathan 2014). It is also loaded with a wide variety of bioactive phytochemicals, such as betalains (Baião et al. 2017) (e.g., betacyanins and betaxanthins), flavonoids, saponins, and polyphenols. Across many food traditions, it is frequently eaten as processed jam, drink, pickle, powder, gel, cooked, and oven-dried (Morgado et al. 2016).

Natural products exhibit limited toxicity in comparison with synthetic drugs; therefore, the process of isolating and characterizing phytochemicals with advantageous effects is currently attracting considerable attention. Beetroot is characterized by a nutritional profile containing significant levels of dietary fiber and a rich concentration of numerous bioactive compounds known as phytochemicals, which consist of both primary and secondary metabolites (1). Primary metabolites are essential for metabolic processes and the overall growth of plants, while (2) secondary metabolites can play a crucial role in protecting against plant damage and diseases. Categorically, alkaloids, phenolics, and terpenoids are the three main classes of phytochemicals with diverse therapeutic properties such as antioxidant, analgesic, antiviral, and antibacterial (Ijaz et al. 2024a).

In addition to its pharmacological properties, beetroot and its phytochemicals have been used to enhance desserts, tomato sauces, wine, jams, soups, and other culinary items by replacing synthetic hues and the adverse reactions they cause. Due to its antioxidant capacity, it would safeguard the health and welfare of the clients. To preserve the beetroot and its phytochemicals, the food industries have adapted different techniques, including drying and freezing, to maintain its antioxidant potential since fresh beetroots degrade quickly because of their high moisture content (Sharma et al. 2012). Beetroot has been included in a variety of dishes as a helpful element. According to reports, the food and meat industries employ many bioactive compounds released when processing vegetables, such as beetroot products and other residues (Jin et al. 2014). Beetroot is well-known for its therapeutic potential and is commonly used to treat a variety of illnesses, such as heart disease, anemia, cancer, and kidney stones (Shedid et al. 2019; Ijaz et al. 2024b; Nogueira Soares et al. 2021). Recent studies have also shown that beetroot benefits blood pressure and cardiovascular health (Cramer et al. 2020). Beetroot supplements substantially increased energy and performance in athletes (Raubenheimer et al. 2017).

A high content of betalains and phenolic compounds highlights beetroot as a highly potent vegetable in terms of antioxidant potential (Ijaz et al. 2023a). The European Union has approved using betalains, a water-soluble pigment, as a food colorant, as E-162. Additionally, beetroot is a valuable source of health insuring compounds that can cure congenital disabilities, prevent anemia, and provide its treatment by providing iron and dietary fibers that are crucial in promoting colon health with fibers. Consequently, adding beetroot by-products to foods like mayonnaise might be less expensive. To create goods with additional value, a variety of food products such as yogurt (Neha et al. 2018; Yadav et al. 2016), ice cream (Evstigneeva et al. 2020), and cakes (Manoharan et al. 2012) utilized beetroot as a natural ingredient. By-products of beetroots can be used as inexpensive culinary ingredients, cutting down on food waste. As a result, plant wastes may be a source of nutraceuticals for functional foods (Lucky et al. 2020). However, a significant amount of beetroot by-products are produced during industrial activities. To get liquefied beetroot that can be atomized, about 40% of the beetroot is wasted. To obtain natural colorants, the liquefied beetroots are often atomized. As shown in Fig. 1, the fibrous mass from extracting the juice has the characteristic purple color of beetroot. This trash is typically used for biofuel or animal feed (Vulić et al. 2012).

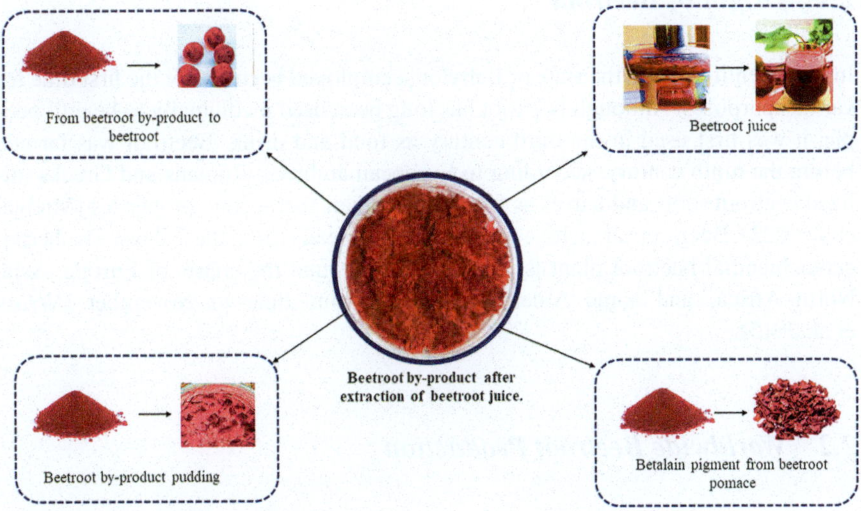

Fig. 1 The visual attributes of the beetroot residue derived from the liquefaction method used to extract beetroot juice

2 Overview of Beetroot

The adaptable vegetable beetroot was used for medical purposes and food and drink until the third century. In the seventh century, the state of Babylonia employed beetroot for the first time in various ways (Chawla et al. 2016). Additionally, it is consumed as a vegetable and supplement (Morgado et al. 2016). It is currently used as food coloring additive E162 in manufacturing industries and ingested regularly as part of a conventional diet in fresh and fermented forms. Every part of this plant has medicinal properties, such as hepatoprotective or cardiovascular health protector properties, anti-inflammatory, diuretic, expectorant, anti-fungal, antioxidant, anti-depressant, carminative, and antimicrobial properties (Ceclu and Nistor 2020). In a study, Georgieva and his team reported betalain and phenolics in beetroot as good antioxidants (Georgiev et al. 2010). This suggested that these compounds may protect against oxidative stressors and that inorganic nitrates may be beneficial in lowering blood pressure (BP) and preventing cardiovascular diseases (CVDs) (Silva et al. 2016). Due to research suggesting that beets, in any form, reduce blood pressure, increase hair growth, and lessen the risk of diabetes, beetroot has become famous as the next "superfood" (Dwivedi et al. 2017). It is thought to have hepatoprotective and anti-carcinogenic effects (Domínguez et al. 2017), enhanced athletic performance (Mudgal et al. 2022), anti-inflammatory and antioxidant capabilities, and wound-healing qualities. A natural food with the highest sugar and nitrate concentration (Chawla et al. 2016). Being a good source of nitrates and sugar that is available naturally, it can help to boost athletes' energy levels.

2.1 *Origin of Beetroot*

In the seventh century, the state of Babylonia employed beetroot for the first time for various purposes. Although beetroot has long been used medicinally to benefit people, it was first used in the third century as food and drink. Beetroot was farmed before the ninth century, according to European archives. Romans and Greeks utilized beetroot roots and leaves as food and for their therapeutic properties (Mudgal et al. 2022). Beetroots are native to the Mediterranean region these days. The herbaceous biennial beetroot plant is widely grown around the coasts of Europe, Asia, North Africa, and some American regions from June to November (Wruss et al. 2015).

2.2 *Worldwide Beetroot Production*

Beetroot is widely available across the world. The output of beetroot was 269,714 million tons globally in 2014. In 2014, Russia produced around 37,844,567 tons and 33,513,369 tons of beetroot in France, respectively. China, Poland, Germany,

Turkey, Egypt, Europe, Ukraine, and the United States are among the top beetroot producers (Razzak et al. 2024). Apart from one of Poland's popular grown root vegetables, Europe also promotes its growth significantly. Data from the Central Statistical Office indicate that Poland produced over 314 thousand tons of beetroot in 2014, an increase of 5.3% from the year before (Sawicki et al. 2016). The four main beetroot cultivars are Crosby Egyptian, crimson globe, early wonder, and Detroit dark green. The last one has dark red meat with smooth and uniform roots. The medium-dark red flesh of the Crimson Globe beetroot has small shoulders. The inside of Crosby Egyptian beetroot is dark purple with hazy zones. Its globe is flat. After seeding, they need 60 days to reach maturity. Early amazing features include flattened roots, green leaves, dark red meat, a crimson, veiny globe, and smooth shoulders (Chawla et al. 2016).

2.3 *Nutrition Profile of Beetroot*

A true superfood, beetroot has a good nutrient content, highlighting folic acid, minerals such as potassium, zinc, salt, phosphorus, calcium, and magnesium, and vitamins A, C, E, K, and B. Moisture (87.58), dietary fiber (2.8), lipids (0.16), protein (1.61), ash (1.08), and carbs (9.56) are all included in the chemical makeup (%). The chemical makeup of beets can vary depending on several factors, including genetics, harvesting conditions, variety types, and ecological circumstances. The distribution of nutritional components in beetroot peel, stem, root, and leaf differs (Sawicki et al. 2016). Because carotenoids and alpha-beta carotene accumulate in plant chloroplasts, beetroot leaves have higher carotenoid content than tubers (Ceclu and Nistor 2020). Beetroot contains micronutrients important for cardiovascular health because they reduce blood pressure and combat cancer (Chhikara et al. 2019a, b). They also include many betalains, phenolic compounds, and antioxidants, all recognized to have various advantageous health benefits (Punia Bangar et al. 2023). Mainly, beetroot contains amino acids such as isoleucine, valine, cysteine, methionine, and threonine, as well as phenolic acids such as ferulic acid, caffeine, and syringic acid, and also a variety of flavonoids with ferulic acid, rhamnocitrin, and rhamnetin abundance (Thiruvengadam et al. 2024).

2.4 *Chemical Makeup and Bioactive Ingredients of By-Product Extracts*

Because of the variety of active chemicals found in beets, including betalain, polyphenols, carotenoids, flavonoids, and nitrates, water-soluble pigments with beneficial physiological and nutritional effects, beets are growing in popularity (Lechner and Stoner 2019). Phytochemicals: flavonoids and phenolics—plants have many phytochemicals (Ijaz et al. 2024c). Consuming beetroot is beneficial to health because it has beneficial bioactive compounds that are highly absorbed and accessible by humans, such as flavonoids, phenolic acids, and antioxidants, in addition to betalains. Researchers have discovered that one of the top ten vegetable species with the best antioxidant properties is betalain, a chemical present in beetroot. In research by Vinson, beets had the greatest total phenolic content (Vinson et al. 1998).

Beetroot has oxalic acid, and its extract contains flavonoids, saponins, and triterpenes. The roots typically contain the least amount of phenolic compounds. Beetle stems, flowers, and leaves are colored because of rich flavonoids. Flavonoids are physiologically active substances with antioxidant qualities that benefit health (Chhikara et al. 2019a, b; Azeredo 2009). Vegetables may lose some of their flavonoid content during preparation. Thus, a thorough assessment of beetroot formulation, including flavonoids, is necessary to maximize the benefits of beetroot eating. Vasconcellos et al. (2016) examined and contrasted the total phenolic content of several beetroot products, including chips, juice, beetroot powder (BP), and cooked beetroot (CB) (Vasconcellos et al. 2016).

The beetroot gel had a higher concentration of phenolic phytochemicals such as flavonoids and saponins and was enhanced with nitrates. Additionally, there was an increase in fiber, carbs, and antioxidant levels. The gel's acceptance rate increased overall, and flavor and texture were well-liked. Higher nitrate concentrations in the gel lower blood pressure and cholesterol. Beetroot had 26 different triterpene saponins, including betavulgarosides I–X. However, it was discovered that beetroot roots did not contain betavulgarosides IX or X. The oxidative degradation of a terminal monosaccharide unit was thought to be involved in the formation of the triterpene oligoglycosides betavulgarosides I–IX, which feature distinct substituents of the acetal and dioxolane types (Punia Bangar et al. 2023).

3 Biological and Functional Properties and Bioactive Compounds from Beetroot By-Products

Vegetables, with their high content of bioactive elements, are an essential component of the human daily diet and offer several nutritional and health benefits. Because of its high nitrate (NO_3) content and reputation as a functional meal, beetroot has drawn the attention of researchers as well as other vegetables. This chemical has considerable nutritional advantages for individuals with cardiovascular disease through endogenous nitric oxide generation (NO) (Dos Santos Baião et al. 2019). Furthermore, betalain pigment, which Caryophyllales order plants mainly generate, has anti-inflammatory, anti-tumor, and anti-hypertensive qualities, among other benefits that make it valuable as a pharmacological agent and nutritional supplement (Mudgal et al. 2022) (Fig. 2).

Growing up in both industrialized and developing nations, metabolic syndrome is associated with several disorders, including diabetes, cardiovascular diseases, and

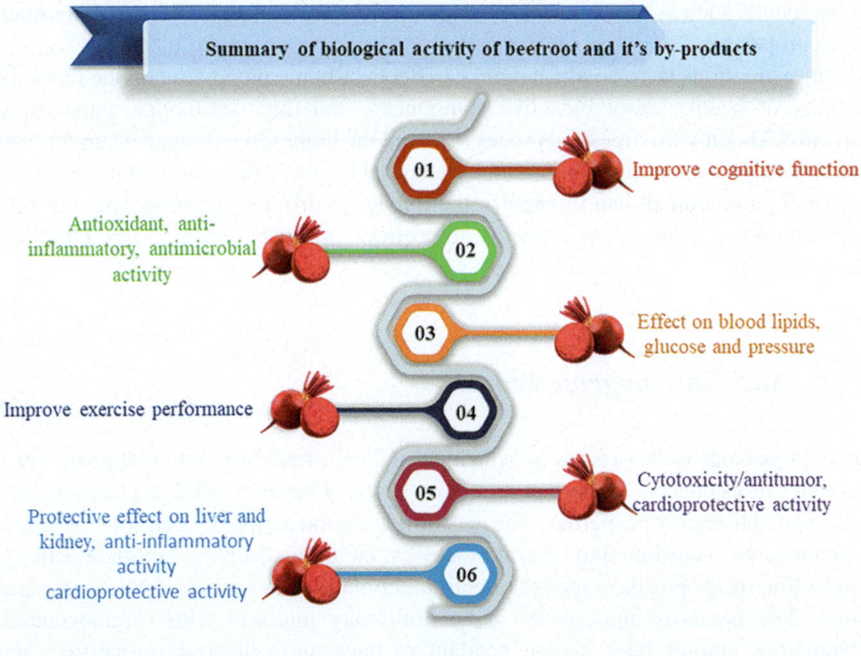

Fig. 2 Biological activities from beetroot by-products

hypertension. Individuals are searching for safer substitutes to enhance their living standards. Vegetables should now be included in the diet regularly (Madadi et al. 2020). Studies suggest that beets can reduce cardiac pressure, enhance athletic performance, and highlight beetroot as a popular "superfood." Beetroot has been identified as a functional food and is thought to have an anti-inflammatory impact. Its nutritional and therapeutic value may be assessed in various medical conditions (Tekin et al., 2023; Neha et al. 2018; Abuajah et al. 2015).

3.1 Antioxidant and Anti-Inflammatory Properties

One of the top ten plants with the greatest potential for antioxidants is beetroot. Betalain's antioxidant action in humans increases low-density lipoproteins, providing oxidation resistance. The thin blood vessel lining may be protected, and betalain extract may decrease inflammation (Miraj 2016). Inflammation controls the body's intrinsic reaction. Shock and septicity are examples of physical and biological stimulants that create disruption of homeostasis and provide an environment for damage due to illness. Prostaglandins are essential to the inflammatory response because inflammatory tissue produces much more of them, which results in redness, swelling, weakness, and temporary discomfort at the inflammation site. Natural compounds such as betalain have prompted scientists to examine its anti-inflammatory properties.

Investigations were conducted on the effects of home processing on the bioavailability of beetroot root bioactive components and their antioxidant capabilities in vitro. The in vitro digestion process showed the highest recovery in a jam for total phenolics (16%) and total antioxidant potential (1.3%). The aforementioned study created comparative data to analyze the impact of different home-processing processes on the antioxidant capacity of beetroot products (Miraj 2016; Guldiken et al. 2016).

3.2 *Anti-Carcinogenic Property*

Carcinogenesis is the process by which normal cells transform into malignant ones, which causes cancer. Numerous studies on beetroot have revealed its chemoprotective and anti-cancer properties. The cancer-prevention action of beetroot's bioactive components is attributed to several processes, including decreased oxidase activity, induction of apoptosis, stoppage of cell inflammation, and excess cytokine generation. The bioactive ingredients in beetroot may interfere with carcinogenesis. Numerous studies have shown betalain to have anti-cancerous properties, and efforts have been made to identify the underlying mechanism. By developing an insulator surrounding wounded tissues, researchers demonstrated that betalain did not affect the normal cell line (Nowacki et al. 2015).

Furthermore, the findings of many studies indicated that a consistent, modest consumption of table beetroot products may positively impact a patient's life expectancy. Betacyanins suppress the growth of human tumor cells and cancer cells in the bladder and ovaries. Beetroot consumption lowers the risk of colon, liver, lung, esophageal, and skin cancers. Anti-Carcinogenic Characteristic Carcinogenesis is the process by which normal cells transform into malignant ones, which causes cancer. Numerous studies on beetroot have revealed its chemoprotective and anti-cancer properties. The cancer-prevention action of beetroot's bioactive components is attributed to several processes, including oxidase activity decrease, induction of apoptosis, stoppage of cell inflammation, and excess cytokine generation (Adhikari et al. 2017).

The bioactive ingredients in beetroot may interfere with carcinogenesis. Numerous investigations have shown betalain to have anti-cancerous properties, and steps have been undertaken to identify the underlying mechanism. By developing an insulator surrounding wounded tissues, researchers demonstrated that betalain did not affect the normal cell line (Nowacki et al. 2015). Furthermore, the findings of many studies indicated that a consistent, modest consumption of table beetroot products may positively impact a patient's life expectancy. Betacyanins suppress human tumors and cancer cells' growth from the bladder and ovaries (Neha et al. 2018; Adhikari et al. 2017). Beetroot consumption lowers the risk of colon, liver, lung, esophageal, and skin cancers.

3.3 Anti-Stress and Anti-Depressive Properties

Beetroots have demonstrated therapeutic promise for mental problems treatment linked to depression and anxiety. Sulakhiya et al. (2016) investigated how beetroot leaf extract affected mice's anxiety, stress, and sadness. Anxiolytic and antidepressant action and potent antioxidant properties were demonstrated in stressed mice. When combined with uridine, which is derived from sugar beetroot, omega-3 can help prevent depression by lowering anxiety and altering mood, which allows the body to relax. Research has indicated that beetroot by-products possess a high concentration of bioactive substances such as polyphenols and betalains, which support its potential as a natural food ingredient and antioxidant. However, beetroot by-products have demonstrated cytotoxic effects on cancer cells and have been connected to a decrease in oxidative stress, suggesting that it may have a role in stress-related illnesses. Furthermore, the ethanolic extract of *Beta vulgaris* leaves has demonstrated potent antioxidant qualities, anxiolytic and anti-depressant effects in stressed animals, and therapeutic potential in neuropsychiatric illnesses. These results illustrate the beetroot by-products' anti-stress and anti-depressive qualities, highlighting their potential as valuable components in the food and pharmaceutical industries (Sulakhiya et al. 2016).

3.4 Anti-Hypertensive Property

Lipoprotein with low density. About one billion people worldwide suffer from hypertension, which is the leading cause of cardiovascular issues. A diet with rich fiber content can reduce arterial pressure. The purpose of the study was to ascertain how RBJ (raw beetroot juice) and CB (cooked beetroot) affected individuals with high blood pressure (Guldiken et al. 2016). The potential of beetroot to counteract hypertension has been demonstrated by its synergistic effect with other nutrients, such as beetroot leaf and stem juice, which have been found to prevent obesity-related health disorders. In experimental rat models, extracts from beetroot have shown anti-hypertensive properties; beetroot juice, in particular, has produced encouraging results. Moreover, beetroot formulations may be useful in treating hypertension due to the cardioprotective properties of dietary nitrate from beetroot and its bioactive phytochemicals, which have been emphasized as a supportive therapy for cardiovascular disorders. These results highlight the importance of beetroot and its by-products as anti-hypertensive agents that can help control hypertension (Adhikari et al. 2017; Sulakhiya et al. 2016).

4 Biological Activities of Beetroot

4.1 Antioxidant Activity

Beetroot has stronger antioxidant activity than other fruits that are said to be rich in antioxidants, such as oranges, pineapple, and prickly pears since it is a rich source of betalain and polyphenols (Koss-Mikołajczyk et al. 2019). It has been discovered that betalain accounts for around half of the antioxidant activity of beetroots overall (Kusznierewicz et al. 2021). Due in large part to their molecular makeup and ability to transfer hydrogen to reactive species, betalains are antiradical compounds with a notable capability to scavenge reactive oxygen species (ROS) (Iqbal et al. 2017). Beetroots contain two forms of betalain pigments, the main ones being betacyanin and betaxanthin. Higher hydroxyl and amine groups are present in yellow-orange betaxanthin than in red-violet betacyanin, often referred to as betanin of beetroot, which gives the extracts greater ability to scavenge radicals.

Nonetheless, acylation and glycosylation offer a variety of betacyanin structures, enhancing beetroots' capacity to function as antioxidants (Rodriguez-Amaya 2019). Beetroot's betanin comprises unsaturated double bonds and aromatic rings with good electron-donating qualities. In addition, it is well-known that beetroot contains phenolic acids (such as caffeic, protocatechuic, salicylic, and p-Coumaric acid), B vitamins (Vitamin B3 and B5), and flavonoids (Kazimierczak et al. 2014; Wruss et al. 2015; Sentkowska and Pyrzyńska 2018). Fertilization of the soil has a major impact on beetroots' antioxidant and antiradical properties. It's interesting to note that less nitrogen-containing fertilizers, such as urea, NH_4NO_3, $Ca(NO_3)_2(NH_4)_2$, and $(NH_4)_2SO_4$, have more antioxidant activity against a variety of in vitro oxidative systems (Babagil et al. 2018).

Furthermore, organically grown beetroots exhibited a much greater overall antioxidant capacity than conventionally grown beetroots. Nonetheless, the scientists concluded that it was impossible to distinguish between beetroots' potential for antioxidants based only on the choice of production methods (conventional and organic) (Carrillo et al. 2019). There are two key factors to consider when comparing the nutritional differences between beetroot varieties: cultivar and growth circumstances (Bárta et al. 2020). For example, recent pot research under net house circumstances examined the impact on biochemical traits, antioxidant enzyme activity, and beetroot development indicators of varying doses of fly ash added to the soil (Shakeel et al. 2020).

The scientists observed that adding 15% fly ash to the soil significantly enhanced the beetroots' antioxidant system (proline concentration and antioxidant activity). This method was also economical while simultaneously improving the beetroots' development metrics. The procedures used for extraction, such as boiling, blanching, fermenting, or drying, determine the retention of beetroot extracts' antioxidant potential. Different extraction methodologies result in extracts with diverse characteristics whose bioactive constituents may or may not be preserved.

Preservation is a typical procedure because extracts are easily handled and may be used again. Since betanins are extremely sensitive to several chemical and physical factors such as preservatives, enzyme activity, copigmentation, different carbohydrates, antioxidants, and their derivatives, in addition to factors like pH, temperature, light intensity, aeration, and water activity (Thippeswamy et al. 2022), different preservation techniques are frequently utilized to modify the antioxidant potential (Ijaz et al. 2024c). For instance, we can change the antioxidant effectiveness of the extracts using a convective airflow drying process that requires high heat for a long duration, which degrades and produces unfavorable modifications in other antioxidant molecules (such as phytochemicals) (Figiel 2010).

One researcher used Instant Controlled Pressure Drop (DIC) technology as a pretreatment for blanching-steaming to solve the issue of pair swell drying. They discovered that by employing this technique, not only did the betalain and other beneficial components of dried beetroot extracts remain maintained, but also its free radical scavenging capacity was enhanced 3.2 times and a 10% higher antioxidant capacity (Sawicki and Wiczkowski 2018). A rise from 47.69 to 76.54% (dry basis) in the anti-oxidative activity of beetroot extracts has also been reported when steam blanching is combined with cold storage. In addition to drying, boiling and fermentation are two more popular technical methods for preserving beetroot extracts.

Comparatively, it was discovered that fermentation preserved the highest level of antioxidant potential in whole beetroot extract compared to boiling. Additionally, it was found that the peel of beetroot was least impacted by processing when comparing the antioxidant activity of the root and its three sections (whole, meat, and peel) (Sawicki and Wiczkowski 2018). Furthermore, according to Thippeswamy et al. (2022), additions such as 1% ascorbic acid, 10% glucose/fructose, sucrose content more than 45%, EDTA at 40 ppm, and sodium benzoate at 350 ppm can act as preservatives in beetroot extract to retain its CUPRAC antioxidant activity (Thippeswamy et al. 2022).

Beyond the root, beetroot's aerial portions have also been thoroughly investigated for their antioxidant capacity. The conclusion drawn from this research is that these unimportant, unusual edible beetroot components may serve as a valuable source of raw materials for producing antioxidant-rich extracts that may be used in both culinary and medical contexts. According to Lasta et al. (2019), extracts with increased antioxidant capacity may be produced efficiently by using green extraction techniques such as supercritical fluid extraction with CO_2 and co-solvents (Lasta et al. 2019). Thus, it can be concluded that various authors have extensively explored the antioxidant potential of beetroot extract (Fu et al. 2020; Hadipour et al. 2020); nevertheless, more in vitro and in vivo studies to demonstrate the bioaccessibility of the antioxidants would be a valuable addition to the current literature.

4.2 Antimicrobial Activity

The fact that beetroots are used as herbs to treat various infectious disorders establishes their antibacterial properties. The considerable antibacterial activity of beetroot extracts evaluated against multiple bacterial strains and their lack of inhibitory effect on fungi and yeasts have been the subject of much research (Kumar and Brooks 2018). According to the scientists' findings, many gram-positive strains, such as *B. cereus*, *S. epidermidis*, *S. aureus*, and *S. pyogenes*, are more susceptible to beetroot extracts than gram-negative strains. Gram-positive bacteria's high sensitivity could be attributed to the polyphenolic compounds (betalains) found in the beetroot extract since they could easily affect their fundamental cell wall development, function, and permeability.

Conversely, gram-negative bacteria were assumed to be more resistant due to their unique periplasmic space, highly hydrophilic lipopolysaccharide and double-layered cell wall. Many extracts contain polyphenolic compounds that can affect bacterial growth and metabolite development by promoting or inhibiting them. The microbial inhibitory property of these beetroot compounds is correlated with extraction efficiency. Several primary antioxidant components were found in beetroot methanol and ethanol extracts, according to Youssef et al. (2021). While ethanol extracts exhibited larger amounts of methyl oxirane, methyl ester, hexadecanoic acid, and n-hexadecanoic acid, 1-dodecanamine and n,n-dimethyl were found to be abundant in methanol extracts (Youssef et al. 2021).

These antioxidant compounds showed a decrease of 66.99–99.79% by methanol extracts and a reduction of 95.6–99.9% by ethanol extracts, suggesting that they have antibacterial and antimycotoxin action against *Alternaria alternata* on potato crops. Similarly, recent research (Salamatullah et al. 2021) found that beetroot peel and pulp extract made with 50% methanol had the greatest bioactive properties and antibacterial activity against all test pathogens (*S. aureus*, *L. monocytogenes*, *E. coli*, *P. aeruginosa*, and *C. albicans*). This extract exhibited more inhibitory zones of gram-negative bacteria than gram-positive bacteria, in contrast to the previously reviewed findings (Kumar and Brooks 2018). The minimum inhibitory dosage varied from 4 to 8 mg/mL to inhibit all bacterial and fungal pathogens.

4.3 Cytotoxic/Anti-Cancer Activity

When it comes to biologically based cancer-prevention therapy, beetroot is the most preferred plant (60.7%) in both mainstream and alternative medical practices (Sárváry and Sárváry 2019). According to a case study, an 80-year-old man's small cell lung cancer spontaneously resolved after 2 years of consistent beetroot juice consumption with no further medicine or therapy. While both beetroot consumption and surgical trauma had a favorable effect on tumor reduction, the precise mechanism is unknown (Song et al. 2021). Beetroots contain mineral elements (like K, Cu, Al, Mn, B, Cu, Zn Ba, Ca, Fe, Mg, and Na) and antioxidant/scavenger-type molecules that affect the redox state and subsequently signal transduction pathways.

Based on current studies, beetroot primarily impedes cell division, induces cell death, and activates autophagy, responsible for its anti-cancer and cytotoxic effects. The existence of anti-cancer action against different cell lines in the extract of beetroots has been thoroughly demonstrated by several publications (Fu et al.

2020; Hadipour et al. 2020; Tan and Hamid 2021). Furthermore, it has been demonstrated that beetroot leaves can also provide anti-cancer properties since they contain flavonoids called apigenin-glycosides, vitexin and vitexin-2-O-xylosidium and rhamnoside (Papi et al. 2013; Farabegoli et al. 2017; da Silva et al. 2020). According to a recent study, the roots and leaves of beet can prevent prostate cancer cell line growth, migration, and colony formation ability. The primary active ingredient in beetroot leaves is apigenin (Mancini et al. 2021). Similar results were demonstrated during p53-modulated apoptosis in HeLa, A549, HCT 116, H1299, and MCF-7 cancer cells by apigenin, which promotes cisplatin cytotoxicity (Liu et al. 2017) and prevents cancer development in hepatic cells through protein kinase B (Yang et al. 2018). On the other hand, Romero et al. (2021) found that extracts were superior to isolated chemicals when they saw the same outcomes for leaf and root beet extracts that lowered the viability efficiency and cloning capacity of the mTOR pathway in HeLa cells. The authors support their claim by emphasizing the bond between the bioactive components in the extracts (Romero et al. 2021).

4.4 Hepatoprotective Activity

Consuming beetroot is now generally acknowledged as a modern dietary strategy to provide protective benefits against oxidative stress to the liver at the highest level of physical activity (Sinaga et al. 2020) and as a potential treatment for liver damage (Butt et al. 2019). Beetroot extracts have demonstrated considerable hepatoprotective efficacy, as evidenced by their capacity evaluated against mice induced with streptozotocin to reduce hepatic oxidative stress, steatosis, inflammation, and dyslipidemia (Al-Harbi et al. 2021). The authors proposed that the preventive mechanism against non-alcoholic fatty liver disease (NAFLD) involves increased endogenous antioxidants and anti-apoptotic Bcl2 and PPARα. Additionally, beetroot regularly consumed by people with NAFLD has enhanced their lipid profile and slowed the pace at which liver transaminases grow (Srivastava et al. 2019; Afzali et al. 2020).

Obese rats given a fat-rich diet along with ethanol extract of beetroot, leaves, and stems (Lorizola et al. 2018) or ultra-filtered beetroot juice (Babarykin et al. 2019) showed comparable hepatoprotective effects. The ability of extracts to reduce hepatic steatosis and liver damage brought on by chemically induced hepatic injury in an animal model is another widely recognized paradigm for researching the hepatoprotective benefits of beetroots. In rats with CCl_4-induced (Butt et al. 2019; Olumese and Oboh 2020) and chlorpyrifos (CPF)-induced hepatotoxicity, providing supplements of dietary beetroot as a nutraceutical drink helps to postpone hepatic stress, improve the antioxidant state of the liver, and minimize lipid peroxidation and inflammation (Iqbal et al. 2018). Additionally, beetroot extracts have demonstrated a hepatoprotective effect in rats treated with silver nanoparticles and persistent nanotoxicity (Albrahim and Alonazi 2020).

The scientists demonstrated that aqueous beetroot extract protected against liver DNA damage, lipid peroxidation, glutathione depletion, and decreased superoxide dismutase and catalase activity.

4.5 Anti-Inflammatory Activity

With a special focus on betalain, multiple in vitro and in vivo trials and clinical studies have been performed to analyze the anti-inflammatory properties of beetroot (Moreno-Ley et al. 2021). Scientists have examined betalains' capacity to reduce inflammatory markers such as TNF-, IL-1, and NF-B and shield against harm caused by xenobiotics like paraquat and quench chemicals generated by reactive oxygen species. The transcriptional factors, such as nuclear erythroid factor 2 (Nrf2) and nuclear erythroid factor 2 (Nrf2), which are transcriptional factors that are considered as master regulators in homeostasis redox, can be activated by beetroot extract. These factors also aid in coordinating the RNA synthesis of many genes involved in cytoprotective, antioxidant, and inflammatory responses (Liu et al. 2019; Bangar et al. 2022; Raish et al. 2019).

A significant impact has been demonstrated on the transcriptional factors of the inflammatory gene (TNF-α, IL-1β, IL-6, and iNOS) in ISP-challenged rat's cardiac tissues by beetroot juice. 150 and 300 mg/kg of beetroot juice has marked significant inhibition and repression of NF-κB (p65) and NF-κB activation, accompanied by a rise in IκBα (Raish et al. 2019). Additionally, the authors noted dose-dependent increases in binding potential of NF-κB, NF-κB DNA, and serum cytokines (TNF-α, IL-6, and IL-10). In inflammatory nociceptive models, acetic acid, phenyl-p-benzoquinone (PBQ), complete Freund's adjuvant (CFA), carrageenan, and the second phase of the formalin test act as a common inducers of TNF-α and IL-1β production.

The inhibitory action of betanin against microglial activation has been verified. The rats' overt pain-like behavior was suppressed by betalains when it came to acetic acid, PBQ, formalin, and CFA. They significantly decreased the mechanical hyperalgesia induced by carrageenan and CFA, which inhibited cytokine release, oxidative stress, and NF-κB activation (106). Additionally, betanin at 500 μM significantly reduced inflammatory mediators (TNF-α, IL-1β, IL-6, and free radicals) and modified MMP, LMP, and ATP in LPS-induced cells (Martinez et al. 2015). Additionally, Kim et al. (2021) discovered 5-Hydroxymaltolas as a molecule separated and generated by serial sample separation from beetroot growth media in fermented beetroot juice, led by antioxidant activity. In LPS-stimulated RAW 264.7 cells, the authors observed a significant reduction in nitric oxide generation and enhanced the suppression of serum cytokines, MAPK activation, and NF-κB translocation (Kim et al. 2021). Similarly, administering chlorpyrifos-induced reprotoxicity rats before chlorpyrifos with powdered beetroot extract demonstrates significantly decreased pro-inflammatory cytokine production, gene expression, and Nos_2 expression (Albasher et al. 2019).

4.6 Cardioprotective Activity

As the risk factors for cardiovascular disease (CVD) are related to three main pathological disorders, including hypertension, hyperlipidemia, and diabetes, beetroot's cardioprotective action involves having a protective effect on people's blood pressure, glucose levels, and lipid profile. Due to COVID-19 forced house confinement, there has been a rise in stress, physical inactivity, and high-calorie food intake. A recent study has demonstrated the benefits of beetroot consumption for preserving endothelium functioning (Ijaz et al. 2023b; Volino-Souza et al. 2020). In addition, researchers examined the impact of modifying the levels of nitrates and other phytochemicals in different beetroot formulations to bolster cardiovascular disease treatments. The stomach's high pH environment exogenously reduced beetroot's high inorganic nitrate (NO^{3-}) content to the beneficial molecule nitric oxide (NO). This NO can control excessive blood pressure by entering the circulation and forcing the endothelium to relax and dilate (112). According to this notion, Jakubcik et al. (2021) tried to comprehend how beetroot-derived NO^{3-} and NO^{2-} do not depend on dosage on plasma levels and cardiovascular parameters. The authors found that consuming NO^{3-} resulted in a notable decrease in mean arterial pressure (MAP) and an increase in plasma NO^{3-} and NO^{2-} (Bahrami et al. 2021; Jakubcik et al. 2021).

Considerable reduction in MAP and systolic and diastolic blood pressure was noticed on day 1 with acute ingestion of NO^{3-} rich beetroot juice; however, no meaningful change was seen until day 28 (Stanaway et al. 2019). In a second research, the same authors also found that older persons experienced a larger drop in diastolic blood pressure and increased plasma nitrate concentrations after consuming 150 mL of beetroot juice (Stanaway et al. 2019). The same was seen for pregnant women, where 97% of them found that supplementing with dietary nitrate in the form of beetroot juice was a suitable dietary intervention (Ormesher et al. 2018). The authors hypothesized that the activity of the oral bacterial nitrate reductase significantly aided the conversion of nitrate to nitrite.

However, Tropea et al. (2020) showed that nitrates in beetroot juice did not affect blood pressure reduction and the considerable improvement of vascular functioning in knockout mice endothelial nitric oxide synthase (Tropea et al. 2020). Dyslipidemia, characterized by increased levels of triglycerides, cholesterol, low-density lipoprotein cholesterol (LDL-C), and decreased levels of high-density lipoprotein cholesterol (HDL-C), is the second known risk factor linked to CVD. It has been demonstrated that eating foods high in betalains, such as beetroot, improves endothelial function and lipid profile because frequent and prolonged consumption of beetroots causes a stabilized form of betalains to accumulate in human physiological fluids (Thongtang et al. 2022).

A related study found that patients receiving treatment in line with current secondary prevention guidelines for coronary artery disease retained less than 1 and 120 μg of betanin in their urine and plasma samples, respectively, after taking beetroot supplements for 2 weeks regularly. These supplements contained 50 mg of betalain or roughly 16.2 mg of betanin. In addition, the authors discovered that betalain supplementation could successfully treat and adjust several atherogenic risk variables, including dyslipidemia, hypertension, hyperglycemia, and hyperhomocysteinemia. The antioxidant, anti-inflammatory, anti-cancer, hepatoprotective, and cardioprotective qualities of bioactive beetroot components have been the subject of several research; yet, the overall benefits of their use are primarily limited to natural colorants and functional drinks. With a dearth of credible clinical evidence, beetroot research is focused and limited to the cellular and animal levels. This is the cause of the underutilization of its potential. The biological activities of the main active compounds in beetroot have been discussed in detail in this part, which highlights a significant gap in the clinical validation of the bioactive profile and activities of these active ingredients. Clarifying associated mechanisms can also be essential to comprehending beetroot's potential for application as a practical, affordable, and safe natural functional food (Rahimi et al. 2019).

5 Nutrients of Beetroot By-Products

Availability of phosphorus, calcium, potassium, sodium, magnesium, copper, iron, zinc, and manganese with a variety of biologically active phytochemicals such as polyphenols, flavonoids, betalains (e.g., betacyanins and betaxanthins), and inorganic nitrate (NO_3) makes the beetroots a valuable source of various minerals and phytochemicals (120). In diverse food cultures, it is consumed as additional juice, powder, bread, gel, cooked, oven-dried, pickled, pureed, or jam-processed (Wruss et al. 2015; Guldiken et al. 2016). 95 kcal of calories per 100 mL, along with 22.6 g of carbs, 0.70 g of proteins, 0.16 g of total fats, 0.91 g of total dietary fiber, and 12 g of total sugars enhanced its nutritional value. Therefore, 100 mL of beetroot juice is anticipated to have 8.8 g of sucrose, 0.86 g of fructose, and 2.5 g of glucose in terms of micronutrient content (Baião et al. 2017). Taken as per 100 g, 1.73–7.85 mg, 10.75–20.36 mg, and 2.02–2.36 mg of the total sugar, vitamin C, and flavonoids are found, respectively, in a variety of commercially produced beetroot juices (Kazimierczak et al. 2016; Guldiken et al. 2016). Approximately 70–100% of the phenolic content of beetroot is comprised of betalains, which are only present in 0.8–1.3 g/L of fresh beetroot juice (approximately 60% betacyanins and 40% betaxanthins) (Wruss et al. 2015) (Fig. 2).

It is considered the primary commercial source of betalains, whether in powdered form, concentrated forms, or as natural hues in confections, dairy, meat, and poultry-derived goods. According to Baiao et al., beetroot gel has the greatest ratio of active polyphenols after in vitro digestion compared to beetroot juice (Baião et al. 2017). Flavonoids undergo alterations during vegetable processing. One of the most significant inorganic compounds in beetroot is NO_3, whose concentration differs ten times across individual types (121). The reported ranges for NO3 content in commercial beetroot juice and powder were 388 ± 19.9 to 3968 ± 252 mg/L and 393 ± 2.23 to 2721 ± 54.4 mg/L, respectively. Despite being comparatively innocuous, nitrate may change its state into NO_2 via bacterial enzymatic mechanisms (NO_3 reductase), which is then non-enzymatically broken down into NO in the oral cavity. The beetroot organs are categorized as follows: seed, fruit, bulb, tuber, root, stem, leaf, and petiole in that order, based on the lowest to greatest NO_3 concentration (Baião et al. 2017). Furthermore, beetroot has a moderately rich oxalic acid composition (Holmes and Assimos 2004), with a typical amount of 94.6–141.6 mg/100 g in raw beetroot and 300–525 mg/L in beetroot juice, respectively. With metal ion chelator property and the production of nephroliths, oxalic acid is considered a health risk, particularly in those at risk for renal disease (Wruss et al. 2015; Webb et al. 2008).

The problem of creating physiologically safe, NO_3-rich beetroot supplements has led to the introduction of several preparation techniques, including freeze-drying beetroot juice to produce beetroot powder. Two further effective forms of beetroot supplements developed recently are pseudoplastic beetroot gels and red round thin beetroot chips. The maximum calorie content (Kcal), total sugar, carbohydrate, and total antioxidant potential (TAP) are found in beetroot chips; however, the lowest total phenolic content (TPC), flavonoids, and saponin level are found in these chips. When compared to other by-products, beetroot gel is not the most popular, but it is the most effective formulation since it has the lowest lipid content and the highest protein content when employed as a pseudoplastic gel for NO3 delivery to athletes (Baião et al. 2017). Sugar beet is produced commercially for sugar production because of its high sucrose content, notwithstanding beetroot's industrial food exploitation (1). The availability of nitrogen affects the sugar's digestion (Fig. 3), particularly during the initial phases of development (Varner 2014; Singh et al. 2019).

Fig. 3 The composition of 100 g of raw beetroot flesh

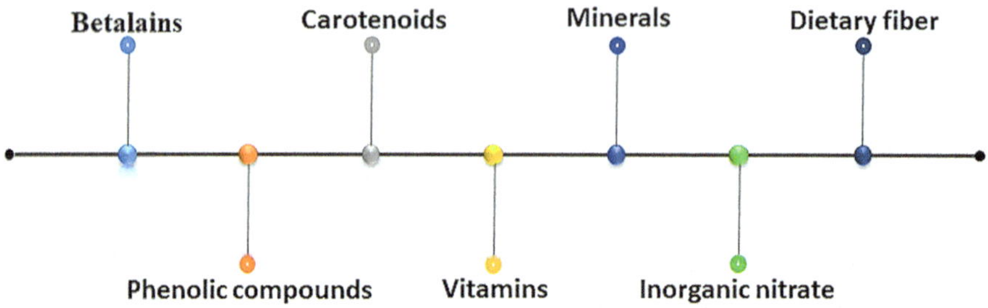

Fig. 4 Bioactive compounds of beetroot by-products

6 Bioactive Compounds of Beetroot By-Products

Beetroot by-products, such as peels, stems, and leaves, are abundant in various bioactive substances that may positively affect health. These substances fall into the primary categories listed below (Fig. 4):

6.1 *Betalains*

Beetroot contains a high amount of water-soluble betalains, which are nitrogenous plant pigments. It is thought that tyrosine serves as a precursor in the synthesis of betalains. The massive accumulation of tyrosine (hydroxylated by tyrosine hydroxylase) produces diphenol/DOPA oxidase activity, catalyzing the dihydroxyphenylalanine (DOPA) conversion into cyclo-DOPA. Thus, the creation of betalamic acid results because of the hydroxylation of tyrosine to DOPA and subsequent breaking of the aromatic ring of DOPA, which is the first step in the synthesis of betalain (Hatlestad et al. 2012). Moreover, the beetroot peel has the highest betalain concentration, although the quantity fluctuates across the entire root. Betalain concentration in beetroot is 120 mg/100 g fresh weight on average, while the ratio of betacyanin to betaxanthin content is 1:3 (Marmion 1992).

Betalamic acid conjugated to cyclo-3,4-dihydroxyphenylalanine (DOPA) with a maximum absorption wavelength of 538 nm gives birth to betacyanins, which have a reddish-violet color. Betacyanins are produced by condensing cyclo-DOPA and betalamic acid to produce non-glycosylated betanidin or isobetanidin chromophores. Tyrosinase and DOPA are the two main enzymes in every step of the biosynthesis process (Aykın-Dinçer et al. 2021). Beetroot has the largest quantities of betacyanins, which are betanin, isobetanin, and neobetanin, in that order. However, the term “betalains” refers to various compounds. When betalamic acid interacts with amino acids or biogenic amines, yellow pigments known as betaxanthins are produced, which have a maximum absorption at 480 nm.

In decreasing sequence, vulgaxanthin I, vulgaxanthin II, indicaxanthin, and miraxanthin are the betaxanthins present in beetroot. The root’s color is determined by the change in the ratios of betacyanins and betaxanthins; a higher concentration of betacyanins gives the root a distinct reddish-purple coloring. Different authors have reported differences in the percentages of betaxanthins and betacyanins in total betalains (Czyzowska et al. 2020; Desseva et al. 2020; Bangar et al. 2022). Certain studies found that betacyanins comprised 80–90% of total betalains, while others found that the percentage ranged from 50 to 70%. Moreover, Domínguez et al. (2020) report that around 90 different betalains (60 betacyanins and 33 betaxanthins) have been found in nature (Domínguez et al. 2020).

Furthermore, different colorants can be changed from yellow to purple-blue colors by combining betaxanthin and betacyanins (Domínguez et al. 2020). With no adverse effects on normal cell lines, betalains have a wide range of therapeutic, hepatoprotective, anti-carcinogenic, and anti-cancer properties (Nowacki et al. 2015). Betalains control homocysteine levels, which release vasodilators and vasoconstrictors and regulate vascular tone, thrombotic activity, platelet function, and delicate stability. According to Miraj (2016), betalain extracts lower inflammation and protect the thin lining of blood vessels (Maraie et al. 2014). Additionally, studies have demonstrated the antiviral, antibacterial, and antiradical qualities of betalains pigments (Slavov et al. 2013).

6.2 Phenolic Compounds

One class of secondary metabolites in plants is called phenolic compounds, which are essential to plant-based foods' quality. Beetroot is rich in phenolics and flavonoids and supports healthy hepatic, renal, hematological, and immune systems (Miraj 2016). The total phenolic content of beetroot juice, powder, chips, and cooked beetroot was determined to be 3.67, 0.51, 0.75, and 2.79 GAE mg/g, in that order (Nemzer et al. 2011; Vasconcellos et al. 2016). The processing methods used by the industry also impact the quantity and quality of phenolic content. Additionally, maintaining antioxidant activity and phenolic component levels depends on storage conditions.

6.3 Phenolic Acids

Large amounts of the two main families of phenolic acids, hydroxybenzoic and hydroxycinnamic acid from beetroot, have been discovered. Research indicates that beetroot's total phenolic acid content is between 50 and 60 mol/g dry weight. The phenolic acids present in beetroot include chlorogenic acid, rutin, vanillic, caffeic acid, epicatechin, catechin hydrate, p-coumaric acid, sinapic acid, syringic acids, protocatechuic proline, ferulic acid, monoterpene dehydrovomifoliol p-coumaroylquinic acid, 2-O-(3,4-dihydroxybenzyl)-2,4,6-trihydroxyphenylacetic acid, and theogallin acid (Kathiravan et al. 2014; Czyzowska et al. 2020; Sawicki et al. 2021). The content of phenolic acids directly impacts the free radical (ROS) scavenging property of beetroot products. Moreover, betalains enhance antioxidant capacities in concert with phenolic acids.

6.4 Flavonoids

Beetroot is a well-known food source of phenolic chemicals, such as flavonoids. Vanillin, rhamnetin, epicatechin, kaempferol, rutin, quercetin, astragalin, apigenin, dihydroisorhamnetin, rhamnocitrin, betagarin, betavulgarin, tiliroside, cochliophilin A, dihydroisorhamnetin, and catechin are among the flavonoids found in beetroot (Platosz et al. 2020). These flavonoid constituents support immune response regulation, mild inflammation reduction, and anti-carcinogenic actions (Iglesias et al. 2015). Flavonoids such as vitexin, vitexin-2-O-xyloside, and vitexin-2-O-rhamnoside have been shown by Slavov et al. (2013) to have a potent anti-proliferative impact on cancer cell lines (Slavov et al. 2013). According to several publications (Slavov et al. 2013; Kumar and Brooks 2018), the most essential flavonoids present in beetroot are rutin, kaempferol, rhamnetin, rhamnocitrin, and astragalin.

6.5 Saponins

Saponins are a defensive tool of plants against illness. A total of 26 triterpene saponins, including betavulgarosides I, II, III, IV, V, VI, VII, VIII, IX, and X, have been found in beetroot (Lechner and Stoner 2019; Mikołajczyk-Bator et al. 2016; Mroczek et al. 2012). Conversely, beetroot leaves contained betavulgarosides IX and X, not roots. It was previously believed that betavulgarosides I–IX, triterpene oligoglycosides with unique acetal and dioxolane-type substituents, were biosynthesized by oxidatively breaking down a terminal monosaccharide unit. Beetroot saponins are effective against various human cancers, such as leukemia, prostate, kidney, lung, breast, colon, and melanoma (Podolak et al. 2010).

6.6 Phenolic Amides

Plants contain many phenolic amide chemicals used in traditional herbal anticoagulant medicines. N-*trans*-Feruloyl 3-methyldopamine, N-*trans*-feruloyltyramine and N-*trans*-feruloylhomovanillylamine, N-*cis*-Feruloyltyramine, N-cis-Feruloyl 3-o-methyldopamine, and N-*trans*-Feruloyltyramine, 5,50,6,60-tetrahydroxy-3,30-biindolyl are the main phenolic amides discovered in beets (Kujala et al. 2002). In addition to the amides already described, early research has identified feruloylputrescine, feruloylserotonin, and feruloylpiperidine.

6.7 Carotenoids

Beetroot is full of carotenoids, powerful antioxidants that help prevent illness. In addition to being photosynthetic accessory pigments, carotenoids act as an oxygen-scavenging radical produced during photosynthesis in chloroplasts, protecting DNA and other biological constituents from damage caused by free radicals. Beetroot's carotenoids have anti-carcinogenic and immune-stimulant properties. The carotene content of beetroot is 1.9 mg/100 g. Carotenoids are mostly found in carrots, with red, yellow, and green capsicums, spinach, and beets following in order of proportion. Beetroot leaves also contain lutein and other oxygenated composites called xanthophyll and β-carotene (Rebecca et al. 2014; Zein et al. 2015).

6.8 Vitamins

Beetroots are high in fat-soluble vitamins (K-phenyladione A-retinol, E-tocopherol) and water-soluble vitamins (B1-thiamine, B2 riboflavin, B3 niacin, B5 pantothenic acid, B9 folates, and B12 cyanocobalamin) (Clifford et al. 2015; Kale et al. 2018). Vitamins help in liver and blood detoxification; they also help treat liver, kidney, and digestive issues; in particular, they help treat fatty livers caused by diabetes, low protein diets, and binge drinking. Additionally, by increasing blood flow to the brain, the beetroot vitamin B component assists in lessening the impacts of dementia and memory loss (Adhikari et al. 2017). A high content of 3.93 mg of vitamin A and 280 mg of K in beetroot leaves is helpful in cancer prevention, cardiovascular health, and lowering blood pressure (Balkrishna et al. 2022). Additionally, beetroot has a high concentration of folic acid, which supports healthy brain function and cancer prevention (Székely et al. 2016).

6.9 Minerals

Beetroot contains natural minerals that support healthy bones, teeth, and tissues, including magnesium (Mn), zinc (Zn), phosphorus (P), copper (Cu), sodium (Na), iron (Fe), potassium (K), and calcium (Ca) (Olumese and Oboh 2020; Kale et al. 2018). Beetroot is used to cure fevers and constipation because its high copper level facilitates the body's iron absorption (Balkrishna et al. 2022). Beetroot leaves are also reportedly a rich source of macro and microminerals (Biondo et al. 2014). Beetroot leaves have a greater concentration of iron (256 mg/kg) than several other vegetables, with (110 mg/kg) in *Senna occidentalis*, (130 mg/kg) in *Chenopodium album*, and *Justicia flava* with (160 mg/kg) (Odhav et al. 2007; Ekholm et al. 2007). Additionally, beetroot leaves have a high content of Cu (13.42 mg/kg), comparable to watercress and rocket (Batool et al. 2019). Similarly, beetroot leaves have a high K content (20,784 mg/kg) that aids in metabolizing proteins and carbs.

6.10 Inorganic Nitrate

The primary food sources of nitrate, between 50 and 85% of the total daily nitrate consumption, are beetroot and leafy green vegetables like spinach and lettuce. According to a 2013 study by Lidder and Webb, beetroot has a high nitrate concentration, ranging from 644 to 1800 mg/kg (Batool et al. 2019). Eating beetroot raised plasma nitrate by 96%. It has been proposed that the cardioprotective benefits of Mediterranean diets may be explained by the high concentration of nitrates in plant foods. Beetroot also has nitrates, which can lower bad cholesterol, normalize blood pressure, guard against ischemia-reperfusion damage, and regulate mitochondrial activity (Guldiken et al. 2016). Dietary nitrate supplementation has been demonstrated to enhance neurovascular coupling in response to visual stimuli and perfusion to brain regions associated with executive function. In response to visual stimuli, the neurovascular coupling, cerebral hemodynamics, and perfusion to brain regions that are related to executive function are enhanced by dietary nitrate supplementation (Guldiken et al. 2016; Presley et al. 2011; Thompson et al. 2014).

6.11 Dietary Fiber

Beetroot is high in dietary fiber (Elleuch et al. 2011). Since beetroot powder makes up about 55% of dietary fiber, it may boost the amount of dietary fiber in cereals and baked products such as bread, cakes, and cookies. Beetroot's bioactive components support several biological functions, including hepatoprotective, anti-carcinogenic, cardioprotective, and antioxidant effects. However, betalain stability is severely hampered by the extraction, processing, and storage techniques, which impact the substance's biological activity. This is why the food and pharmaceutical sectors use beetroot bioactive components sparingly. To stabilize betalain, studies must be done on investigating and improving more recent extraction, processing, and storage techniques. Furthermore, it's critical to comprehend the kinetics of beetroot phenolic compound degradation to investigate the compounds' potential for various applications. Additional clinical research might help determine the connection between betalains' stability and health-promoting properties (Ingle et al. 2017; Kohajdová et al. 2018; Sahni and Shere 2016; Singh et al. 2016).

7 Health Promising Potential of Beetroot By-Products

In the sphere of public health, sophisticated lifestyles have gained many converts, which contributes to major chronic illnesses, including diabetes, heart disease, and cancer. Because of this, it's now essential to include fruits and vegetables in one's diet daily. Several scientific studies have shown that beetroot improves clinical outcomes for various disorders (Sawicki et al. 2019).

7.1 Bioavailability

A food ingredient must be accessible in vivo, which implies that after ingestion, its bioactive components pass through the digestive system and end up in the plasma in sufficient amounts for cells to employ them (Wootton-Beard and Ryan 2011). Furthermore, food ingredients must maintain their chemical makeup through several stages of digestion, each of which presents a unique metabolic challenge for the molecule and affects the degree and rate of absorption to establish a systemic circulatory system and carry out a specific advantageous function (Rein et al. 2013; Ting et al. 2014).

It has been determined that betalains, the primary active ingredient in beetroot, and inorganic nitrate are potential sources of bioactive chemicals for human consumption (Van Velzen et al. 2008; Kanner et al. 2001). Participants' urine, examined after 300 mL of beetroot juice consumption, contained 0.5–0.9% of the betacyanins (betanin, isobetanin) they had taken, according to Kanner et al. (2001). This implies that people may absorb even very low betacyanins (Powers and Jackson 2008). Additionally, they found that the greatest urine exclusion ratio of betacyanins, an absorption indication, happened 2–4 h after ingestion, despite significant inter-individual variability throughout this period (Kanner et al. 2001).

7.2 Oxidative Stress

Supplementing with beetroot may be a helpful strategy for boosting innate antioxidant defenses, which will shield biomolecules from oxidative damage. A cell's physiological state is considered to be in a redox equilibrium phase under normal metabolic conditions, or, to put it another way, an equilibrium between (antioxidant) reducing and (pro-oxidant) oxidizing agents occurs (Powers and Jackson 2008). RONS is essential for various physiological and metabolic processes at deficient concentrations, such as gene expression, apoptosis, cell division, and muscle contraction. Overexposure of cells to RONS generated intracellularly (e.g., inflammation, aberrant cellular metabolism) or extracellularly (e.g., xenobiotics, UV rays) can disrupt the antioxidant capacity of cells, resulting in an imbalance in redox cell function and, ultimately, oxidative stress.

Tesoriere et al. (2008) state that several in vitro studies have shown that betalain pigments, in particular, protect cellular constituents from oxidative damage (Tesoriere et al. 2008). Numerous physiologically active phenolics, such as rutin, epicatechin, and caffeic acid, are found in beetroot and are all potent antioxidants (Tesoriere et al. 2008). Similarly, it has been demonstrated that nitrite and other NO sources, including beetroot, inhibit the generation of radicals and actively scavenge potentially hazardous RONS, such as hydrogen peroxide, implying the possibility of nitrate's antioxidant qualities (Lundberg et al. 2011; Wink et al. 2011). The researchers also found that betanin was the most effective lipid peroxidation inhibitor, the most common betalain in beetroot (300–600 mg/kg). Betanin's exceptional electron-providing ability and ability to defeat highly reactive radicals that assault cell membranes are the sources of its potent antioxidant action (Kanner et al. 2001).

7.3 *Inflammation*

Most of the time, inflammation is a helpful process that controls our natural response to physiological stresses, including injury, infection, and other organisms that might damage the body and interfere with homeostasis (Monteiro and Azevedo 2010). However, the host may have adverse consequences from the immune response. Inadequate immune system renewal and persistence of the invasive element can lead to chronic inflammation and chronic cell dysfunction (Kundu and Surh 2012). It has been shown that betalains and beetroot extracts work well as anti-inflammatory agents. They interact with pro-inflammatory communication pathways to partially mediate their anti-inflammatory effects (El Gamal et al. 2014). Beetroot's anti-inflammatory and antioxidant qualities have aroused interest in its potential use in aberrant immune cell activity conditions. In animal research, betacyanin extracts have also been demonstrated to have chemopreventive effects on malignant skin, lung, and liver cells. These effects have been observed in human pancreatic, skin, breast, and prostrate carcinogenic cells (Kapadia et al. 2013). Beetroot's anti-inflammatory and antioxidant qualities have sparked attention to its possible use in conditions marked by abnormal immune cell activity. Research indicates that betalain extracts from beetroot may help mitigate the consequences of prolonged inflammation, which has been linked to the formation of malignant tumors (Das et al. 2013).

7.4 *Endothelial Function*

The nitrate content of beetroot is converted to nitrite, which, with further reduction, produces NO (Hobbs et al. 2012). By controlling thrombotic activity, vascular tone, platelet function, and the delicate balance between the release of vasodilating substances (prostacyclin, NO) and vasoconstricting agents (thromboxane, endothelin), the endothelium maintains vascular homeostasis (Davignon and Ganz 2004). Endothelial dysfunction is primarily caused by reduced NO bioavailability with aging, as NO modulates several endothelial functions (Lidder and Webb 2013). Atherosclerosis and hypertension have advanced due to endothelial dysfunction, which has been identified as a major possible cause of many cardiovascular illnesses. Like beetroot, which is high in nitrate, betalains and caffeic acid are bioactive components that might help control cardiovascular disease (Joris and Mensink 2013; Machha and Schechter 2011).

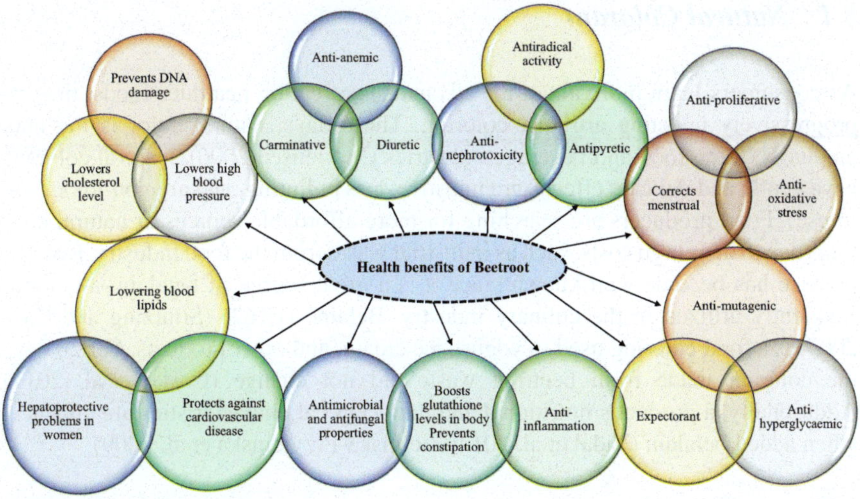

Fig. 5 Health benefits of beetroot

7.5 *Cognitive Function*

One of the primary pathologic processes that coexist with the decline in cognitive performance associated with aging is impaired cerebral blood circulation (Bond et al. 2013). The aging process has been related to a decrease in cerebral perfusion, which has been connected to several neurological conditions that impair cognitive function, such as Alzheimer's disease, brain damage, and clinical dementia (Poels et al. 2008). One of the leading causes and detrimental effects of cerebral hypoperfusion's onset and development is a disruption in neurovascular function, which is partially mediated by reduced NO activity. Therefore, a NO originator such as beetroot may be able to counteract deficiencies in cognitive function and enhance cerebral blood flow. Gilchrist et al. (2014) reported that compared to the control group, older (67 years) type-2 diabetics who received 250 mL of beetroot juice for 2 weeks showed significant improvement in simple reaction time. On the other hand, no effect was seen in different cognitive tasks involving decision-making, rapid processing, morphology, and spatial memory (Gilchrist et al. 2014) (Fig. 5).

8 Application of Beet By-Product in the Beetroot and Beverage Industry

Due to its numerous valuable properties, red beetroot pomace has grown in popularity in the food and beverage industry. It is used as a food colorant preservative to make various functional foods.

8.1 Natural Colorant

As consumers learn more about the advantages of eating healthier foods, they are progressively rejecting artificial coloring. These days, natural color is the main emphasis of the food and beverage industries (Mortensen 2006). Natural colorants have gentle and delicate effects, are nutritive, less polluting, and are non-toxic (safe to use). Food producers are searching for more affordable sources of natural colorants due to their high costs, such as industrial waste from the food industry. Beetroot pomace has become well-known since it is a great source of betalain, a colorant frequently utilized in the culinary industry. Betaine (E162) (Stintzing and Carle 2007) is a food coloring used in yogurt, ice cream, and other products. For 20 days, the colorant made from beetroot waste did not change (Costa et al. 2017). Additionally, a product's nutritional and nutraceutical characteristics are enhanced when added betalain (Vidal et al. 2014; Zielińska-Przyjemska et al. 2009).

8.2 Natural Preservative

Variable preservatives and techniques are utilized to extend the shelf life of improved-quality products (Ozaki et al. 2020). Because these substances function as potent antibacterial agents that damage the product and enhance its sensory attributes, synthetic nitrate and nitrite are frequently utilized in the food processing industry. Essentially, their potential for harm and their significant correlation with health problems prompt food makers to focus more on natural preservatives (Petkoska et al. 2021). Beetroot pomace now naturally produces nitrates and nitrites. In addition to being high in nitrate and nitrite, beetroot pomace is rich in many other beneficial ingredients that, in addition to maintaining food quality, provide functional qualities (Jin et al. 2014). It is very common to find biodegradable packaging with an antibacterial component. The antibacterial efficacy of edible packaging is enhanced by adding beetroot extract (Domínguez et al. 2020; Tran et al. 2017).

8.3 Functional Foodstuff

Beetroot pomace provides intriguing technical properties for the food industry, such as enriching pastries, cakes, and pasta (Costa et al. 2017). It is also a good source of phytonutrients and dietary fiber. Owing to advances in nutritional science, bakery products such as cakes, cookies, and biscuits are now categorized as junk food that contains empty calories. Therefore, value addition is required to improve the nutritional quality of these widely consumed snack foods and make them higher in fiber and essential nutrients. Most pomace powders also function as organic flavorings, colorants, and fragrances without changing the physiognomies of the products they are added to (Tran et al. 2017; Manopriya and Arivuchudar 2019). Because beetroot pomace powders improve emulsion consistency and enhance water and oil retention, they can be used as a low-cost, non-caloric bulking component instead of wheat, sugar, or fat to improve food functionality (Elleuch et al. 2011). About 62.75% of the powder beetroot pomace contains crude fiber. The fiber content of the cookies increased with the inclusion of beetroot pomace powder. When beetroot pomace is added in large enough quantities, the quantity of crude fiber, moisture, protein, and ash increases, but the amount of carbohydrates decreases (Sahni and Shere 2016).

8.4 Uses in Fortification and Value-Added Food Products

Beetroot is a treasure trove of bioactive chemicals with related health advantages, including the potential for both antioxidant and antibacterial activity (Chhikara et al. 2019a, b). Juice from beetroot has become incredibly popular as a healthy tonic snack. Beetroot may be processed into various formulations that provide a valuable and innovative way to obtain dietary nitrate and its bioactive components (Baião et al. 2017). Food producers can explore multiple methods of creating nutritious food items through value addition, increasing their market potential and economic value (Pant and Chinwan 2014). Convenience is an important marketing technique in the food sector.

Additionally, processing aids in extending shelf life to address seasonality and perishability (Dhiman et al. 2022). As was previously said, beetroot may be processed into various creative value-added products that enable customers to benefit from its medicinal properties (Chhikara et al. 2019a, b). A well-liked beverage, beetroot juice can be purchased as a pasteurized or unpasteurized fresh-squeezed juice. Since betalains are lost during heat treatment, freshly squeezed juice is recommended since it preserves most flavor and aroma. But juice must be pasteurized to be stable and edible (Dos Santos et al. 2018). Fresh, clean beetroots are ground to get juice, which is subsequently cleared to get rid of pulp and other impurities. Because of the color change and nutritional loss following heat treatment, this juice may also be used as a colorant in some goods like ice cream, candies, and other confections (Chandran et al. 2014). After standardizing the amount of black pepper and ginger variation beetroot RTD, they were thermally processed for 30 min at 90–100 °C. They were then poured into glass bottles that had been sterilized, which helped to maintain their physicochemical qualities for 2 weeks at room temperature (Dhiman et al. 2020).

Dhineshkumar and Ramasamy combined 150 g of skim milk with 850 mL of distilled water to prepare beetroot-flavored yogurt. Following 15 min of pasteurization at 80–85 °C, the temperature was lowered to 42–45 °C. To halt fermentation, starting culture was added, mixed, and chilled at 4 °C for 12 h at the same temperature. The pH of the ideal yogurt ranged from 6.5 to 7.8. Yogurt's mineral content and nutritional value rose when it added beetroot juice (Dhineshkumar and Ramasamy

2016). Similarly, Yadav et al. (2016) prepared yogurt with a beetroot foundation and high sensory ratings using dried beetroot powder. Yogurt's protein, carbohydrate, lipid, and ash contents rose as the concentration of beetroot powder increased, enhancing the food's nutritional value (Yadav et al. 2016). A natural colorant in strawberry-flavored ice cream was beetroot juice. To make this dish, combine butter (80% fat), skim milk powder (95% solids not fat), and milk (5% fat and 9.5% solids not fat). After heating the milk to between 55 and 60 °C, butter, stabilizer, and powdered skim milk were added. After homogenizing the mixture at 2500 and 500 psi, it was pasteurized for 30 s at 80 °C, chilled overnight, reheated, and blended with beetroot juice (Manoharan et al. 2012).

Beetroot puree was utilized by Evstigneeva et al. (2020) to make ice cream with a beetroot basis. In addition to adding nice color, beetroot added valuable nutrients to the mixture. A delicate paste was made from clean, peeled beetroots to manufacture beetroot candies in a different investigation. This mixture was brought to a boil with 65% sugar and 3% pectin while constantly stirring. After 55 min of heat desiccation, the paste was mixed with 0.5% citric acid when the appropriate consistency was reached. The thick mixture was rolled into desired-sized and shaped sweets once it had cooled to room temperature (Evstigneeva et al. 2020). Beetroot candies may be a healthier alternative to the artificially flavored sweets sold in stores (Fatma et al. 2016).

It was also discovered that beetroot enhanced the nutritional content of apple jam during preparation. To extract the coloring and antioxidants, beetroot was cooked in water. After that, 1900 g of apples and 700 g of sugar were cooked in 1.5 L of water previously used to boil beets. Ten cinnamon sticks were introduced throughout this procedure but were removed before moving on. The product was then crushed and cooked again to eliminate any leftover surplus water and get the right consistency. The final product had a high AA concentration and a low-sugar content (Guiné et al. 2016). The low-sugar jam used 60% beetroot pulp and 0.5% strawberry flavoring. It could be stored for up to 6 months under refrigeration without chemical preservatives (Perumpuli et al. 2018). In a different study, Wang et al. (2020) discovered that while processing beetroot in a jam lowers its TPC, TBC, and AA, it increases the compounds' bioaccessibility when compared to fresh beetroot. This finding may result from molecular structural changes brought about by jam processing. TPC, TFC, TBC, and AA dropped by 33.9, 10.6, 7.9, and 9.7–65.7%, respectively, with storage (Wang et al. 2020).

Additionally, beetroot may be used to make jelly by combining 61% sugar, fruit extract, 0.5% citric acid, and 2% pectin. The mixture was heated while stirring continuously until the TSS reached 65 °Brix. A couple of the process's drawbacks include the creation of hazy jelly as a result of surface scum and the failure to set when there is less than 2% pectin. Overall, acceptance of the finished product was high (Chaudhari and Nikam 2015). The wine was also made with beetroot juice. *Saccharomyces cerevisiae* was used to carry out anaerobic fermentation. Overall acceptance of the resulting wine was equivalent to red grape wine made commercially (Mo et al. 2020). The length of fermentation affects the beetroot wine's taste and physicochemical characteristics. The technique to halt fermentation, such as pasteurization or cold shock, might affect the end product's color (Otegbayo et al. 2020). Beetroot microcapsules were created by homogenizing beetroot juice with gum Arabic solution and spray-drying the mixture (Pitalua et al. 2010). As a natural red food coloring, root powder may be used in dry mixes (such as soups and Indian curry blends), desserts, jams, jellies, and more (Gokhale and Lele 2011).

High temperatures cause beetroot powder manufactured with conventional methods, including spray-drying, to become discolored due to thermal deterioration. Therefore, the finest approaches for producing powder are new ones. Kowalski and Szadzińska (2014) found that intermittent drying, as an alternative to stationary convective drying, produced substantially superior quality goods with more color retention, higher betanin concentration, and less product degradation. In addition, the non-stationary methods of beetroot drying followed by osmotic and blanching pretreatment produce dried goods with improved color, a larger concentration of naturally occurring pigments that are beneficial to health, and increased durability (Kowalski and Szadzińska 2014).

Combining many approaches is highly beneficial as it compensates for the drawbacks of each approach when used separately. Figiel (2010) combined convective and vacuum-microwave approaches and found that introducing vacuum-microwave finish drying early on can significantly shorten the time it takes for beetroot cubes to dry. Additionally, this assisted in keeping the quality comparable to freeze drying (Figiel 2010). Beetroot pulp (0–40%) was added to wheat flour by Chhikara et al. (2019a, b) to create noodles with better nutritional value, bioactive chemicals, and color intensity. The amount of beetroot integrated rose along with the betalain, TPC, and AA levels. Considering the sensory, phytochemicals, cooking, and nutritional aspects, 30% beetroot-integrated noodles are the best (Chhikara et al. 2019a, b). Based on the aforementioned research, it can be inferred that beetroot has the potential to be processed into a wide variety of products that offer diverse nutritional benefits to customers. Value addition will also raise the product's market potential and profitability. To fully realize its potential, a more thorough study of product development techniques must be done (Fig. 6).

8.5 Non-Food Applications of Beetroot By-Products

Even though beetroot was used medicinally in ancient Greece and Rome, there has been sporadic scientific interest in the plant in recent decades. As a result, during the last 20 years, some studies have demonstrated that beetroot, among the top ten vegetables in potency, is a potent source of natural antioxidants (Sawicki et al. 2019). Numerous demographic groups now have altered health conditions due to modern lifestyle choices, including diabetes, cancer, and heart disease. Fruits and vegetables must now be included in the diet regularly (Wootton-Beard and Ryan 2011). They are highly recommended as dietary medicine since they are abundant in bioactive chemicals with potent antioxidant potential. Beetroot-root has been shown in vitro and pre-clinical studies to have the following health benefits: anti-cancer,

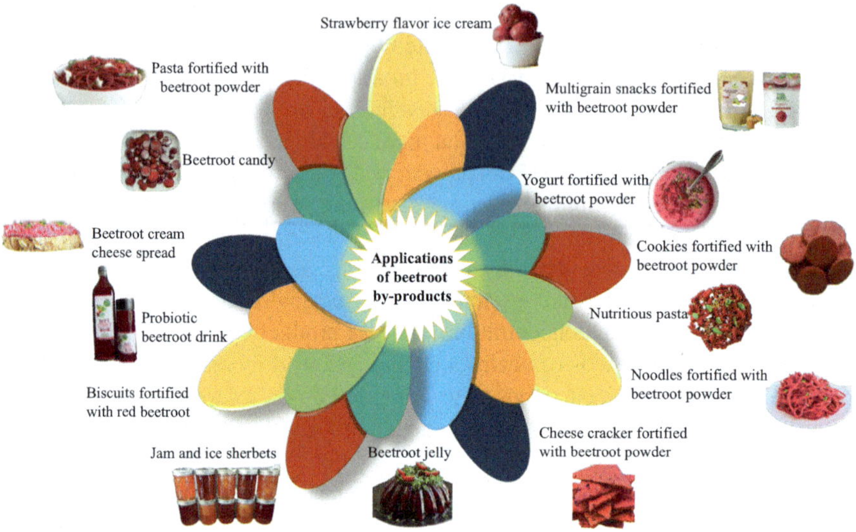

Fig. 6 Valorization of by-products for non-health purposes

antioxidant, anti-depressant, anti-oxidative stress, anti-mutagenic, anti-carcinogenic, radioprotective properties; anti-hypertensive, antibacterial, anti-hypercholesterolemia, anti-proliferative, hematopoietic, anti-hyperglycemic, diuretic, anti-nephrotoxicity, anti-inflammatory, hepatoprotective, and immuno-modulatory properties (Lu et al. 2009; Murthy and Manchali 2012).

Beetroot-root eating can also help ward against age-related illnesses (Ravichandran et al. 2013; Kapadia and Rao 2012). Beetroot may be eaten in various ways to get all its health benefits, including raw, juiced, salad, boiled, baked, dried, or as chips or other combinations. The leaves are a great source of minerals and vitamins A, K, C, and E. These nutrients can help treat a variety of illnesses and conditions, including cancer, dandruff, gastric ulcers, kidney problems, liver toxicity, hepatitis, jaundice, diarrhea, and vomiting. The primary bioactive ingredient that beetroots may contain is betalain, which is a member of the family of red and yellow pigments. Two types of pigments are found in betalain: betaxanthin, a yellow-to-orange pigment, and betacyanins, a violet pigment. The antioxidant activity of betanin is three times more than that of catechin and ten times greater than that of tocopherol. When cytochrome c tests linoleate peroxidation, the chemicals get cationized, increasing their affinity for membranes, a very advantageous property for antioxidants (Kapadia and Rao 2012; Kanner et al. 2001).

Because of their pharmacological actions as antioxidants, anti-cancer, anti-inflammatory, hepatoprotective (Georgiev et al. 2010), anti-lipidemic, and antibacterial agents, betalains, particularly betacyanins, are significant for human health. In vitro, they can suppress the growth of cells in human tumors and cervical, ovarian, and bladder cancer cells (Neha et al. 2018; Adhikari et al. 2017). Beetroot consumption decreases the risk of esophageal, colon, liver, lung, and skin tumors. Beetroot extract is a natural pigment widely used as a colorant in medications and cosmetics because of its potential to prevent cancer. Additionally, it can work in concert with anti-cancer medications like doxorubicin (Adriamycin) to lessen treatment-related drug toxicity (Kapadia and Rao 2012). Consuming beetroot (in the form of juice or supplemented bread) has positive physiological benefits that may enhance the course of treatment for several diseases, including type 2 diabetes, hypertension, atherosclerosis, and neurodegenerative illnesses (Edziri et al. 2019; Dos Santos Baião et al. 2019). Beetroot's B vitamins increase blood flow to the brain, which helps to lessen the effects of dementia and memory loss (Adhikari et al. 2017). Consuming beetroot as a nitrate source offers a natural way to raise the availability of nitric oxide (NO) in vivo, which may help prevent and treat diseases, including endothelial function and hypertension (Dos Santos Baião et al. 2019).

Betalains improve oxidation resistance, which benefits metmyoglobin and low-density lipoproteins in humans. In addition to reducing oxidative damage to lipids, betalains and other phenolic compounds can help lessen blood vessel, joint, and bone inflammation. In times of inflammation, this rebuilding benefits individuals with osteoporosis and asthma. This explains why betalains have anti-inflammatory and antiradical properties, which aid in regulating illnesses linked to human oxidative stress (Ravichandran et al. 2013). Beetroot is a viable therapeutic intervention in several clinical diseases connected to inflammation and oxidative stress (Netzel et al. 2005). According to Netzel et al. (2005), a single intake of beetroot juice increased the amount of antioxidant chemicals, such as betalains, excreted in the urine (Netzel et al. 2005). It has elevated levels of tyrosine, iron, folic acid, vitamin C, and red betalains, which are antioxidants. Some people produce red urine, or beeturia, due to their inability to metabolize red betanin. The "free iron" redox cycle (cytochrome c), H_2O_2-activated metmyoglobin, and lipoxygenase, respectively, were shown to catalyze lipid peroxidation in membranes or linoleate emulsions that benin from beetroots was discovered to block (Kanner et al. 2001). Obesity is a result of both inflammation and oxidative damage. One dietary element that has antioxidant potential is betalain pigments, found in beetroot. Zielińska-Przyjemska et al. (2009) looked into the impact of beetroot juice and chips on oxidative metabolism and apoptosis in the neutrophils of obese people in vitro. Beets are rich in fiber and vitamin C. They also include potassium and manganese, two essential minerals. Because it minimizes the muscles' need for the body's energy source, adenosine triphosphate, potassium is necessary for normal nerve and muscle function (Zielińska-Przyjemska et al. 2009).

Constipation is avoided as waste is moved through the intestines with fiber. Similarly, the colon is guarded by beets' antioxidant components against cancer. Beetroot's hemostatic, carminative, emmenagogue and kidney-protecting qualities are among its purported medicinal uses. Peeling et al. (2015) researched the effect of beetroot on blood's oxygen-carrying capacity, aiming to promote high levels of athletic performance, particularly in athletes who race or kayak. Consuming beetroot also prevents macular degeneration, which is linked to carotenoids and is known to lower the incidence of cataract development (Peeling et al. 2015).

The vitamins and nutrients aid in the liver and blood detoxification process. They can also treat disorders of the kidneys, liver, and digestive tract, especially those involving the accumulation of fatty deposits in the liver brought on by diabetes, protein shortages, or alcoholism. Natural elements found in beetroot provide bone strength (Neha et al. 2018; Adhikari et al. 2017). Beetroot juice is sometimes taken as a home treatment for erectile dysfunction. Some significant findings from a significant pilot trial on beetroot-based cereal bars and their effect on people at risk for cardiovascular illnesses were reported. It examined phenolic chemicals' antioxidant capacity and stability, demonstrating their reliance on the physicochemical circumstances inside the various digestive system segments. The duodenal digestion is unaffected by the total amount of phenolic chemicals present in beetroot formulations. Some phenolic compounds may change into other compounds and be vulnerable to the duodenal fluid's alkaline pH. The study concluded that eating beetroot-cereal bars encouraged the bioconversion of NO_3^- to NO, which is a factor that reduces the development of cardiovascular disease (Dos Santos Baião et al. 2019) (Fig. 7).

Medical research has looked at the possible health advantages of beetroot by-products, including powder and extract. These leftovers include many vitamins, minerals, and antioxidants, which may help the heart, reduce inflammation, and fight cancer. Beetroot by-products have been found in certain trials to help decrease blood pressure, boost cognitive function, and improve athletic performance. Additionally, they could be helpful in the treatment of diseases, including diabetes, obesity, and liver disease. Additionally, beetroot is being investigated as a possible therapeutic intervention for several oxidative stress and inflammation-linked

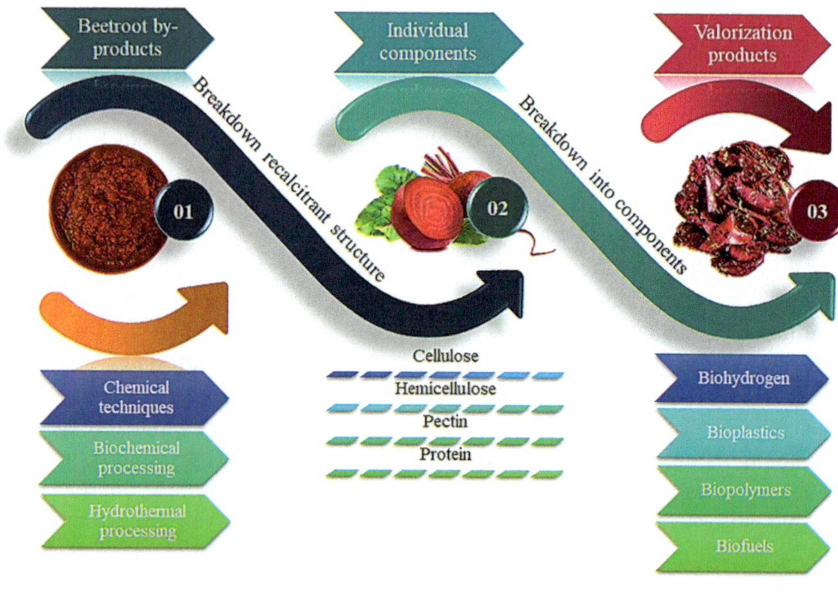

Fig. 7 Beetroot by-products: applications in medicine

clinical disorders. The betalain component of beetroot has potent anti-inflammatory, antioxidant, and chemopreventive properties in vitro and in vivo (Peeling et al. 2015; Clifford et al. 2015).

By-products of beetroot are well recognized for their beneficial effects on several pathologic and clinical outcomes. Beetroot juice supplementation is a cost-effective strategy for acute and chronic conditions that may show promise in managing blood pressure and vascular function, diabetes and insulin hemostasis, renal health, and potential effects on microbiota abundance. The non-significant variation in systolic and diastolic blood pressure was a secondary result and physiological reaction of microbiota abundance regulation. Additionally, some research has indicated that beetroot juice has reno-protective qualities, which are linked to a lower death rate and positive modifications to kidney function parameters in individuals with renal diseases. Similarly, it has been demonstrated that consuming beetroot juice consistently lowers the blood glucose peak and successfully delays the postprandial glycemic response. The impact of reducing blood pressure significantly has been observed in normotensive patients; it tends to be more pronounced in hypertensive persons and in adults who are overweight (Mirmiran et al. 2020; Chen et al. 2021).

8.6 Beetroot By-Products for Diabetes

Alpha-lipoic acid is an antioxidant found in beets. This substance may improve insulin sensitivity and decrease blood glucose levels. Alpha-lipoic acid's impact on diabetic neuropathy symptoms. Researchers discovered that giving alpha-lipoic acid supplements orally and intravenously reduced the symptoms of peripheral and autonomic neuropathy in people with diabetes. Nonetheless, most of the dosages used in these investigations were far greater than those found in beetroot. Based on the research that is currently available, the consequences of lower food dosages remain unclear (Salehi et al. 2019).

8.7 Heart Health and Blood Pressure

A study including 68 hypertensive individuals investigated the effects of consuming 250 mL of beetroot juice daily. After consumption, the researchers discovered that doing this considerably reduced blood pressure. They propose that the high nitrate content of the beetroot juice was the cause of this anti-hypertensive impact. High nitrate vegetable consumption is suggested as an efficient and inexpensive means of managing hypertension. However, stopping a prescription blood pressure medication without consulting a doctor is never a good idea. For cardiovascular disorder (CVD), high blood pressure is regarded as the leading risk factor. Reducing it with dietary modifications and other strategies can help stop heart attacks, strokes, heart failure, and other potentially fatal CVD consequences (Kapil et al. 2015).

8.8 Digestion and Regularity

The amount of fiber in one cup of beetroot is 3.81 g. Enough fiber intake is necessary for gut health and easy digestion. Depending on an individual's age and gender, a single cup of beets can supply over 8.81% of their daily fiber needs, according to the United States Department of Agriculture (USDA). One strategy to enhance dietary fiber intake is to include beetroot in one's diet. Beetroot by-products, like pulp and peel, have good qualities for regularity and digestion. According to studies, beetroot peel is full of bioactive substances, including phenolics and betalains, that can improve the composition and activity of the gut microbiota and support gastrointestinal health.

Furthermore, when subjected to gastrointestinal proteases, the proteins isolated from beetroot leaves showed good digestibility, indicating simple breakdown and absorption in the digestive system. Anaerobic digestion of sugar beets and potato by-products has also been investigated; specific methane yields suggest that agricultural leftovers, including beetroot components, may be efficiently recycled by anaerobic processes. Overall, the advantageous bioactive components and digestibility properties of beetroot by-products present a viable path for improving regularity and digestion (McKillop et al. 2021).

8.9 Exercise and Athletic Performance

Supplementing with beetroot juice has been shown in certain studies to increase the quantity of oxygen that muscles take in when they are doing out. Researchers discovered that giving seasoned cyclists large dosages of beetroot juice enhanced their time trial performance. The same year, another research looked at 12 active female volunteers in their leisure time. On the other hand, the researchers found that supplementing with beetroot juice enhanced the athletes' performance. Thus, more investigation is required to validate the beneficial effects of beetroot on exercise performance (Rokkedal-Lausch et al. 2019).

8.10 Cancer Prevention

It was discovered that some of the chemicals in beets can prevent cells from undergoing malignant mutations. Among these substances is betalain, a pigment and antioxidant that gives beets their red or yellow hue. Beets may have some role in lowering cancer risk, but further study is required before medical experts suggest them in favor of other conventional cancer risk reduction techniques. Because betalains, polyphenols, flavonoids, and nitrates are significant sources of bioactive chemicals with anti-inflammatory, chemopreventive, and antioxidant characteristics, beetroot and its by-products have shown remarkable potential in cancer prevention and therapy. Research on a variety of cancer cell lines, including human carcinoma of the uterine (HeLa) cells, has demonstrated the anti-cancer benefits of natural chemicals produced from beetroot extracts, such as betanin and apigenin, via altering cell proliferation, migration, apoptosis, and the mTOR pathway. Beetroot peel powder has also been shown to have higher antioxidant activity, total phenolic content, and improved sensory and textural qualities when added to value-added products like mayonnaise. These results highlight the potential of beetroot by-products in developing functional and health-promoting food formulations (Lechner and Stoner 2019).

9 Beetroot By-Product in Microbiology

The beetroot matrix was utilized as a lyophilized powder in a dried or fresh form. A 1012 CFU/g Lactobacillus plantarum MIUG BL3 culture inoculum was sprayed on the vegetative tissue. The viability of the lactic acid bacteria (LAB) in each product was assessed for 21 days using microbiological culture techniques. The amount of betalains and the viability of LAB were connected with the antioxidant activity of the food items produced. The polyphenolic compound content revealed a significant concentration of bioactive substances, ranging from 225.7 to 1314 mg/L. According to the confocal laser scanning microscope study, many live probiotic cells were found in all variations, particularly in the fresh beetroot-root cubes. The high Lb. plantarum (CFU per gram) of the food items after 21 days of refrigeration was ascribed to the nutrients and bio-compounds of the vegetal matrix, which in some way preserved the survival of the bacterial cells by providing protection. Because of their positive qualities throughout the conducted studies, the food items obtained and enhanced with probiotic LAB might be considered novel functional food products (Barbu et al. 2020).

9.1 Beetroot By-Products Used to Remove Heavy Metal Toxicity

Chronic poisoning with cadmium (Cd) raises the quantity of Cd in the blood, liver, and kidney, which triggers oxidative processes in these organs. This suppresses the immune response in chickens, as seen by changes in humoral and cell immunity indices. In Cd-exposed hens, oxidative stress indicators, biochemical markers, and immunological indices were nearly restored to baseline levels after 10 days of ultrafiltration beetroot juice. When hens were exposed to Cd ingested fractionated beetroot juice, it inhibited the oxidative effects of the heavy metal and had immune-modulating (Vasiljeva et al. 2018). By activating the intrinsic and extrinsic apoptotic pathways, the growth of cancer cell lines can be inhibited by beetroot phytochemicals with significant antioxidant and anti-inflammatory properties. A substantial amount of data also suggests that these phytochemicals have an impact on the downregulation of angiogenesis-controlling genes, such as hypoxia-inducible factor 1A and vascular endothelial growth factor A, as well as pro-survival genes, such as baculoviral inhibitor of apoptosis repeat-containing 5 and catenin beta^{-1} (Javed et al. 2023; Ninfali et al. 2017). Higher red pigment content beetroot types, which function as free-radical-scavenging reducing agents and phase II enzyme-inducing activators, are more effective. Lee et al. (2014) reported on the anti-proliferative properties of betanin and betalain against HepG2 cancer cell lines in vitro. One noteworthy finding is that when given at extremely low doses, the study's beetroot extract demonstrated a chemopreventive impact against liver cancer (Kapadia et al. 2003). The beetroot extract has a cytotoxic effect on human prostate and breast cancer cell lines similar to that of doxorubicin (Adriamycin). The chemical structure of the aforementioned red-colored anti-cancer antibiotic is strikingly similar to that of betanin (Kapadia et al. 2011), with a planar arrangement of an aromatic chromophore coupled to a sugar molecule. Its examination against mice's skin has demonstrated an in vivo anti-tumor promoting action, and lung bioassays have demonstrated a noteworthy tumor inhibitory impact (Kapadia et al. 1996). In rats given N-nitrosomethylbenzylamine, drinking water colored like beetroot inhibits the development of esophageal cancer (Lechner et al. 2010; Ijaz et al. 2023a, b).

10 Conclusion

Beetroot by-products have diverted attention in passing years due to their high content of bioactive phytochemicals, which are known to provide various health benefits. These compounds, present in the leaves, stems, and peel of beets, are recognized for their antioxidant, anti-inflammatory, and anti-cancer properties. Consequently, beetroot by-products are being investigated for their potential uses in functional foods, nutraceuticals, and pharmaceuticals. Among the essential bioactive compounds in beetroot by-products are betalains, a group of red and yellow pigments unique to beets that exhibit potent antioxidant properties. Antioxidants are crucial in shielding the body from oxidative stress and damage induced by free radicals, which are linked to chronic diseases like heart disease, cancer, and diabetes. By incorporating betalain-rich beetroot by-products into the diet, individuals may enhance their antioxidant intake and decrease the risk of developing these diseases. Besides their antioxidant characteristics, beetroot by-products contain other bioactive compounds like phenolics, flavonoids, and carotenoids. These compounds have demonstrated anti-inflammatory effects, which can aid in reducing inflammation in the body and lowering the risk of chronic inflammatory conditions such as arthritis and inflammatory bowel disease.

Moreover, some research studies have indicated that these bioactive compounds may possess anti-cancer properties, suggesting their potential role in cancer prevention and treatment. The bioactive phytochemicals present in beetroot by-products have been associated with various health advantages. For instance, certain research studies have indicated that these substances could aid in reducing blood pressure and enhancing cardiovascular health, thus proving beneficial for individuals dealing with hypertension or heart conditions. Moreover, beetroot by-products have exhibited antimicrobial characteristics, which can defend against harmful bacteria and pathogens, potentially lowering the likelihood of infections. Due to the numerous health benefits of bioactive phytochemicals in beetroot by-products, there is a growing interest in integrating these compounds into functional foods and nutraceuticals. Functional foods have additional health benefits beyond the basics, while nutraceuticals have medicinal or health advantages derived from food sources. By including beetroot by-products in these products, manufacturers may develop innovative and nutritious choices for consumers seeking to enhance their health and overall well-being.

Additionally, the use of beetroot by-products in the pharmaceutical sector is under investigation. Several studies have proposed that bioactive compounds found in beetroot by-products could be utilized for potential therapeutic purposes in various diseases such as cancer, diabetes, and neurodegenerative disorders. Through a deeper examination of the bioactive properties of these compounds and their mechanisms of action, scientists may have the opportunity to create novel pharmaceutical products that capitalize on the health advantages of beetroot by-products. Ultimately, beetroot by-products serve as a valuable reservoir of bioactive phytochemicals with promising health benefits. These compounds, which encompass betalains, phenolic compounds, flavonoids, and carotenoids, provide antioxidant, anti-inflammatory, and potentially anti-cancer attributes. Researchers and manufacturers could exploit these bioactive compounds' therapeutic capabilities to enhance human health and well-being by integrating beetroot by-products into functional foods, nutraceuticals, and pharmaceuticals. Although further investigation is necessary to fully comprehend the health advantages of beetroot by-products and their potential applications across various industries, the outlook appears encouraging for these sustainable and nutritious sources of bioactive phytochemicals.

References

Abuajah CI, Ogbonna AC, Osuji CM (2015) Functional components and medicinal properties of food: a review. J Food Sci Technol 52:2522–2529

Adhikari A, Saha A, Indu R, Sur TK, Das AK (2017) Evaluation of anti-inflammatory effect of beetroot extract in animal models. Int J Basic Clin Pharmacol 6:2853–2858

Afzali N, Ebadi SS, Afzali H, Sharif MR, Vazirian M, Ebadi SA, Shahkarami V, Rahimi H (2020) Effect of Beta vulgaris extract on liver enzymes in patients with non-alcoholic fatty liver disease: a randomized clinical trial. Hepat Mon 20(7)

Albasher G, Almeer R, Al-Otibi FO, Al-Kubaisi N, Mahmoud AM (2019) Ameliorative effect of Beta vulgaris root extract on chlorpyrifos-induced oxidative stress, inflammation and liver injury in rats. Biomolecules 9(7):261

Albrahim T, Alonazi MA (2020) Role of beetroot (Beta vulgaris) juice on chronic nanotoxicity of silver nanoparticle-induced hepatotoxicity in male rats. Int J Nanomed 3471–3482

Al-Harbi LN, Alshammari GM, Al-Dossari AM, Subash-Babu P, Binobead MA, Alhussain MH, AlSedairy SA, Al-Nouri DM, Shamlan G (2021) *Beta vulgaris* L.(beetroot) methanolic extract prevents hepatic steatosis and liver damage in T2DM rats by hypoglycemic, insulin-sensitizing, antioxidant effects, and upregulation of PPARα. Biology 10(12):1306

Aykın-Dinçer E, Güngör KK, Çağlar E, Erbaş M (2021) The use of beetroot extract and extract powder in sausages as natural food colorant. Int J Food Eng 17(1):75–82

Azeredo HM (2009) Betalains: properties, sources, applications, and stability—a review. Int J Food Sci Technol 44(12):2365–2376

Babagil A, Tasgin E, Nadaroglu H, Kaymak HC (2018) Antioxidant and antiradical activity of beetroot (Beta vulgaris L. var. conditiva Alef.) grown using different fertilizers. J Chem 2018(1):7101605

Babarykin D, Smirnova G, Markovs J, Vasiljeva S, Basova N, Simanis R, Viksna L (2019) Therapeutic effect of fractionated by ultrafiltration red beetroot (Beta vulgaris L.) juice in rats with food-induced fatty liver. Eur J Biol Res 9(1):1–9

Bahrami LS, Arabi SM, Feizy Z, Rezvani R (2021) The effect of beetroot inorganic nitrate supplementation on cardiovascular risk factors: A systematic review and meta-regression of randomized controlled trials. Nitric Oxide 115:8–22

Baião DDS, da Silva DV, Del Aguila EM, Paschoalin VMF (2017) Nutritional, bioactive and physicochemical characteristics of different beetroot formulations. Food Addit 6(6):21–43

Balkrishna A, Joshi B, Srivastava A, Bisht S (2022) Agriculture food crops diversity of India: a scenario. Indian Ecol Soc 49:1011–1022

Bangar SP, Sharma N, Sanwal N, Lorenzo JM, Sahu JK (2022) Bioactive potential of beetroot (Beta vulgaris). Food Res Int 158:111556

Barbu V, Cotârleț M, Bolea CA, Cantaragiu A, Andronoiu DG, Bahrim GE, Enachi E (2020) Three types of beetroot products enriched with lactic acid bacteria. Foods 9(6):786

Bárta J, Bártová V, Šindelková T, Jarošová M, Linhartová Z, Mráz J, Bedrníček J, Smetana P, Samková E, Laknerová I (2020) Effect of boiling on colour, contents of betalains and total phenolics and on antioxidant activity of colourful powder derived from six different beetroot (Beta vulgaris L. var. conditiva) cultivars. Pol J Food Nutr Sci 70(4):377–385

Batool R, Aziz E, Salahuddin H, Iqbal J, Tabassum S, Mahmood T (2019) Rumex dentatus could be a potent alternative to treatment of microbial infections and of breast cancer. J Tradit Chin Med 39(6)

Biondo PBF, Boeing JS, Barizão ÉO, Souza NED, Matsushita M, Oliveira CCD, Boroski M, Visentainer JV (2014) Evaluation of beetroot (Beta vulgaris L.) leaves during its developmental stages: a chemical composition study. Food Sci Technol 34:94–101

Bond Jr V, Curry BH, Adams RG, Asadi MS, Millis RM, Haddad GE (2013) Effects of dietary nitrates on systemic and cerebrovascular hemodynamics. Cardiol Res Pract 2013(1):435629

Butt MS, Randhawa MA, Shahid M (2019) Hepatoprotective effects of red beetroot-based beverages against CCl4-induced hepatic stress in Sprague Dawley rats. J Food Biochem 43(12):e13057

Carrillo C, Wilches-Pérez D, Hallmann E, Kazimierczak R, Rembiałkowska E (2019) Organic versus conventional beetroot. Bioactive compounds and antioxidant properties. Lwt 116:108552

Ceclu L, Nistor OV (2020) Red beetroot: composition and health effects—a review. J Nutr Med Diet Care 6(1):1–9

Chandran J, Nisha P, Singhal RS, Pandit AB (2014) Degradation of colour in beetroot (Beta vulgaris L.): a kinetics study. J Food Sci Technol 51:2678–2684

Chaudhari SN, Nikam MP (2015) Development and sensory analysis of beetroot jelly. Int J Sci Res 4(10):827–830

Chawla H, Parle M, Sharma K, Yadav M (2016) Beetroot: A health promoting functional food. Inventi Rapid Nutraceuticals 1(1):0976–3872

Chen L, Zhu Y, Hu Z, Wu S, Jin C (2021) Beetroot as a functional food with huge health benefits: Antioxidant, antitumor, physical function, and chronic metabolomics activity. Food Sci Nutr 9(11):6406–6420

Chhikara N, Kushwaha K, Jaglan S, Sharma P, Panghal A (2019a) Nutritional, physicochemical, and functional quality of beetroot *(Beta vulgaris* L.) incorporated Asian noodles. Cereal Chem 96(1):154–161

Chhikara N, Kushwaha K, Sharma P, Gat Y, Panghal A (2019b) Bioactive compounds of beetroot and utilization in food processing industry: A critical review. Food Chem 272:192–200

Clifford T, Howatson G, West DJ, Stevenson EJ (2015) The potential benefits of red beetroot supplementation in health and disease. Nutrients 7(4):2801–2822

Costa APD, Hermes VS, Rios ADO, Flôres SH (2017) Minimally processed beetroot waste as an alternative source to obtain functional ingredients. J Food Sci Technol 54:2050–2058

Cramer MN, Hieda M, Huang M, Moralez G, Crandall CG (2020) Dietary nitrate supplementation does not influence thermoregulatory or cardiovascular strain in older individuals during severe ambient heat stress. Exp Physiol 105(10):1730–1741

Czyzowska A, Siemianowska K, Sniadowska M, Nowak A (2020) Bioactive compounds and microbial quality of stored fermented red beetroots and red beetroot juice. Pol J Food Nutr Sci 70(1)

Das S, Williams DS, Das A, Kukreja RC (2013) Beet root juice promotes apoptosis in oncogenic MDA-MB-231 cells while protecting cardiomyocytes under doxorubicin treatment. J Exp Second Sci 2:1–6

da Silva LGS, Morelli AP, Pavan ICB, Tavares MR, Pestana NF, Rostagno MA, Simabuco FM, Bezerra RMN (2020) Protective effects of beet (Beta vulgaris) leaves extract against oxidative stress in endothelial cells in vitro. Phytother Res 34(6):1385–1396

Davignon J, Ganz P (2004) Role of endothelial dysfunction in atherosclerosis. Circulation 109(23_suppl_1):III–27

Desseva I, Stoyanova M, Petkova N, Mihaylova D (2020) Red beetroot juice phytochemicals bioaccessibility: An in vitro approach. Pol J Food Nutr Sci 70(1)

Dhiman A, Chawla D, Thakur D, Chauhan D, Bharara C, Singh A (2020) Development and optimization of ready to serve (RTS) beetroot drink. Asian J Dairy Food Res 39(1):49–57

Dhiman A, Suhag R, Thakur D, Gupta V, Prabhakar PK (2022) Current status of Loquat (Eriobotrya japonica Lindl.): Bioactive functions, preservation approaches, and processed products. Food Rev Int 38(sup1):286–316

Dhineshkumar V, Ramasamy D (2016) Studies on development of yogurt flavoured with beetroot juice (Beta vulgaris L.). Int J Adv Res Biol Sci 3(2):108–117

Domínguez R, Cuenca E, Maté-Muñoz JL, García-Fernández P, Serra-Paya N, Estevan MCL, Herreros PV, Garnacho-Castaño MV (2017) Effects of beetroot juice supplementation on cardiorespiratory endurance in athletes. A systematic review. Nutrients 9(1):43

Domínguez R, Maté-Muñoz JL, Cuenca E, García-Fernández P, Mata-Ordoñez F, Lozano-Estevan MC, Veiga-Herreros P, da Silva SF, Garnacho-Castaño MV (2018) Effects of beetroot juice supplementation on intermittent high-intensity exercise efforts. J Int Soc Sports Nutr 15:1–12

Domínguez R, Munekata PE, Pateiro M, Maggiolino A, Bohrer B, Lorenzo JM (2020) Red beetroot. A potential source of natural additives for the meat industry. Appl Sci 10(23):8340

Dos Santos CD, Ismail M, Cassini AS, Marczak LDF, Tessaro IC, Farid M (2018) Effect of thermal and high pressure processing on stability of betalain extracted from red beet stalks. J Food Sci Technol 55:568–577

Dos Santos Baião D, d'El-Rei J, Alves G, Neves MF, Perrone D, Del Aguila EM, Paschoalin VMF (2019) Chronic effects of nitrate supplementation with a newly designed beetroot formulation on biochemical and hemodynamic parameters of individuals presenting risk factors for cardiovascular diseases: A pilot study. J Funct Foods 58:85–94

Dwivedi N, Singh A, Jaiswal M, Agrahari K (2017) Standardization and development of beetroot-based product. Int J Home Sci 3(2):26–30

Edziri H, Jaziri R, Haddad O, Anthonissen R, Aouni M, Mastouri M, Verschaeve L (2019) Phytochemical analysis, antioxidant, anticoagulant and in vitro toxicity and genotoxicity testing of methanolic and juice extracts of Beta vulgaris L. S Afr J Bot 126:170–175

Ekholm P, Reinivuo H, Mattila P, Pakkala H, Koponen J, Happonen A, Hellström J, Ovaskainen ML (2007) Changes in the mineral and trace element contents of cereals, fruits and vegetables in Finland. J Food Compos Anal 20(6):487–495

El Gamal AA, AlSaid MS, Raish M, Al Sohaibani M, Al-Massarani SM, Ahmad A, Hefnawy M, Al-Yahya M, Basoudan OA, Rafatullah S (2014) Beetroot (Beta vulgaris L.) extract ameliorates gentamicin-induced nephrotoxicity associated oxidative stress, inflammation, and apoptosis in rodent model. Mediat Inflamm 2014(1):983952

Elleuch M, Bedigian D, Roiseux O, Besbes S, Blecker C, Attia H (2011) Dietary fibre and fibre-rich by-products of food processing: Characterisation, technological functionality and commercial applications: A review. Food Chem 124(2):411–421

Evstigneeva T, Iakovchenko N, Kuzmicheva N, Skvortsova N (2020) Applying beetroot as food ingredient in ice cream production. Agron Res 18(S3):1662–1672

Farabegoli F, Scarpa ES, Frati A, Serafini G, Papi A, Spisni E, Antonini E, Benedetti S, Ninfali P (2017) Betalains increase vitexin-2-O-xyloside cytotoxicity in CaCo-2 cancer cells. Food Chem 218:356–364

Fatma S, Sharma N, Singh SP, Jha A, Kumar A (2016) Fuzzy analysis of sensory data for ranking of beetroot candy. Int J Food Eng 2(1):26–30

Figiel A (2010) Drying kinetics and quality of beetroots dehydrated by combination of convective and vacuum-microwave methods. J Food Eng 98(4):461–470

Fu Y, Shi J, Xie SY, Zhang TY, Soladoye OP, Aluko RE (2020) Red beetroot betalains: Perspectives on extraction, processing, and potential health benefits. J Agric Food Chem 68(42):11595–11611

Georgiev VG, Weber J, Kneschke EM, Denev PN, Bley T, Pavlov AI (2010) Antioxidant activity and phenolic content of betalain extracts from intact plants and hairy root cultures of the red beetroot Beta vulgaris cv. Detroit dark red. Plant Foods Hum Nutr 65:105–111

Gilchrist M, Winyard PG, Fulford J, Anning C, Shore AC, Benjamin N (2014) Dietary nitrate supplementation improves reaction time in type 2 diabetes: development and application of a novel nitrate-depleted beetroot juice placebo. Nitric Oxide 40:67–74

Gokhale SV, Lele SS (2011) Dehydration of red beet root (*Beta vulgaris*) by hot air drying: Process optimization and mathematical modeling. Food Sci Biotechnol 20(4):955

Guiné R, Roque AR, Gonçalves F, Correia P (2016) Development of an innovative jam based on beetroot. J Food Sci Res 1(2):49–53

Guldiken B, Toydemir G, Nur Memis K, Okur S, Boyacioglu D, Capanoglu E (2016) Home-processed red beetroot (*Beta vulgaris* L.) products: Changes in antioxidant properties and bioaccessibility. Int J Mol Sci 17(6):858

Hadipour E, Taleghani A, Tayarani-Najaran N, Tayarani-Najaran Z (2020) Biological effects of red beetroot and betalains: A review. Phytother Res 34(8):1847–1867

Hatlestad GJ, Sunnadeniya RM, Akhavan NA, Gonzalez A, Goldman IL, McGrath JM, Lloyd AM (2012) The beet R locus encodes a new cytochrome P450 required for red betalain production. Nat Genet 44(7):816–820

Hobbs DA, Kaffa N, George TW, Methven L, Lovegrove JA (2012) Blood pressure-lowering effects of beetroot juice and novel beetroot-enriched bread products in normotensive male subjects. Br J Nutr 108(11):2066–2074

Holmes RP, Assimos DG (2004) The impact of dietary oxalate on kidney stone formation. Urol Res 32:311–316

Iglesias R, Citores L, Di Maro A, Ferreras JM (2015) Biological activities of the antiviral protein BE27 from sugar beet (Beta vulgaris L.). Planta 241:421–433

Ijaz S, Iqbal J, Abbasi BA, Ullah Z, Yaseen T, Kanwal S, Mahmood T, Sydykbayeva S, Ydyrys A, Almarhoon ZM, Sharifi-Rad J (2023a) Rosmarinic acid and its derivatives: Current insights on anticancer potential and other biomedical applications. Biomed Pharmacother 162:114687

Ijaz S, Iqbal J, Abbasi BA, Tufail A, Ullah Z, Sharifi-Rad J, Yaseen T, Uddin S (2023b) Biofortification: lessons from the past and strategies for future food security. In: Legumes biofortification. Springer, Cham, pp 521–545

Ijaz S, Iqbal J, Abbasi BA, Tufail A, Yaseen T, Uddin S, Usman K, Ullah R, Bibi H, Inam P, Sagindykova E (2024a) Current stage of preclinical and clinical development of guggulsterone in cancers: Challenges and promises. Cell Biol Int 48(2):128–142

Ijaz S, Iqbal J, Abbasi BA, Kanwal S, Tavafoghi M, Ahmed MZ, Mahmood T (2024b) Investigation of bioactive constituents and evaluation of different in vitro antimicrobial, antioxidant, and cytotoxicity potentials of different Portulacaria afra extracts. J King Saud Univ Sci 36(2):103033

Ijaz S, Iqbal J, Abbasi BA, Yaseen T, Rehman S, Kazi M, Ansari L, Mahmood T (2024c) Role of OMICS-based technologies in plant sciences. OMICs Based Tech Global Food Sec 45–66

Ingle M, Thorat SS, Kotecha PM, Nimbalkar CA (2017) Nutritional assessment of beetroot (*Beta vulgaris* L.) powder cookies. Asian J Dairy Food Res 36(3)

Iqbal J, Abbasi BA, Mahmood T, Kanwal S, Ali B, Shah SA, Khalil AT (2017) Plant-derived anticancer agents: A green anticancer approach. Asian Pac J Trop Biomed 7(12):1129–1150

Iqbal J, Abbasi BA, Batool R, Mahmood T, Ali B, Khalil AT, Kanwal S, Shah SA, Ahmad R (2018) Potential phytocompounds for developing breast cancer therapeutics: nature's healing touch. Eur J Pharmacol 827:125–148

Jakubcik EM, Rutherfurd-Markwick K, Chabert M, Wong M, Ali A (2021) Pharmacokinetics of nitrate and nitrite following beetroot juice drink consumption. Nutrients 13(2):281

Javed R, Ijaz S, Hameed H, Nazish M, Sharif MS, Afreen A, Alarjani KM, Elshikh MS, Mehboob S, Abdul Razak S, Waheed A (2023) Phytochemical-mediated biosynthesis of silver nanoparticles from Strobilanthes glutinosus: Exploring biological applications. Micromachines 14(7):1372

Jin SK, Choi JS, Moon SS, Jeong JY, Kim GD (2014) The assessment of red beet as a natural colorant, and evaluation of quality properties of emulsified pork sausage containing red beet powder during cold storage. Korean J Food Sci Anim Resour 34(4):472

Joris PJ, Mensink RP (2013) Beetroot juice improves in overweight and slightly obese men postprandial endothelial function after consumption of a mixed meal. Atherosclerosis 231(1):78–83

Kale RG, Sawate AR, Kshirsagar RB, Patil BM, Mane R (2018) Studies on evaluation of physical and chemical composition of beetroot (*Beta vulgaris* L.). Int J Chem Stud 6(2):2977–2979

Kanner J, Harel S, Granit R (2001) Betalains a new class of dietary cationized antioxidants. J Agric Food Chem 49(11):5178–5185

Kapadia GJ, Rao GS (2012) Anticancer effects of red beet pigments. In: Neelwarne B (ed) Red beet biotechnology: food and pharmaceutical applications. Springer, Boston, MA, pp 125–154

Kapadia GJ, Tokuda H, Konoshima T, Nishino H (1996) Chemoprevention of lung and skin cancer by Beta vulgaris (beet) root extract. Cancer Lett 100(1-2):211–214

Kapadia GJ, Azuine MA, Sridhar R, Okuda Y, Tsuruta A, Ichiishi E, Mukainake T, Takasaki M, Konoshima T, Nishino H, Tokuda H (2003) Chemoprevention of DMBA-induced UV-B promoted, NOR-1-induced TPA promoted skin carcinogenesis, and DEN-induced phenobarbital promoted liver tumors in mice by extract of beetroot. Pharmacol Res 47(2):141–148

Kapadia G, Azuine MA, Subba Rao G, Arai T, Iida A, Tokuda H (2011) Cytotoxic effect of the red beetroot (Beta vulgaris L) extract compared to doxorubicin (Adriamycin) in the human prostate (PC-3) and breast (MCF-7) cancer cell lines. Anticancer Agents Med Chem 11(3):280–284

Kapadia GJ, Rao GS, Ramachandran C, Iida A, Suzuki N, Tokuda H (2013) Synergistic cytotoxicity of red beetroot (Beta vulgaris L.) extract with doxorubicin in human pancreatic, breast and prostate cancer cell lines. J Complement Integr Med 10(1):113–122

Kapil V, Khambata RS, Robertson A, Caulfield MJ, Ahluwalia A (2015) Dietary nitrate provides sustained blood pressure lowering in hypertensive patients: a randomized, phase 2, double-blind, placebo-controlled study. Hypertension 65(2):320–327

Kathiravan T, Nadanasabapathi S, Kumar R (2014) Standardization of process condition in batch thermal pasteurization and its effect on antioxidant, pigment and microbial inactivation of Ready to Drink (RTD) beetroot (Beta vulgaris L.) juice. Int Food Res J 21(4)

Kazimierczak R, Hallmann E, Lipowski J, Drela N, Kowalik A, Püssa T, Matt D, Luik A, Gozdowski D, Rembiałkowska E (2014) Beetroot (Beta vulgaris L.) and naturally fermented beetroot juices from organic and conventional production: metabolomics, antioxidant levels and anticancer activity. J Sci Food Agric 94(13):2618–2629

Kazimierczak R, Siłakiewicz A, Hallmann E, Srednicka-Tober D, Rembiałkowska E (2016) Chemical composition of selected beetroot juices in relation to beetroot production system and processing technology. Notulae Botanicae Horti Agrobotanici Cluj-Napoca 44(2):491–498

Kim SL, Choi HS, Ko YC, Yun BS, Lee DS (2021) 5-hydroxymaltol derived from beetroot juice through lactobacillus fermentation suppresses inflammatory effect and oxidant stress via regulating NF-kB, MAPKs pathway and NRF2/HO-1 expression. Antioxidants 10(8):1324

Kohajdová Z, Karovičová J, Kuchtová V, Lauková M (2018) Utilisation of beetroot powder for bakery applications. Chem Pap 72:1507–1515

Koss-Mikołajczyk I, Kusznierewicz B, Wiczkowski W, Sawicki T, Bartoszek A (2019) The comparison of betalain composition and chosen biological activities for differently pigmented prickly pear (Opuntia ficus-indica) and beetroot (Beta vulgaris) varieties. Int J Food Sci Nutr 70(4):442–452

Kowalski SJ, Szadzińska J (2014) Kinetics and quality aspects of beetroots dried in non-stationary conditions. Dry Technol 32(11):1310–1318

Kujala TS, Vienola MS, Klika KD, Loponen JM, Pihlaja K (2002) Betalain and phenolic compositions of four beetroot (Beta vulgaris) cultivars. Eur Food Res Technol 214:505–510

Kumar S, Brooks MSL (2018) Use of red beet (Beta vulgaris L.) for antimicrobial applications—a critical review. Food Bioprocess Technol 11:17–42

Kundu JK, Surh YJ (2012) Emerging avenues linking inflammation and cancer. Free Radic Biol Med 52(9):2013–2037

Kusznierewicz B, Mróz M, Koss-Mikołajczyk I, Namieśnik J (2021) Comparative evaluation of different methods for determining phytochemicals and antioxidant activity in products containing betalains—verification of beetroot samples. Food Chem 362:130132

Lasta HFB, Lentz L, Mezzomo N, Ferreira SRS (2019) Supercritical CO2 to recover extracts enriched in antioxidant compounds from beetroot aerial parts. Biocatal Agric Biotechnol 19:101169

Lechner JF, Stoner GD (2019) Red beetroot and betalains as cancer chemopreventative agents. Molecules 24(8):1602

Lechner JF, Wang LS, Rocha CM, Larue B, Henry C, McIntyre CM, Riedl KM, Schwartz SJ, Stoner GD (2010) Drinking water with red beetroot food color antagonizes esophageal carcinogenesis in N-nitrosomethylbenzylamine-treated rats. J Med Food 13(3):733–739

Lee EJ, An D, Nguyen CT, Patil BS, Kim J, Yoo KS (2014) Betalain and betaine composition of greenhouse-or field-produced beetroot (*Beta vulgaris* L.) and inhibition of HepG2 cell proliferation. J Agric Food Chem 62(6):1324–1331

Lidder S, Webb AJ (2013) Vascular effects of dietary nitrate (as found in green leafy vegetables and beetroot) via the nitrate-nitrite-nitric oxide pathway. Br J Clin Pharmacol 75(3):677–696

Liu R, Ji P, Liu B, Qiao H, Wang X, Zhou L, Deng T, Ba Y (2017) Apigenin enhances the cisplatin cytotoxic effect through p53-modulated apoptosis. Oncol Lett 13(2):1024–1030

Liu X, Huang J, Li J, Mao Q, He J (2019) Effects of liraglutide combined with insulin on oxidative stress and serum MCP-1 and NF-κB levels in type 2 diabetes. J Coll Physicians Surg Pak 29(3):218–221

Lorizola IM, Furlan CP, Portovedo M, Milanski M, Botelho PB, Bezerra RM, Sumere BR, Rostagno MA, Capitani CD (2018) Beet stalks and leaves (Beta vulgaris L.) protect against high-fat diet-induced oxidative damage in the liver in mice. Nutrients 10(7):872

Lu X, Wang Y, Zhang Z (2009) Radioprotective activity of betalains from red beets in mice exposed to gamma irradiation. Eur J Pharmacol 615(1-3):223–227

Lucky AR, Al-Mamun A, Hosen A, Toma MA, Mazumder MAR (2020) Nutritional and sensory quality assessment of plain cake enriched with beetroot powder. Food Res 4(6):2049–2053

Lundberg JO, Carlström M, Larsen FJ, Weitzberg E (2011) Roles of dietary inorganic nitrate in cardiovascular health and disease. Cardiovasc Res 89(3):525–532

Machha A, Schechter AN (2011) Dietary nitrite and nitrate: a review of potential mechanisms of cardiovascular benefits. Eur J Nutr 50:293–303

Madadi E, Mazloum-Ravasan S, Yu JS, Ha JW, Hamishehkar H, Kim KH (2020) Therapeutic application of betalains: A review. Plants 9(9):1219

Mancini MCS, Ponte LGS, Silva CHR, Fagundes I, Pavan ICB, Romeiro SA, da Silva LGS, Morelli AP, Rostagno MA, Simabuco FM, Bezerra RMN (2021) Beetroot and leaf extracts present protective effects against prostate cancer cells, inhibiting cell proliferation, migration, and growth signaling pathways. Phytother Res 35(9):5241–5258

Manoharan A, Ramasamy D, Dhanalashmi B, Gnanalashmi KS, Thyagarajan D (2012) Organoleptic evaluation of beetroot juice as natural color for strawberry flavor ice cream. Indian J Med Healthc 1(1):34–37

Manopriya M, Arivuchudar R (2019) Cakes enriched with Beta vulgaris pomace powder: development, sensory evaluation, proximate composition and total antioxidant evaluation. Adalya J 8(11):688–694

Maraie NK, Abdul-Jalil TZ, Alhamdany AT, Janabi HA (2014) Phytochemical study of the Iraqi Beta vulgaris leaves and its clinical applications for the treatment of different dermatological diseases. World J Pharm Pharm Sci 3:5–19

Marmion DM (1992) Handbook of US colorants: foods, drugs, cosmetics, and medical devices. Wiley, New York

Martinez RM, Longhi-Balbinot DT, Zarpelon AC, Staurengo-Ferrari L, Baracat MM, Georgetti SR, Sassonia RC, Verri WA, Casagrande R (2015) Anti-inflammatory activity of betalain-rich dye of Beta vulgaris: effect on edema, leukocyte recruitment, superoxide anion and cytokine production. Arch Pharm Res 38:494–504

McKillop K, Harnly J, Pehrsson P, Fukagawa N, Finley J (2021) Fooddata central, USDA's updated approach to food composition data systems. Curr Dev Nutr 5:596

Mikołajczyk-Bator K, Błaszczyk A, Czyżniejewski M, Kachlicki P (2016) Identification of saponins from sugar beet (*Beta vulgaris*) by low and high-resolution HPLC–MS/MS. J Chromatogr B 1029:36–47

Miraj S (2016) Chemistry and pharmacological effect of beta vulgaris: A systematic review. Pharm Lett 8(19):404–409

Mirmiran P, Houshialsadat Z, Gaeini Z, Bahadoran Z, Azizi F (2020) Functional properties of beetroot (Beta vulgaris) in management of cardio-metabolic diseases. Nutr Metab 17:1–15

Mo E, Ji E, Ca O (2020) Proximate chemical compositions and sensory properties of wine produced from beetroot (Beta vulgaris). Chem Sci Rev Lett 9(33):58–66

Monteiro R, Azevedo I (2010) Chronic inflammation in obesity and the metabolic syndrome. Mediat Inflamm 2010(1):289645

Moreno-Ley CM, Osorio-Revilla G, Hernández-Martínez DM, Ramos-Monroy OA, Gallardo-Velázquez T (2021) Anti-inflammatory activity of betalains: A comprehensive review. Hum Nutr Metab 25:200126

Morgado M, de Oliveira GV, Vasconcellos J, Monteiro ML, Conte-Junior C, Pierucci APTR, Alvares TS (2016) Development of a beetroot-based nutritional gel containing high content of bioaccessible dietary nitrate and antioxidants. Int J Food Sci Nutr 67(2):153–160

Mortensen A (2006) Carotenoids and other pigments as natural colorants. Pure Appl Chem 78(8):1477–1491

Mroczek A, Kapusta I, Janda B, Janiszowska W (2012) Triterpene saponin content in the roots of red beet (Beta vulgaris L.) cultivars. J Agric Food Chem 60(50):12397–12402

Mudgal D, Singh S, Singh BR (2022) Nutritional composition and value added products of beetroot: A review. J Curr Res Food Sci 3(1):01–09

Murthy KNC, Manchali S (2012) Anti-diabetic potentials of red beet pigments and other constituents. In: Neelwarne B (ed) Red beet biotechnology: food and pharmaceutical applications. Springer, Boston, MA, pp 155–174

Neha P, Jain SK, Jain NK, Jain HK, Mittal HK (2018) Chemical and functional properties of Beetroot (Beta vulgaris L.) for product development: A review. Int J Chem Stud 6(3):3190–3194

Nemzer B, Pietrzkowski Z, Spórna A, Stalica P, Thresher W, Michałowski T, Wybraniec S (2011) Betalainic and nutritional profiles of pigment-enriched red beet root (Beta vulgaris L.) dried extracts. Food Chem 127(1):42–53

Netzel M, Stintzing FC, Quaas D, Strass G, Carle R, Bitsch R, Bitsch I, Frank T (2005) Renal excretion of antioxidative constituents from red beet in humans. Food Res Int 38(8-9):1051–1058

Ninfali P, Antonini E, Frati A, Scarpa ES (2017) C-glycosyl flavonoids from Beta vulgaris cicla and betalains from Beta vulgaris rubra: antioxidant, anticancer and antiinflammatory activities—A review. Phytother Res 31(6):871–884

Nogueira Soares R, Machado-Santos AP, Barros-Santos E, Vieira De Oliveira G, Murias JM, Alvares TS (2021) Acute supplementation with beetroot juice improves endothelial function in HIV-infected individuals. Appl Physiol Nutr Metab 46(3):213–220

Nowacki L, Vigneron P, Rotellini L, Cazzola H, Merlier F, Prost E, Ralanairina R, Gadonna JP, Rossi C, Vayssade M (2015) Betanin-enriched red beetroot (Beta vulgaris L.) extract induces apoptosis and autophagic cell death in MCF-7 cells. Phytother Res 29(12):1964–1973

Odhav B, Beekrum S, Akula US, Baijnath H (2007) Preliminary assessment of nutritional value of traditional leafy vegetables in KwaZulu-Natal, South Africa. J Food Compos Anal 20(5):430–435

Olumese FE, Oboh HA (2020) Aqueous beetroot juice extract improves renal function and some biochemical parameters in carbon tetrachloride-induced toxicity in Sprague Dawley rats. Ann Trop Pathol 11(2):171–175

Ormesher L, Myers JE, Chmiel C, Wareing M, Greenwood SL, Tropea T, Lundberg JO, Weitzberg E, Nihlen C, Sibley CP, Johnstone ED (2018) Effects of dietary nitrate supplementation, from beetroot juice, on blood pressure in hypertensive pregnant women: A randomised, double-blind, placebo-controlled feasibility trial. Nitric Oxide 80:37–44

Otegbayo BO, Akwa IM, Tanimola AR (2020) Physico-chemical properties of beetroot (Beta vulgaris l.) wine produced at varying fermentation days. Sci Afr 8:e00420

Ozaki MM, Munekata PE, de Souza Lopes A, do Nascimento MDS, Pateiro M, Lorenzo JM, Pollonio MAR (2020) Using chitosan and radish powder to improve stability of fermented cooked sausages. Meat Sci 167:108165

Pant S, Chinwan D (2014) Value addition of food products. Int J Eng Sci Res 4(12):946–949

Papi A, Farabegoli F, Iori R, Orlandi M, De Nicola GR, Bagatta M, Angelino D, Gennari L, Ninfali P (2013) Vitexin-2-O-xyloside, raphasatin and (−)-epigallocatechin-3-gallate synergistically affect cell growth and apoptosis of colon cancer cells. Food Chem 138(2–3):1521–1530

Peeling P, Cox GR, Bullock N, Burke LM (2015) Beetroot juice improves on-water 500 m time-trial performance, and laboratory-based paddling economy in national and international-level kayak athletes. Int J Sport Nutr Exerc Metab 25(3):278–284

Perumpuli PABN, Fernando G, Kaumal M, Arandara M, Silva S (2018) Development of low sugar vegetable jam from beetroot (Beta vulgaris L.): Studies on Physicochemical Sensory and Nutritional Properties. Int J Theor Appl Sci 10(2):22–27

Petkoska AT, Daniloski D, D'Cunha NM, Naumovski N, Broach AT (2021) Edible packaging: Sustainable solutions and novel trends in food packaging. Food Res Int 140:109981

Pitalua E, Jimenez M, Vernon-Carter EJ, Beristain CI (2010) Antioxidative activity of microcapsules with beetroot juice using gum Arabic as wall material. Food Bioprod Process 88(2–3):253–258

Platosz N, Sawicki T, Wiczkowski W (2020) Profile of phenolic acids and flavonoids of red beet and its fermentation products. Does long-term consumption of fermented beetroot juice affect phenolics profile in human blood plasma and urine? Pol J Food Nutr Sci 70(1)

Podolak I, Galanty A, Sobolewska D (2010) Saponins as cytotoxic agents: a review. Phytochem Rev 9:425–474

Poels MM, Ikram MA, Vernooij MW, Krestin GP, Hofman A, Messen WJ, Van Der Lugt A, Breteler MM (2008) Total cerebral blood flow in relation to cognitive function: the Rotterdam Scan Study. J Cereb Blood Flow Metab 28(10):1652–1655

Powers SK, Jackson MJ (2008) Exercise-induced oxidative stress: cellular mechanisms and impact on muscle force production. Physiol Rev 88(4):1243–1276

Presley TD, Morgan AR, Bechtold E, Clodfelter W, Dove RW, Jennings JM, Kraft RA, King SB, Laurienti PJ, Rejeski WJ, Burdette JH (2011) Acute effect of a high nitrate diet on brain perfusion in older adults. Nitric Oxide 24(1):34–42

Punia Bangar S, Singh A, Chaudhary V, Sharma N, Lorenzo JM (2023) Beetroot as a novel ingredient for its versatile food applications. Crit Rev Food Sci Nutr 63(26):8403–8427

Rahimi P, Mesbah-Namin SA, Ostadrahimi A, Abedimanesh S, Separham A, Jafarabadi MA (2019) Effects of betalains on atherogenic risk factors in patients with atherosclerotic cardiovascular disease. Food Funct 10(12):8286–8297

Raish M, Ahmad A, Ansari MA, Alkharfy KM, Ahad A, Khan A, Ali N, Ganaie MA, Hamidaddin MAA (2019) Beetroot juice alleviates isoproterenol-induced myocardial damage by reducing oxidative stress, inflammation, and apoptosis in rats. 3 Biotech 9(4):147

Raubenheimer K, Hickey D, Leveritt M, Fassett R, de Zevallos O, Munoz J, Allen JD, Briskey D, Parker TJ, Kerr G, Peake JM, Pecheniuk NM (2017) Acute effects of nitrate-rich beetroot juice on blood pressure, hemostasis and vascular inflammation markers in healthy older adults: a randomized, placebo-controlled crossover study. Nutrients 9(11):1270

Ravichandran K, Saw NMMT, Mohdaly AA, Gabr AM, Kastell A, Riedel H, Cai Z, Knorr D, Smetanska I (2013) Impact of processing of red beet on betalain content and antioxidant activity. Food Res Int 50(2):670–675

Razzak MA, Sharif MK, Naz T, Rauf MA, Shahid F, Shahzad R, Saleem R, Aslam T, Inam A (2024) Evaluating the bioactive compounds of beetroot and their pharmacological activities in promoting health. Eur J Health Sci 10(1):13–30

Rebecca LJ, Sharmila S, Das MP, Seshiah C (2014) Extraction and purification of carotenoids from vegetables. J Chem Pharm Res 6(4):594–598

Rein MJ, Renouf M, Cruz-Hernandez C, Actis-Goretta L, Thakkar SK, da Silva Pinto M (2013) Bioavailability of bioactive food compounds: A challenging journey to bioefficacy. Br J Clin Pharmacol 75(3):588–602

Rodriguez-Amaya DB (2019) Update on natural food pigments-A mini-review on carotenoids, anthocyanins, and betalains. Food Res Int 124:200–205

Rokkedal-Lausch T, Franch J, Poulsen MK, Thomsen LP, Weitzberg E, Kamavuako EN, Karbing DS, Larsen RG (2019) Chronic high-dose beetroot juice supplementation improves time trial performance of well-trained cyclists in normoxia and hypoxia. Nitric Oxide 85:44–52

Romero SA, Pavan ICB, Morelli AP, Mancini MCS, da Silva LGS, Fagundes I, Silva CHR, Ponte LGS, Rostagno MA, Bezerra RMN, Simabuco FM (2021) Anticancer effects of root and beet leaf extracts (Beta vulgaris L.) in cervical cancer cells (HeLa). Phytother Res 35(11):6191–6203

Sahni P, Shere DM (2016) Physico-chemical and sensory characteristics of beet root pomace powder incorporated fibre rich cookies. Int J Food Ferment Technol 6(2):309–315

Salamatullah AM, Hayat K, Alkaltham MS, Ahmed MA, Arzoo S, Husain FM, Al-Dossari AM, Shamlan G, Al-Harbi LN (2021) Bioactive and antimicrobial properties of oven-dried beetroot (pulp and peel) using different solvents. Processes 9(4):588

Salehi B, Berkay Yılmaz Y, Antika G, Boyunegmez Tumer T, Fawzi Mahomoodally M, Lobine D, Akram M, Riaz M, Capanoglu E, Sharopov F, Martins N (2019) Insights on the use of α-lipoic acid for therapeutic purposes. Biomolecules 9(8):356

Sárváry A, Sárváry A (2019) Use of complementary and alternative medicine among breast cancer patients in Hungary: A descriptive study. Complement Ther Clin Pract 35:195–200

Sawicki T, Wiczkowski W (2018) The effects of boiling and fermentation on betalain profiles and antioxidant capacities of red beetroot products. Food Chem 259:292–303

Sawicki T, Bączek N, Wiczkowski W (2016) Betalain profile, content and antioxidant capacity of red beetroot dependent on the genotype and root part. J Funct Foods 27:249–261

Sawicki T, Martinez-Villaluenga C, Frias J, Wiczkowski W, Peñas E, Bączek N, Zieliński H (2019) The effect of processing and in vitro digestion on the betalain profile and ACE inhibition activity of red beetroot products. J Funct Foods 55:229–237

Sawicki T, Wiczkowski W, Hrynkiewicz M, Bączek N, Hornowski A, Honke J, Topolska J (2021) Characterization of the phenolic acid profile and in vitro bioactive properties of white beetroot products. Int J Food Sci Technol 56(2):629–638

Sentkowska A, Pyrzyńska K (2018) Zwitterionic hydrophilic interaction liquid chromatography coupled to mass spectrometry for analysis of beetroot juice and antioxidant interactions between its bioactive compounds. LWT 93:641–648

Shakeel A, Khan AA, Hakeem KR (2020) Growth, biochemical, and antioxidant response of beetroot (Beta vulgaris L.) grown in fly ash-amended soil. SN Appl Sci 2:1–9

Sharma KD, Karki S, Thakur NS, Attri S (2012) Chemical composition, functional properties and processing of carrot—a review. J Food Sci Technol 49(1):22–32

Shedid SM, Abdel-Magied N, Saada HN (2019) Role of betaine in liver injury induced by the exposure to ionizing radiation. Environ Toxicol 34(2):123–130

Silva DVTD, Silva FDO, Perrone D, Pierucci APTR, Conte-Junior CA, Alvares TDS, Aguila EMD, Paschoalin VMF (2016) Physicochemical, nutritional, and sensory analyses of a nitrate-enriched beetroot gel and its effects on plasmatic nitric oxide and blood pressure. Food Nutr Res 60(1):29909

Sinaga FA, Hasibuan R, Risfandi M, Sinaga RN (2020) Hepatoprotective effects of beetroot juice at maximum physical activity. J Phys Conf Ser 1462(1):012007. IOP

Singh B, Hathan BS (2014) Composição química, propriedades funcionais e processamento de beterraba-uma revisão. Int J Sci Eng Res 5(1):679–684

Singh JP, Kaur A, Shevkani K, Singh N, Singh B (2016) Physicochemical characterisation of corn extrudates prepared with varying levels of beetroot (Beta vulgaris) at different extrusion temperatures. Int J Food Sci Technol 51(4):911–919

Singh IP, Mittal S, Bhalla V, Srivastava V, Singh K, Chopra M, Manchanda HK, Chhibber M, Kaur N (2019) Invited talks. In: 10th national conference on recent advances in chemical and environmental sciences

Slavov A, Karagyozov V, Denev P, Kratchanova M, Kratchanov C (2013) Antioxidant activity of red beet juices obtained after microwave and thermal pretreatments. Czech J Food Sci 31(2):139–147

Song SH, Ha CW, Kim C, Seong GM (2021) Complete spontaneous remission of small cell lung cancer in the absence of specific treatment: A case report. Thorac Cancer 12(19):2611–2613

Srivastava S, Siddiqi Z, Singh T, Bala L (2019) Beetroot supplementation on non-alcoholic fatty liver disease patients. Curr Res Nutr Food Sci J 7(1):96–101

Stanaway L, Rutherfurd-Markwick K, Page R, Wong M, Jirangrat W, Teh KH, Ali A (2019) Acute supplementation with nitrate-rich beetroot juice causes a greater increase in plasma nitrite and reduction in blood pressure of older compared to younger adults. Nutrients 11(7):1683

Stintzing FC, Carle R (2007) Betalains—emerging prospects for food scientists. Trends Food Sci Technol 18(10):514–525

Sulakhiya K, Patel VK, Saxena R, Dashore J, Srivastava AK, Rathore M (2016) Effect of Beta vulgaris Linn. leaves extract on anxiety-and depressive-like behavior and oxidative stress in mice after acute restraint stress. Pharmacogn Res 8(1):1

Székely D, Illés B, Stéger-Máté M, Monspart-Sényi J (2016) Effect of drying methods for inner parameters of red beetroot (Beta vulgaris L.). Acta Universitatis Sapientiae. Alimentaria 9(1):60–68

Tan ML, Hamid SBS (2021) Beetroot as a potential functional food for cancer chemoprevention, a narrative review. J Cancer Prev 26(1):1

Tekin İ, Özcan K, Ersus S (2023) Optimization of ionic gelling encapsulation of red beet (Beta vulgaris L.) juice concentrate and stability of betalains. Biocatal Agric Biotechnol 51:102774

Tesoriere L, Fazzari M, Angileri F, Gentile C, Livrea MA (2008) In vitro digestion of betalainic foods. Stability and bioaccessibility of betaxanthins and betacyanins and antioxidative potential of food digesta. J Agric Food Chem 56(22):10487–10492

Thippeswamy B, Joshi A, Sethi S, Dahuja A, Kaur C, Tomar BS, Varghese E (2022) Chemical additives for preserving the betalain pigment and antioxidant activity of red beetroot. Sugar Tech 24(3):890–899

Thiruvengadam M, Chung IM, Samynathan R, Chandar SH, Venkidasamy B, Sarkar T, Rebezov M, Gorelik O, Shariati MA, Simal-Gandara J (2024) A comprehensive review of beetroot (Beta vulgaris L.) bioactive components in the food and pharmaceutical industries. Crit Rev Food Sci Nutr 64(3):708–739

Thompson KG, Turner L, Prichard J, Dodd F, Kennedy DO, Haskell C, Blackwell JR, Jones AM (2014) Influence of dietary nitrate supplementation on physiological and cognitive responses to incremental cycle exercise. Respir Physiol Neurobiol 193:11–20

Thongtang N, Sukmawan R, Llanes EJB, Lee ZV (2022) Dyslipidemia management for primary prevention of cardiovascular events: Best in-clinic practices. Prev Med Rep 27:101819

Ting Y, Jiang Y, Ho CT, Huang Q (2014) Common delivery systems for enhancing in vivo bioavailability and biological efficacy of nutraceuticals. J Funct Foods 7:112–128

Tran TN, Athanassiou A, Basit A, Bayer IS (2017) Starch-based bio-elastomers functionalized with red beetroot natural antioxidant. Food Chem 216:324–333

Tropea T, Renshall LJ, Nihlen C, Weitzberg E, Lundberg JO, David AL, Tsatsaris V, Stuckey DJ, Wareing M, Greenwood SL, Sibley CP (2020) Beetroot juice lowers blood pressure and improves endothelial function in pregnant eNOS−/− mice: importance of nitrate-independent effects. J Physiol 598(18):4079–4092

Van Velzen AG, Sips AJ, Schothorst RC, Lambers AC, Meulenbelt J (2008) The oral bioavailability of nitrate from nitrate-rich vegetables in humans. Toxicol Lett 181(3):177–181

Varner AS (2014) Modeling and optimization of the dehydration of beets for use as a value-added food ingredient. Doctoral dissertation, University of Georgia

Vasconcellos J, Conte-Junior C, Silva D, Pierucci AP, Paschoalin V, Alvares TS (2016) Comparison of total antioxidant potential, and total phenolic, nitrate, sugar, and organic acid contents in beetroot juice, chips, powder, and cooked beetroot. Food Sci Biotechnol 25:79–84

Vasiljeva S, Smirnova G, Basova N, Babarykin D (2018) Cadmium-induced oxidative damage and protective action of fractioned red beet (*Beta vulgaris*) root juice in chickens. Agron Res 16:1517–1526

Vidal PJ, López-Nicolás JM, Gandía-Herrero F, García-Carmona F (2014) Inactivation of lipoxygenase and cyclooxygenase by natural betalains and semi-synthetic analogues. Food Chem 154:246–254

Vinson JA, Hao Y, Su X, Zubik L (1998) Phenol antioxidant quantity and quality in foods: vegetables. J Agric Food Chem 46(9):3630–3634

Volino-Souza M, Oliveira GVD, Conte-Junior CA, Alvares TS (2020) COVID-19 quarantine: impact of lifestyle behaviors changes on endothelial function and possible protective effect of beetroot juice. Front Nutr 7:582210

Vulić J, Čanadanović-Brunet J, Ćetković G, Tumbas V, Djilas S, Četojević-Simin D, Čanadanović V (2012) Antioxidant and cell growth activities of beet root pomace extracts. J Funct Foods 4(3):670–678

Wang T, Liu L, Rakhmanova A, Wang X, Shan Y, Yi Y, Liu B, Zhou Y, Lü X (2020) Stability of bioactive compounds and in vitro gastrointestinal digestion of red beetroot jam: Effect of processing and storage. Food Biosci 38:100788

Webb AJ, Patel N, Loukogeorgakis S, Okorie M, Aboud Z, Misra S, Rashid R, Miall P, Deanfield J, Benjamin N, MacAllister R (2008) Acute blood pressure lowering, vasoprotective, and antiplatelet properties of dietary nitrate via bioconversion to nitrite. Hypertension 51(3):784–790

Wink DA, Hines HB, Cheng RY, Switzer CH, Flores-Santana W, Vitek MP, Ridnour LA, Colton CA (2011) Nitric oxide and redox mechanisms in the immune response. J Leukoc Biol 89(6):873–891

Wootton-Beard PC, Ryan L (2011) A beetroot juice shot is a significant and convenient source of bioaccessible antioxidants. J Funct Foods 3(4):329–334
Wruss J, Waldenberger G, Huemer S, Uygun P, Lanzerstorfer P, Müller U, Höglinger O, Weghuber J (2015) Compositional characteristics of commercial beetroot products and beetroot juice prepared from seven beetroot varieties grown in Upper Austria. J Food Compos Anal 42:46–55
Yadav M, Masih D, Sonkar C (2016) Development and quality evaluation of beetroot powder incorporated yogurt. Int J Sci Eng Technol 4(4):582–586
Yang J, Pi C, Wang G (2018) Inhibition of PI3K/Akt/mTOR pathway by apigenin induces apoptosis and autophagy in hepatocellular carcinoma cells. Biomed Pharmacother 103:699–707
Youssef NH, Qari SH, Behiry SI, Dessoky ES, El-Hallous EI, Elshaer MM, Kordy A, Maresca V, Abdelkhalek A, Heflish AA (2021) Antimycotoxigenic activity of beetroot extracts against Alternaria alternata mycotoxins on potato crop. Appl Sci 11(9):4239
Zein H, Hashish AE, Ismaiel G (2015) The antioxidant and anticancer activities of Swiss chard and red beetroot leaves. Curr Sci Int 4(4):491–498
Zielińska-Przyjemska M, Olejnik A, Dobrowolska-Zachwieja A, Grajek W (2009) In vitro effects of beetroot juice and chips on oxidative metabolism and apoptosis in neutrophils from obese individuals. Phytother Res 23(1):49–55

Bioactive Phytochemicals from Carrot (*Daucus carota*) By-Products

Zeliha Üstün Argon, Zinar Pınar Gümüş, Süleyman Doğu, and Turan Akdağ

Contents

Z. Üstün Argon (✉) · S. Doğu
Medicinal and Aromatic Plants Program, Meram Vocational School, Necmettin Erbakan University, Konya, Turkey

Medicinal and Cosmetic Plants Application and Research Center, Necmettin Erbakan University, Konya, Turkey
e-mail: zargon@erbakan.edu.tr; sdogu@erbakan.edu.tr

Z. P. Gümüş
Central Research Testing and Analysis Laboratory Research and Application Center, Ege-Matal, Ege University, Izmir, Turkey
e-mail: zinar.pinar.gumus@ege.edu.tr

T. Akdağ
Medicinal and Aromatic Plants Program, Meram Vocational School, Necmettin Erbakan University, Konya, Turkey
e-mail: takdag@erbakan.edu.tr

M. F. Ramadan (ed.), *Bioactive Phytochemicals in By-products from Leaf, Stem, Root and Tuber Vegetables*, https://doi.org/10.1007/978-3-031-80700-8_10

Abstract Carrot (*Daucus carota* subsp. *sativus*) is a root vegetable with total production, including turnips, reaching approximately 41 million tons in 2020. Carrots have high nutritional value and provide a rich source of bioactive compounds that significantly benefit human health, including carotenoids, vitamins, and minerals. The popularity of carrot consumption is on the rise, attributed to its recognition as a natural antioxidant source that may help prevent chronic diseases, along with its industrial advantages. Following the harvest, carrots that fail to meet the consumer market's quality standards and size and shape requirements are discarded and often repurposed as cattle feed. This plant is also utilized across various industries, with carrot peels, tops (green leafy parts), pulp, seed oil, pomace, and residue after juice extraction considered by-products. The primary bioactive phytochemicals in *Daucus carota* by-products include carotenoids, polyacetylenes, phenolic compounds, dietary fiber, and minerals, predominantly extracted from carrot peels and tops. Due to these valuable components, *Daucus carota* by-products are preferred in food, animal feed, pharmaceutical and cosmetic industries, and agriculture. The energy sector also utilizes carrot by-products for various purposes.

Keywords *Daucus carota* · Carrot by-products · Bioactive complex · Polyphenols · Phytochemicals · Human health

1 Introduction and Economic Values of Carrot By-Product

The food processing industries have produced billions of tons of waste every year. After processing the raw materials, food products are reached to the customers through food chains. In each stage of production and distribution of the foods, wastes such as pods, leaves, seeds, husks, and roots can be an essential issue from environmental perspectives and economic losses since they do not meet the market standards (Sharma et al. 2012; Ahmad et al. 2019; Vaz et al. 2022). The Food and Agriculture Organization of the United Nations (FAO) defines food waste as "As processed/semi-processed/raw food and the related inedible parts discarded from the human food supply chain in the household, food service and retail sectors. These discarded food or inedible parts are employed as landfilling materials, composting, controlled combustion, co/anaerobic digestion, litter, land applications, compost/aerobic digestion." In the UNEP Food Waste Index Report of 2021, it has been stated that in 2019, 931 million tons of food waste was generated, and 61% of it came from households, 26% was from food service, and 13% was from retail (FAO 2021) and the economic value of this loss was about 1 trillion dollars. Evaluating the amount of food waste and its economic value with sustainable development goals such as protecting the planet, ensuring peace and prosperity, and ending poverty, contributing to sustainable food production and improving food safety demands a new perspective to utilize food waste (Sharma et al. 2012; Ying et al. 2021; Kim 2023). The European Union (EU) circular economy framework also mentions agri-food by-products and aims to minimize them by focusing on reducing, reusing, and

Fig. 1 *Daucus carota* (Freepik, 2023)

recycling raw materials. This approach enhances the value and extends the useful life of resources within the economy by developing new products from these materials (Ferreira et al. 2021) (Fig. 1).

Since the nutritional value of some of the wastes is high, they are called "by-products" and can be used as raw materials or ingredients in food or other industries. Post-harvest of agricultural products is the first stage where the waste or, with the different terms, by-products are generated. The roots, peels, seeds, and stems are separated from the plants in the first stage of the process, and the amount of these parts is reported to be more than 50%. Composting is one solution to make food waste helpful in improving the soil quality, and fruits, vegetables, eggshells, and nuts can be used for this purpose. Using the wastes as livestock feed was another solution because of the high nutritional value of the materials. With the recent technological studies, however, evaluating these products' chemical composition and availability in large amounts brings the possibility of utilizing these organic wastes as value-added products. Recovering food loss through re-designed ingredients will also positively impact society, the economy, and the environment (Ying et al. 2021; Vaz et al. 2022).

The loss of fruits and vegetables is an important issue. Since fruits and vegetables are a noteworthy source of protein, carbohydrates, fiber, fats, vitamins, minerals, phytochemicals such as flavonoids, carotenoids, phenolics, and technologically relevant compounds, their recovery and conversion into health-promoting functional and nutritional products is essential for a wide range of different purposes including bakery products, beverages, extruded snacks (Ahmad et al. 2019; Ferreira et al. 2021; Ying et al. 2021; Vaz et al. 2022).

The carrot (*Daucus carota* L) is one of the most popular and most commonly consumed vegetables, especially in the meal kit industry (Kim 2023). This root is grown worldwide, especially in China, the United States, Ukraine, Uzbekistan, and Russia (FAO 2024) and is accepted as one of the main sources of dietary carotenoids, fiber, phenolic compounds, minerals, and vitamins. Due to their health

Fig. 2 Carrot and carrot juice

benefits, such as antioxidant properties, anti-cancer activities, and being a precursor of vitamin A, carrots and their products have been preferred and produced more in recent years (Sharma et al. 2012) (Fig. 2).

Large quantities of harvested carrots (25–25%) are discarded annually in different parts of the world because of their poor quality, irregular color, size, and shape. These discarded carrots are primarily used as animal feed. Moreover, the production of carrot puree and juice results in several waste products, including carrot peels, which can be reclaimed for their bioactive compounds. These processing by-products and discarded carrots can be converted into valuable substances such as carotenoids, phenolic compounds, dietary fiber fractions, and bioethanol. Different studies have shown that the leaf extract of carrots has been used for anticarcinogenic studies, and other wastes have been employed for the feed of pigs, cows, and rabbits to enhance dry matter and nutritional composition. Additionally, carrots can be used to create anthocyanin-rich concentrates for the pigment industry, with the leftover pomace being extracted to produce high-value phenolic compounds for use in functional food ingredients (Ahmad et al. 2019; Eliopoulos et al. 2022; Kim 2023).

Carrot juice production generates significant waste, leaving 30–50% of the raw material as pomace. Challenges such as limited space in processing facilities, inadequate water supply for managing waste, and environmental concerns often arise, leading to the loss of valuable nutrients and biomass. To address these issues, new technologies and innovative applications are needed for by-product recovery.

Pomace is rich in fiber, uronic acids, carotenoids, and neutral sugars. However, up to 50% of the carotene can be lost in the pomace, typically used for livestock feed, compost, or manure (Sharma et al. 2012).

In this chapter, with the recent studies, carrot (*Daucus carota* L) by-products will be evaluated for their chemical compositions, biological and functional properties, food and non-food applications of extracts and bioactive compounds from by-products, and valorization of by-products for non-health purposes.

2 Chemical Composition and Bioactive Compounds of Extracts from By-Products

The carrot (*Daucus carota*) is a root vegetable, part of the Apiaceae (Umbelliferae) family, and is widely used raw, cooked, or in juices. Carrots, *Daucus carota*, are a root vegetable full of nutritious and health-promoting components. Carrots have a high content of anthocyanidins, carotenoids, and fiber (Clementz et al. 2019). The chemical composition of carrots is quite rich and contains many nutrients. The general chemical composition of carrots includes water (85–90%), protein (0.6–1.1%), a very small amount of fat (0.2–0.3%), and carbohydrates (7–10%). Most carbohydrates are sugars (glucose, fructose, sucrose) and fiber. The carrot roots contain many times more sucrose than either glucose or fructose. Vitamin A (β-carotene) is the best-known carrot component and an important antioxidant. In addition to vitamin A, carrots contain vitamins C, K, B (B1, B2, B3, B6) and E. Carrots, including potassium, manganese, calcium, magnesium, and phosphorus, are rich in various bioactive compounds. These components have positive effects on health and can help prevent diseases. The bioactive components of carrots are carotenoids (beta-carotene, alpha-carotene, lutein and zeaxanthin), polyacetylenes (falcarinol and falcarindiol), and phenolic compounds (chlorogenic acid, caffeic acid, *p*-hydroxybenzoic acids, *p*-coumaric acid, and numerous cinnamic acid derivatives). Chlorogenic acid represents 42.2 and 61.8% of the total phenolic compounds detected in different carrot tissues (Zhang and Hamauzu 2004; Gonçalves et al. 2010). Carrots also have a unique combination of three flavonoids: kaempferol, quercetin, and luteolin (Ching and Mohamed 2001; Lila 2004; Horbowicz et al. 2008).

Its most notable feature is its high beta-carotene content; this compound is converted to vitamin A in the body and supports healthy vision, strengthens the immune system, and protects skin health. In addition, other carotenoids, such as alpha-carotene, lutein, and zeaxanthin, are particularly beneficial for eye health. Carrots are also a good source of vitamins. Vitamin A, which is necessary for vision, skin, and immune system health, is made by converting beta-carotene in the body. Vitamin K1 is essential for blood clotting and bone health, vitamin B6 plays a role in protein and fat metabolism and supports brain function, and vitamin C supports the immune system with its antioxidant properties and protects skin health. Minerals are an essential part of carrots and important for muscle, heart, antioxidant enzymes, bone health, and various biochemical reactions. Soluble and insoluble fiber support digestive health, while antioxidants fight free radicals and prevent cell damage. Fiber also regulates cholesterol levels and helps control blood sugar. Falcarinol and falcarindiol from polyacetylenes have anti-inflammatory, antimicrobial, and anti-cancer properties. Chlorogenic acid and p-coumaric acid from phenolic compounds have antioxidant properties and may help regulate blood sugar levels and prevent cell damage. Chemical composition is among the most important factors determining carrots' nutritional value and health benefits (Sharma et al. 2012).

Carrots can vary depending on their color and the region in which they are grown. Carrots come in many colors; the most common is orange, but there are also tropical carrots with white, yellow, red, purple, and black skins (Singh et al. 2018). While orange carrots are rich in beta-carotene, purple carrots are rich in anthocyanins. Red carrots contain lycopene, and yellow carrots contain lutein. White carrots contain few pigments. Carrot seed essential oil is a common fragrance ingredient in cosmetics and perfumes and a flavor component in several food categories. It is also a major source of the sesquiterpene alcohol carotol. Carrot essential oil is a common fragrance ingredient in cosmetics and perfumes. These differences diversify the nutritional value of carrots and their effects on health. Carrots are a versatile vegetable with many health benefits due to their chemical composition. Regular consumption can positively affect vision, skin, heart, and digestive health. Including carrots in a balanced diet is beneficial for overall health and well-being. This rich chemical composition makes carrots a nutritious and valuable vegetable in terms of health (da Silva Dias 2014).

Carrots are a versatile vegetable served raw, cooked, baked, roasted, sautéed, pickled, or made into carrot juice. The carrot processing industry generates significant amounts of by-products. The carrot root consists of three parts: stem (83.19%), head (5.01%), and peel (14.19%). Carrot peels, the main residue from the industrial use of carrots, pose a significant environmental problem. However, the extraction of beneficial bioactive compounds provides alternative application routes for this bio-waste in functional foods, drugs, pharmaceuticals, and cosmetics (Nguyen and Scarlett 2016). Carrot waste contains high levels of dietary fiber and phenolic compounds (Vodnar et al. 2017; Singh et al. 2006).

An important component of carrot peel is pectin, a soluble dietary fiber composed of d-galacturonic acid units, L-rhamnose, L-arabinose, d-galactose, and other monosaccharides. Pectin (E440), a universally used additive in functional foods (frozen foods, jellies, jams, or low-calorie foods), has thickening and gelling properties, making it a possible choice to replace hydrocolloids (Encalada et al. 2019). Other valuable components of carrot peel are carotenoids (α- and β-carotene) (Clementz et al. 2019).

In their study, Bao and Chang attributed the fact that carrot products are a good source of fiber to the presence of 4–5% protein, 8–9% reducing sugars, 5–6% minerals, and 37–48% total fiber (on a dry weight basis) in carrot pulp (Bao and Chang 1994).

Carrot pulp is a good source of vitamins, dietary fiber, and carotenoids, so that it can be used as a valuable bioactive and functional compound. Dried carrot pulp powder, which can be stored for a long time, can modify the textural properties of foods and their nutritional potential. Increased consumption of antioxidant-rich fruits and vegetables helps reduce the risk of cancer and cardiovascular disease. Carrots are a rich source of calcium pectate, an exceptional pectin fiber with cholesterol-lowering properties.

While carrots (*Daucus carota*) are generally known as vegetables with high nutritional value, the by-products generated during carrot processing, namely carrot peel, pulp and residues, contain valuable bioactive compounds. These by-products' chemical composition and bioactive compounds are an important area of research for the food, health, and cosmetic industries. Several studies investigate the chemical composition and bioactive compounds of extracts obtained from carrot by-products. Carrot juices and blends are among the most popular soft drinks (Stoll et al. 2003a). Despite significant improvements in processing techniques, such as depolymerizing enzymes and puree decanter technology, many valuable compounds, such as carotenes, uronic acids, and natural sugars, remain in the pulp. Various attempts have been made to use carrot pulp in food products such as bread, cakes, and sauces and in the production of functional beverages. It has been shown that the pigments of spray-dried carrot pulp waste are susceptible to deterioration during storage, depending on the storage time and temperature.

Due to the low yields associated with carrot juice production, up to 50% of the raw material remains as pulp, and this pulp contains large amounts of valuable compounds such as carotenoids, dietary fiber (Nocolle et al. 2003), uronic acids, and neutral sugars (Stoll et al. 2003b). The total carotene content of the pulp can be up to 2 g/kg dry matter, depending on the processing conditions (Singh et al. 2006). Carrot pulp contains 17 and 31–35% of the total α-carotene in fresh, uncooked and cooked carrots, respectively (Bao and Chang 1994). Nawirska and Kwasniewska (2005) reported the composition of fiber components of carrot pulp (dry weight basis) as pectin (3.88%), hemicellulose (12.3%), cellulose (51.6%), and lignin (32.1%).

Nguyen and Scarlett (2016) determined the mass fraction of carrot root and the effects of four different solvents (methanol, water, ethanol, and hexane) on carrot peel's total phenolic and saponin content and antioxidant capacity. Thus, they indicated the solvent system's importance for effectively extracting bioactive compounds from carrot peel. The results showed that carrot root consisted of stem, head, and peel with fresh weight mass fractions of 83.19%, 5.01%, and 14.19%, respectively. Among the four solvents tested, methanol obtained the highest extraction efficiency (54.02% dry weight), total phenolic content (9.02 mg GAE/g dry weight), and antioxidant capacity (DPPH radical scavenging capacity (87.0 g TE/100 g), copper ion reducing antioxidant capacity and iron reducing antioxidant capacity) from carrot peel. At the same time, water extracted the highest saponin content (272.9 mg EE/g dry weight) and had the maximum ABTS radical scavenging capacity (Nguyen and Scarlett 2016).

In their studies, Shyamala and Jamuna investigated the chemical composition and antioxidant potential of carrot (Daucus carota) pulp waste. The protein content of carrot pulp waste extracts was reported to be 6.21 mg/100 g, while total polyphenols were higher in methanol extracts (220–250 mg/100 g) than in ethanol and aqueous extracts. Antioxidant activity determined by the DPPH method showed 40% activity in methanol extracts of carrot pulp waste (20 mg) (Shyamala and Jamuna 2010).

In the study of Chantaro et al., carrot peel, a food industry by-product, was investigated as a raw material for producing antioxidant dietary fiber powder. In terms of antioxidant activity, the effect of blanching and hot air drying on drying kinetics and physicochemical properties of dietary fiber powder, thermal degradation during both blanching and drying caused a decrease in β-carotene and phenolic compound contents, thus leading to the loss of antioxidant activity of the final product. In the study, the amount of dietary fiber in carrot peel was reported as 45.4 ± 0.4 g/100 g, carbohydrate as 33.0 ± 0.7 g/100 g, lipid as 1.5 ± 0.1 g/100 g, β-carotene as 20.4 mg GAE/g DW, polyphenols as 1371 mg GAE/g DW, antioxidant activity coefficient as 90.56–94.67 mg GAE/g (Chantaro et al. 2008).

Smigielski and his colleagues studied the effect of ultrasound-assisted maceration on the bioactivity, quality, and yield of essential oil obtained from spent carrot seeds (Daucus carota). The main components of the oil were reported to be carotene (34.84–35.71%), sabinene (10.16–9.84%), α-pinene (8.02–7.79%), and daukol (3.73–4.51%). Monoterpenes constitute 40.6% of the oil composition, while sesquiterpenes constitute 47.3% (Smigielski et al. 2014). In the study by Özcan and Chalchat, volatile oil components were determined in carrot seeds from Konya, Turkey, to investigate their potential use. They identified carotol, daucene, (Z, Z)-α-farnesene, germacrene D, *trans*-α-bergamotene and β-selinen, daukol, and copaenol (Özcan and Chalchat 2007).

In the study conducted by Encalada et al. in 2019, they developed pectin-enriched fractions that can be used in functional food formulation due to the antioxidants (α- and β-carotenes, lutein, α-tocopherol) extracted with carrot residues and their gel-forming effect. In this study, α-carotene (52 mg/100 g), β-carotene (80 mg/100 g), lutein (6.4 mg/100 g), and α-tocopherol (7.1 mg/100 g) contents were determined in carrot powder (Encalada et al. 2019).

In their study, Tang and Chen investigated the extraction of carotenoids from carrot pulp waste and reported that the carotenoids were 111.73 μg/g (Tang and Chen 2000).

Vodnar et al. evaluated the antioxidant, antimutagenic, and antimicrobial activities of the leading Romanian agro-industrial wastes (apple peels, carrot pulp, white and red grape peels, red beet peels, and pulp). Linoleic acid was the main fatty acid in all analyzed samples, while carrot extracts were reported to have no antimicrobial activity. The HPLC-DAD-ESI-MS method identified phenolic compounds such as those found in the extracts of carrot fresh waste. These are caffeic acid 4-O-glucoside, 3-caffeoylquinic acid, phloridzin (phloretin 2-O-glucoside), phloretin 2-O-xylosyl-glucoside, rutin (quercetin 3-O-rutinoside), quercitrin (quercetin 3-O-rhamnoside), quercetin 3,4-O-diglucoside (Vodnar et al. 2017).

In the study by Özcan and Chalchat, the main fatty acids of carrot seeds identified by gas chromatography were petroselinic (59.35%), linoleic (11.82%), palmitic (10.01%), and stearic (2.41%) acids (Özcan and Chalchat 2007).

In the study by Kamiloglu et al., polyphenol-rich black carrot and its by-products (peel and pulp) were used to modulate the inflammatory response in tumor necrosis factor a (TNF-a)-treated endothelial cells. Anthocyanins and phenolic acids were detected as the major polyphenols in black carrot, peel and pomace samples. The following anthocyanins were identified: cyanidin-3-xylosyl-glucosyl-galactoside, cyanidin-3-xylosyl-galactoside, cyanidin-3-xylosyl-sinapoyl-glucosyl-galactoside, cyanidin-3-xylosyl-feruloyl-glucosyl-galactoside, cyanidin-3-xylosyl-coumaroyl-glucosyl-galactoside. Neochlorogenic, chlorogenic, and caffeic acids were identified as the three major phenolic acids (Kamiloglu et al. 2017).

Mustafa et al. used a pressurized liquid extraction technique by selecting ethanol as an eluent solvent and evaluated α- and β-carotene content in two types of frozen carrot by-products: soft, soggy carrots, and orange carrots. They reported that the amounts of β-carotene obtained were 12.3–22.9 mg/100 g and 8.1–19.3 mg/100 g fresh weight, respectively (Mustafa et al. 2012).

Solyom et al. investigated the dielectric properties of by-products using microwave heating. This study focused on the effect of temperature and moisture content on the dielectric constant (30) and loss (300) of carrot waste, apple pulp, pineapple peel, and used coffee grounds at 2.45 GHz. In conclusion, the authors suggested that when considering food-waste processing by microwave technology, the dielectric behavior of fresh food material should not be considered without considering the compositional similarity between fresh food and food waste (Solyom et al. 2020).

A process for the recovery of carrot pulp hydrolysate, which involves fine grinding, enzymatic hydrolysis, finishing, homogenization, and concentration techniques, was developed by Stoll et al. The total carotene content (α- and β-carotene) was reported to be 64 mg/kg (Stoll et al. 2003b)

In their study, Tanska et al. investigated the mineral composition of dried carrot pulp to test the effect of 5, 7.5 and 10% addition of dried carrot pulp on selected wheat dough and bread properties. They determined the elements Na, K, P, Ca, Mg, Cu, Mn, Fe, and Zn to be 3.2, 18.6, 1.8, 3.0, 1.1, 4.0, 10.8, 30.5, and 29.4 mg/g, respectively (Tanska et al. 2007).

In the study by Agcam et al., the optimal ultrasonic processing conditions were used to maximize the extraction of anthocyanin pigments from black carrot pulp, and the yield of five different anthocyanin compounds from black carrot pulp was maximized. The most abundant anthocyanin compound was cyanidin-3-xyloside-galactoside-glucoside-ferulic acid (60.85–74.22 mg/L). The potential of ultrasonic technology for extracting valuable by-products from the fruit and vegetable juice industry was emphasized (Agcam et al. 2017).

Therefore, the by-product obtained after carrot juice extraction represents a promising source of compounds with bioactive properties that can be explored in developing food ingredients and dietary supplements (Moure et al. 2001; Schieber et al. 2001; Sharma et al. 2012). Adding value to waste directly benefits both processors and consumers for sustainable agriculture.

3 Biological and Functional Properties of Extracts and Bioactive Compounds from By-Products

Carrots (*Daucus carota*) are widely recognized for their rich nutritional profile and health benefits. However, the processing of carrots for human consumption produces various by-products, such as peels, pulp, and residues. Generally considered waste, these by-products contain important bioactive compounds with important biological and functional properties (Jayesree et al. 2021; Clementz et al. 2019). This section examines the biological and functional properties of carrot extracts and bioactive compounds obtained from carrot by-products, highlighting their potential applications and benefits. Carrot by-products can be used in various areas and provide economic and environmental benefits (Sharma et al. 2012). The utilization of carrot by-products reduces the amount of waste, allows the emergence of new products, and contributes to sustainability goals (Schieber 2017). Food applications are the most important products that can be made from carrot co-products in different areas (Guerra et al. 2023).

As food products, carrot pulp and puree can be used in soups, sauces, smoothies, and snacks. Carrot pulp can produce flour or added fiber foods (Kaisangsri et al. 2016). Carrot peel or pulp can make energy bars, crackers, or healthy snacks. Carrot by-products can enrich foods with fiber, vitamins, and antioxidants (Mustafa et al. 2012; Wan et al. 2021). For example, carrot pulp can be added to baked goods or beverages to enhance nutritional value (Stoll et al. 2003a). Carrot pulp and extracts can be used as vitamin and mineral supplements. Dietary supplements can be produced as natural health products, especially those rich in vitamins and antioxidants. These extracts contain high concentrations of vitamins, minerals, and antioxidants. Natural colors derived from carrots can be used as color additives in the food industry. Natural colorants derived from carotenoids can be used in various foods to enhance visual appeal without compromising health, providing a natural alternative to synthetic colorants.

As a plant-based alternative, carrot pulp also has the potential to be used in the production of plant-based meatballs and meat alternatives. There are also several applications other than food. Small pieces and pulps of carrot by-products can be used as a nutritious animal feed (Giller et al. 2022). It can be particularly suitable for sheep, cattle, and other livestock. Carrot peel and pulp can be composted and turned into a natural fertilizer. This fertilizer can be used in gardens and agricultural areas. Carrot by-products can be used for biogas production.

Biofuel is a process used to generate energy from organic waste (Szymanska-Chargot et al. 2017). These by-products, which also have the potential to be used in cosmetics and personal care products, can be used for skin and hair care. Carrot pulps, rich in beta-carotene and antioxidants, can be useful in skin care products, especially in anti-aging creams and masks. In addition, carrot extracts and oils can be used in hair masks and shampoos. Using carrot co-products in various industries offers both environmental and economic benefits.

The biological properties of carrot co-products determine the nutritional value, health effects, and potential uses of these co-products. The biological properties of carrot by-products are related to the dietary fiber, vitamins, minerals, carotenoids and antioxidants, polyphenols, and flavonoids found in their chemical content (Chantaro et al. 2008; Salari et al. 2024). Carrot by-products contain both soluble and insoluble fiber. Soluble fiber can lower cholesterol and balance blood sugar. Insoluble fiber regulates the digestive system and prevents constipation. It contains important vitamins, especially vitamin A (in the form of beta-carotene), vitamin C, and vitamin K. Vitamin A is necessary for vision, skin health, and the immune system. Vitamin C has antioxidant properties and supports the immune system. Vitamin K is important for blood clotting and bone health. Carrot flesh contains the most minerals, including potassium, calcium, and magnesium. Potassium regulates blood pressure, while calcium and magnesium are essential for bone and muscle health. Beta-carotene, a powerful antioxidant among carotenoids and antioxidants, fights free radicals and is converted to vitamin A in the body. It also contributes to skin and eye health. In addition, alpha-carotene and lutein have similar antioxidant properties and support eye health. While the phenolic components of carrot by-products have antioxidant and anti-inflammatory properties, flavonoids also show antioxidant, anti-inflammatory, and anti-cancer properties (Kamiloglu et al. 2017). The chemical compounds it possesses increase the antioxidant activity of carrot by-products and provide the ability to fight free radicals. Carrot by-products fight oxidative stress caused by free radicals because they contain potent antioxidants that can reduce cell damage and slow aging. Carrot by-products can support cardiovascular health through several mechanisms. The high potassium content helps regulate blood pressure, while the antioxidants and dietary fiber contribute to reducing cholesterol levels. This combination of nutrients can help lower the risk of heart disease and stroke. Studies investigating the antimicrobial properties of carrot by-products suggest that extracts from carrot by-products have antimicrobial properties that may help combat pathogenic bacteria and fungi. This property may be useful in food preservation and natural health products (Vodnar et al. 2017; Singh et al. 2006).

In summary, among the biological effects of carrot by-products, anti-inflammatory effects, immune system support, and digestive health can be counted as the most important features. It can improve the general health status, strengthen the immune system, protect against diseases, help the digestive system work regularly, and support intestinal health. Food safety and processing techniques are very sensitive to preserving chemical content and biological properties. Carrot by-products may lose some nutrients depending on processing and storage conditions. Appropriate processing and storage methods are important to preserve chemical and biological properties. The biological properties of carrot by-products make them valuable from a health and commercial perspective. Understanding and evaluating these properties can help develop sustainable use and waste reduction strategies.

Functional uses for carrot by-products include nutritional enhancement, supplements, natural food additives, animal feed, compost, and fertilizer. As a nutritional enhancer, carrot by-products can enrich foods with additional fiber, vitamins, and antioxidants. Bioactive compounds derived from carrot by-products can be used in dietary supplements. Natural colorants derived from carotenoids in carrot by-products can be used as food additives, providing a natural alternative to synthetic colorants. These products can be processed into animal feed, providing a nutrient-rich option that supports animal health. Carrot by-products can be composted to produce natural fertilizers. This process reduces waste and provides a valuable resource to improve soil quality and support sustainable agriculture.

In their study, Esposto et al. extracted bioactive compounds from carrot by-products and evaluated their chemical stability after encapsulation in chitosan-coated liposomes or using sodium tripolyphosphate for chitosan complexation. Thus, they are promising as a biotechnological product that can be applied in the food industry or any aqueous matrix (Esposto et al. 2022).

The study by Kirbas et al. investigated the effects of using different fiber sources (apple pulp powder, carrot pulp powder, and orange pulp powder) on dough rheology and quality properties of gluten-free rice flour-based cakes. Gluten-free cake doughs were formulated by replacing different amounts of rice flour with fruit pulp powders. It was found that carrot by-products could have a different application in the formulation of gluten-free cakes (Kırbaş et al. 2019).

The antimicrobial activity of fermented by-products obtained using tomato, carrot, and melon by-products was tested by Ricci et al. to extend the shelf life of food products, which is one of the areas of research in the food industry. The authors concluded that lactic acid fermentation of plant by-products could be a good strategy to obtain useful antimicrobials in food biopreservation.

Encalada et al. evaluated the performance of a carrot waste fraction enriched with orange pectin as a food additive to stabilize a model spreadable chia oil-in-water emulsion (20:80 w/w). They reported that carrots, considered an effective natural additive for functional food preservation, could be successfully enhanced (Encalada et al. 2020).

The effects of pasta samples developed by replacing hard wheat semolina with finger millet flour, pearl millet flour, and carrot pulp powder on cooking quality, color, and texture in pasta production were evaluated by Gull et al. The authors reported that millet flour and carrot pulp allowed partial replacement of hard wheat semolina in pasta with acceptable properties (Gull et al. 2015).

Tanska and colleagues investigated the effect of adding dried carrot pulp to bread regarding carotenoids, fiber and mineral components and evaluated these breads rheologically and organoleptically (Tanska et al. 2007).

Carrot co-products are a valuable and often overlooked source of essential nutrients and bioactive compounds. Their antioxidant, anti-inflammatory, anti-cancer, cardiovascular, skin health, immune support, and digestive health properties make them a beneficial addition to the diet. By utilizing carrot co-products, we improve our health and contribute to sustainable food practices by reducing waste. The functional use of carrot co-products is an emerging area with significant potential. Industries can create value-added products while promoting sustainability by utilizing the bioactive compounds in these by-products. Future research could explore the optimization of extraction methods to maximize yields of bioactive compounds and develop innovative applications in health, nutrition, and environmental management.

4 Food and Non-Food Applications of Extracts and Bioactive Compounds From Carrot By-Products

Carrot is one of the most popular vegetables in the world due to its high nutritional value and health effects. Producing carrot juice is one of the main processes, which leaves 30–50% of the raw material as pomace with important nutrients such as uronic acids, fiber, neutral sugar, and carotenoids. 50% of the carotene is lost within the pomace and is usually used for livestock feed, compost, or manure. Carrot shreds can be dehydrated for further processes. Carrot pomace has also been an ingredient for food products such as dressings, bread, high-fiber biscuits, toffees, cake, pickles, and beverages to enrich or build formulations (Sharma et al. 2012).

Most studies used the carrot by-products in powder form with the optimal properties. Different studies focused on the optimization of process conditions. Upadhyay et al. (2008) studied to determine the optimal drying temperature of the carrot pomace, and it was found that 65 °C was the optimal temperature in terms of the retention of ascorbic acid and β-carotene amounts. Chen and Tang (1998) also determined that the isomerization of carotenoids from the carotenoid powder of carrot waste fell under the high storage temperatures of around 45 °C.

The literature has important parameters for fresh vegetables and fruits, but these results cannot be implied for the by-products; therefore, the parameters for the food-waste processes should also be determined. Solyom et al. (2020) studied the microwave heating parameters for food by-products, including carrots. The effects of temperature and moisture level on the dielectric constant and the loss of raw material have been focused on. The moisture changed from 31.6 to 90.4%. The temperature was between 27.1 and 68.3 °C due to determined dielectric properties. For the carrot peel, penetration depth was found to be decreased when the temperature was increased for all moisture applications.

Vegetable residues, including carrots, were powdered after several processes such as dehydration, freeze-drying, hot air drying, and milling to provide fine powders. Powdered carrots were evaluated for antioxidant, physicochemical, and technological properties at the end of the production processes and during storage. Storage time increased the moisture content and decreased the quality of powders and antioxidants. Compared to other residues, carrot powder showed the highest color differences, luminosity, and solubility index. During the fourth month of the storage, color differences decreased in carrot powders. As a result, the powder obtained from carrot by-products can be used as an ingredient to enrich foods and as a processing aid (Bas-Bellver et al. 2023)

Wheat bread is one of the food products in which carrot pomace is used to fortify the bread with its fiber, carotenoids, and minerals such as zinc, potassium, manganese, and iron. Tanska et al. (2007) added carrot pomace to wheat bread in 5, 7.5, and 10%, and Ambroizak (1998) also tested a 5.5% addition of dried carrot. The results showed that rheological and organoleptic specifications improved, especially adding 5% pomace. Kumar and Kumar (2012) and Sharma and Kumar (2017) added carrot pomace by 2.5–5–7.5–10% w/w to small-sized wheat bread and buns. Adding 2.5 and 5% of carrot pomace resulted in better water absorption, bread volume, elasticity of bread crust.

Carrot pomace powder has been tested in gluten-free cakes with some other pomaces, such as apple and orange, by Kırbaş et al. (2019) for celiac patients. The results showed that adding pomace powders improved the rheological properties of batters and the quality parameters of the fiber-enriched gluten-free cakes. Carrot is one of the important ingredients of gluten-free products for celiac patients. Studies showed that adding 5–15% orange and black carrot flour to the formulations improved the gluten-free pasta's nutritional, technological, and sensory properties (Arslan Bayrakçı 2020).

Different studies tested the addition of carrot, black carrot pomace, and pomace powders into cookies and biscuits formulations. The inclusion amount changed from 0 to 30% w/w. The enrichment in fiber, carotenoid, and phenolic content, antioxidant capacity, and positive effects on color, texture, and moisture were reported (Kumar and Kumar 2011; Turksoy et al. 2011; Hernández-Ortega et al. 2013). The biscuits with the carrot pomace and pomace powder inclusion showed better spread ratios and changes in chemical parameters such as fat, crude fiber, β-carotene, and protein were observed (Kumari and Grewal 2007; Gayas et al. 2012; Sharma and Kumar 2017). The effects of by-products of carrots on brittle bakery products generally have positive impacts, affecting the overall acceptance, color, and antioxidant values with the other process specifications. Wheat rolls with the addition of carrot pomace powders had increased water absorption, stability, and higher acceptance (Kohajdová et al. 2012). In another study about ready-to-eat snacks, Ying et al. (2021) evaluated a formulation including carrot powder, carrot pomace powder, broccoli, and pomace powders. The results indicated that including pomace powders in food products will improve fiber content, decrease carbohydrate levels, increase shelf life, and reduce food waste.

Carrot pomace powder has also been tried in pasta products. One study replaced the powdered by-product with refined wheat flour at 5–15–25% w/w. Adding carrot pomace improved the appearance of the pasta in terms of color. The flour and taste of the pasta were also developed, and the optimum level of carrot pomace inclusion was determined to be 15% w/w (Mishra and Bhatt 2018). Another study mixed carrot pomace with millet flour in pasta ingredients up to 25% w/w. The results showed that up to 15% of inclusions developed the fiber and carotenoid content and gave optimal results for overall acceptance, and additions of up to 25% improved the appearance (Gull et al. 2015).

Ferreira et al. (2015) evaluated some fruits and vegetables such as carrots, spinach, rocket, watermelon, passion fruit, orange, mint, taro, lettuce, courgetti, and cucumber for their pomaces. In the first step, the juice was extracted, and the pomace was dried. The flour was tested in different formulations of the biscuits and cereal bars in various amounts, changing between 20 and 35%. Since the mixture has high fiber and carbohydrate levels, the water holding capacity was better than the oil samples. Mineral and fiber content were also higher in biscuits and cereal bars. Other parameters such as chemical, sensorial, and microbiological properties were also found positive which shows that flour is an alternative product in the food sector, which decreases biowaste.

Production of toffees with the carrot pomace in the formula has been tried. The results showed that these toffees had more soluble solids, total reducing and non-reducing sugar (Jagtap et al. 2000). In their study, Sharma et al. (2012) mention the processes of different food applications of carrot wastes such as kheer, carrot powder, carrot yogurt, toffees, and carrot pomace powder in detail. These results show that carrot waste can be successfully applied to food formulations. Esposto et al. (2022) evaluated carrot by-products for extracting total phenolic compounds and carotenoids. Sodium tripolyphosphate (TPP) and chitosan encapsulated these two components in liposomes. TPP particles more successfully protected carotenoid extracts from degradation, and liposomes were effectively coated by chitosan. Therefore, this coated by-product could be an essential material for food formulations.

Essential oil of *Daucus carota* L. subsp. *sativus* is commonly extracted from the cultivated root and used as a flavoring agent in perfumery and food products. This oil also effectively prevented bacteria and fungus activities (Staniszewska et al. 2005). Green leaves or tops of the *D. carota* are also classified as biowastes. Chiboub et al. (2019) evaluated the carrot tops for essential oil extraction and, if possible, for recovery of some other value-added products. The extracted essential oil was evaluated for antimicrobial, antifungal activities and phytochemical properties. The results showed that, since carrot essential oil is of high quality and yield, it can be used as an important ingredient for food, cosmeceuticals, and natural medicines.

Meat products also have potential for improvement through the incorporation of carrot by-products. Chicken sausages were examined with dried carrot pomace and wheat bran for fresh and frozen products. The study aimed to increase the fiber content, and including 3–6 and 9% w/w carrot pomace developed the fiber content. The best results were observed at 6%, with the highest acceptance and 15 days of shelf life (Yadav et al. 2018). For poultry products, the application of carrot pomace and other ingredients resulted in increased fiber content and overall acceptance. Adding 6% w/w of the pomace to the formulations was found optimal for the meat products (Eliopoulos et al. 2022). Another study on dried fermented sausage used carrot dietary fiber as a fat replacer. Including 1.5% w/w helped develop the product's sensorial properties (Garcia et al. 2002).

Dairy products represent a new segment where carrot by-products can be utilized. Carrot pomace beads were added to the yogurt with 2.5–5% w/w and the carotene level and antioxidant activity increased significantly. Physicochemical and microbiological properties of the yogurt were stable during the 28 days storage period at 4 °C (Šeregelj et al. 2021). Different ice cream types were enriched by adding carrot puree (CP) between 0 and 30% and probiotic bacteria *Bifidobacterium animalis* subsp. *Lactis* to be able to prepare a product that can be consumed mainly by children. Considering the results, a carrot puree addition up to 10% along with the probiotic bacteria improved the amount of probiotics in the ice cream, color characteristics, and general acceptability. Also, the formulation will help children to have carrots' nutrients, which are not easy to consume due to their hard texture (Tasbasi 2022).

One of the carrot by-products, the peels, was used in a jam formulation. A jam of carrot peels was found to have a higher phenolics content (87.4 mg/100 g) than apple pomace, banana, and mandarin peels. Other properties such as total flavonoids, DPPH anti radical scavenger activity, and lightness were found at 35.9 mg CAT/100 g, 1.8 μg/mL, 29.46, respectively. The taste, odor, and color parameters also give acceptable results (Hussein et al. 2015)

Beverages are one of the main areas where the carrot by-products can be used. Pérez et al. (2023) developed a juice formula by using broccoli stalk juice (82.5%), carrot by-product juice (17.5%), extra virgin olive oil (1%), lemon juice (0.25%), black pepper, salt and garlic powder (0.1% each). The aim of the study was to create a new beverage formula with the addition of antimicrobial and antioxidant seasonings. Various treatments were applied, such as high hydrostatic pressure (HHP), ultrasound (US), pasteurization, and their combinations. The antioxidant contents and sulfur components have been preserved by using non-thermal applications and the shelf life was extended. Microbiological counts were reduced in HHP treatments, and the developed beverage was recommended as a good source for nutritional intakes such as carotenoids and glucosinolates and a potential alternative for using plant by-products.

Another study evaluated cucumber, tomato, apple, and carrot pomace to obtain sugars, hemicellulose, lignin, cellulose, and pectic polysaccharides. The pomaces were treated with water, acidic and alkali solutions and an oxidative reagent. Carrot pomace is identified with lignocellulosic material, and the chemical composition was determined in the 100 g dry pomace as cellulose (10.0 g), hemicellulose (5.73 g), and lignin (2.5 g) and neutral detergent fiber (18.24 g). Polysaccharide fractions of carrot pomace were found to be 12.21–43.77–130.49 and 186.48 mg GalA/g cell wall material of pomace for water soluble pectins fraction, chelator soluble pectins fraction, diluted alkali-soluble pectins fractions and the total, respectively. The finest microfibrils amount was 3% for carrots, and those were thinner than 15 nm. In the case of cellulose isolated from carrot pomaces, the Segal crystallinity index ranged at 68.73%, the crystallinity degree was 56.51%, and the largest crystallites of celluloses belonged to carrot among the other plants in the study (Szymanska-Chargot et al. 2017).

Turkoglu (2017) evaluated the protein content of the carrot and its skins along with potatoes. The raw materials were freeze-dried, and the protein content of the whole carrot and its skin was determined to be 7.741% (dry basis, db) and 9.593% (wb). The effects of protein isolate powders were evaluated in the batter of traditional "lokma" sweets. The moisture content of the water activity was within secure limits. The oil retention was decreased, and the panelists found the product acceptable for consumption (Turkoglu 2017).

Due to their dietary fiber, carrot pulp has also been investigated in Turkish Delight production along with carrot and orange pulp. The study aimed to determine the possibility of using biowaste as an ingredient, increasing the fiber and the nutrients while decreasing the sugar content of the delight. The samples were evaluated for physical, microbiological, chemical, and textural properties at different storage periods. All types of added pulps have improved nutritional value, dietary fiber, and process properties, and they have reduced sugar. However, the samples adding 6 and 9% of carrot pulp had higher acceptability (Hanoglu 2022).

Another study evaluated the possible use of off-grade apple and carrot pomace in producing fruit sheets and must. Varying amounts of apple and carrot pomace mixtures were tested, and the results showed that the formula with 80% apple and 20% carrot was the best option for the suitable fruit sheets. The process parameters such as moisture content, pH, water activity, brix value, ash value, HMF, total phenolic content, antioxidant value and total sugar value were followed and to be able to enrich the formulation and decrease the sugar level of lemon peel pomegranate and grape seed are also added to the formula. The studied parameters show that the formula with carrot pomace and other ingredients not only helped to improve the properties but also created a functional food with an alternative way to use by-products (Yıldırım 2022).

Fermented left-overs of black carrots are another by-product of carrots and were evaluated for the traditional shalgam drink process. The residue was dried and added to the tarhana formulation, a traditional fermented product in Turkey. The formula for adding the 1,2,3, and 5% black carrot powder has been examined regarding microbiology, physiochemistry, and sensorial parameters. The results imply that water and foam holding and emulsion capacity were increased. The mineral content of the formula improved with the addition of carrot pomace in terms of Mg, K, Ca, P, Na, and S. In tarhana, 38 different flavor components were determined and ethyl hexanoate was found with the highest amount. Adding 5% of black carrot pomace resulted in the most preferred color, smell, taste, acidity, texture, and overall acceptability in tarhana (Altiparmak 2023).

Carrot by-products are also evaluated for health-focused purposes. In a study, black carrot and its by-products are rich in polyphenols; therefore, these have been tested for modulating the inflammatory response in tumor necrosis factor "treated endothelial cells after gastrointestinal digestion and in a co-culture of intestinal Caco-2 and endothelial EA.hy926 cell model." The results showed that anthocyanins and phenolic acids of polyphenol-rich black carrots may exert their effects by inhibiting the inflammatory cascade in endothelial cells (Kamiloglu et al. 2017).

The health industry can also benefit from carrot by-products due to their rich therapeutic compounds. Carotenoid content in carrots is a very important component in health and nutrition. These valuable materials, which can remain in the pomace in high amounts could otherwise be considered waste. Pressurized liquid extraction (PLE) can extract the carotenoids from the carrot by-products. Mustafa et al. (2012) evaluated the extraction process and found that temperature and extraction time were two important factors in the yield of carotenoids. The study showed that, along with a dispersing agent, after 20 min of extraction, 80% of the carotenoids can be recovered.

5 Valorization of By-Products for Non-Health Purposes

Daucus carota is a plant that has been used in many ways as a vegetable, juice, or ingredient for meals. As shown in previous parts of this text, the by-products could be used for many different purposes, such as food applications. This section will mention the use of carrot by-products for non-health purposes.

Carrot by-products have been examined for soap production. In the study of Łusiak et al. (2024), carrot and apple pomace have been used in various amounts of 5–15% in soap formulations. Properties such as cutting force, hardness, and penetration work were found to be improved compared to the control group, especially at the molecular level.

Converting the carrot waste into 2,3-butanediol (BDO) with a microbial path can be another way to reduce biowaste and improve value-added products. In a study, *Paenibacillus polymyxa* DSM 365 was used to valorize the carrot waste to 2,3-butanediol (BDO). The study showed that 100 g of dried carrot discard and the microbial agent, 18.8 g BDO, 2.5 g ethanol, and 2.1 g acetoin could be obtained. Since there was no need for hydrothermal pre-treatment, the profitability could increase (López-Linares et al. 2023).

Sharma and Kumar (2017) mentioned using carrot residues in different fields. Occasionally, carrot pomace was used as pig feed mixed in the regular diet at 20–25% w/w. The effects of feeding carrots and apples and their by-products to piglets during weaning were studied. The impact on intestinal microbiota and morphology, inflammation markers, diarrhea incidence, and oxidative stress was evaluated, and essential components, pectins and polyphenols, were considered to enhance intestinal health. Not just as a nutritious alternative for piglets but as a part of the circular economy in the European Union and the world, these by-products are inexpensive and natural sources of valuable compounds (Pistol et al. 2024)

Furthermore, some studies mentioned that carrot waste is an appropriate additive for cow feed due to its metabolizable energy. Compared to the effects of apple and citrus by-products on milk production, carrot by-products were found to be more effective for increasing metabolizable energy than in affecting milk production. Dried pomaces of the apple, white grape, pomegranate, aronia, red grape, orange, beetroot, carrot, and tomato were evaluated for short-term fermentation characteristics by adding them to a cow diet. Carrot pomace increased the gas formation due to its digestible organic matter amount (Giller et al. 2022).

Rabbit diets were also tested with the addition of dried carrots. The waste of carrots was added to the diet of the rabbits by up to 50% but showed no negative effects on animal health, growing parameters, or economic efficiency. Some studies compared the tomato (6% w/w) and carrot waste (50% w/w) in rabbit diets to determine the contributions of the wastes to enriching the nutritional value. At the end of the study tomato helped the overall acceptability of the cooked meat and carrot had no negative impact (Eliopoulos et al. 2022).

The food industry by-product carrot pulp was examined for silage production. Five different formulations with different additions were tested for organic matter, i.e., digestibility, nutritional value, and metabolic energy level. The results showed that all the silage formulations with carrots were found to be "high quality" and "good quality," according to the Fleig scores (Kandemir 2022). Another study examined barley shochu, carrot, mandarin, and orange juice residues for silage development. The silages' nutritional value was higher than the controls due to the addition of crude fiber and protein (Onodera et al. 1997).

Argentinean cockroaches, yellow mealworms, black soldier flies, and house crickets are insects that can be used as food or feed. Therefore, their diets, which can support them in growing faster than average, could be considered economically important. In a study, these four insects were fed with different by-products such as beet molasses, potato steam peelings, spent grains , beer yeast, bread and cookie remains. Carrot was used as a water source. Considering the purpose of the carrot usage in this study, the pomace was also re-evaluated for a similar purpose in another study. The results of this study proved that the insects can grow with the by-products effectively for both human and animal consumption (Oonincx et al. 2015).

Different by-products from the food industry were evaluated for pigment production. Black carrot, apple, pomegranate, and red beet pulps were used as raw material along with filamentous fungi *Aspergillus carbonarius*. Fungal pigment production parameters (fermentation conditions, incubation period, pulp particle size, initial substrate pH) have been evaluated for optimal process conditions, and melanin was found to be the main product (Arikan et al. 2020).

By-products also can be evaluated as an essential tool for biopreservation. The potential of carrot, melon, and tomato were used in a study as substrates for lactic acid fermentation. Lactobacillus strains, including *Lactobacillus plantarum*, *L. paracasei*, *L. rhamnosus*, *L. genus*, and *L. casei* were used for the fermentation. The fermented by-products were extracted and tested as an antimicrobial agent for pathogenic *Escherichia coli*, *Salmonella* spp., *Staphylococcus aureus*, *Bacillus cereus*, and *Listeria monocytogenes*. All the extracts showed high antimicrobial activity against all the pathogens except *E. coli*, which was affected less. So, vegetable by-products were effective natural antimicrobials from natural sources, especially for food preservation (Ricci et al. 2019).

Carrot pomace has also been evaluated for biofuel production. After hydrolyzation of the carrot pomace, the sample was fermented to ethanol using *Saccharomyces cerevisiae*. The study examined various types of acids and concentrations and found the most effective combination as 0.76% sulfuric acid when the ethanol value reached 2.16 g/L. Also this study found that pH 5 was the level at which the highest ethanol was reached (1.81 g/L). The determination of the most productive inoculation rate resulted in 20% (v/v) and the highest ethanol production (3.55 g/L) reached when 45.80% (w/v) carrot pomace was used (Siyakus 2013).

Biogas production has also been evaluated using solid carrot juice waste. The study assessed pH, C/N ratio, hydraulic retention time, and waste. At the end of the study for the fed-batch and semi-continuous systems, the daily biogas production was 2.3 L and 2.1 L, respectively. Also, there was an inverse proportion between hydraulic retention and biogas production efficiency (Gülsen 2014).

In another study, carrot pulp was evaluated for bioethanol production. For fermentation *Saccharomyces cerevisiae*, *Pichia stipites* and for identification *Candida boidinii* were used. When the raw material was used in the amount of 100–150 g/L the highest ethanol production was determined. Adding peptone, yeast extraction and different inorganic components to the medium also increased the ethanol yield. The highest bioethanol concentration for *S. cerevisiae, P. stipitis, and C. boidinii* was 27.48 g/L at 48th hour and 11.32 g/L at 96th hour, respectively (Çakır 2022).

The ashes from different plants (including carrot) were added into the formulation of glaze along with transparent frit, potassium feldspar, sodium feldspar, ulexite, iron colored with copper oxide, cobalt oxide, tin oxide, and manganese oxide. Glaze formulations were applied on vitrified clay at a temperature of 1200 °C. The results succeeded for all the multiplied trials and three-dimensional applications (Ereshova 2021).

Carrot pomace is also being explored for use in vegetable-based bioplastics. In a study, carrot pomace, a waste product, can be transformed entirely into biofilms. The material's structure and configuration were identified using techniques such as NMR, FTIR, DMTA, and DSC. The mechanical properties of the material were excellent. The composition of the bioplastic films included cellulose (52%), pectin (43%), hemicellulose (3%), and aliphatic polyesters (4%). Pectin and sugars acted as plasticizers, with pectin enhancing the material's flexibility. These process parameters could be applied to various scales of manufacturing environmentally friendly products (Perotto et al. 2020).

6 Conclusion

The food processing sector produces billions of tons of waste each year, including inedible components like pods, leaves, seeds, and husks, creating environmental and economic issues due to their exclusion from market standards. The FAO reports that this waste, encompassing processed and raw food, is frequently disposed of through landfilling or composting. In 2019 alone, 931 million tons of food waste was recorded, leading to approximately $1 trillion in economic losses. Tackling this problem is vital for sustainable development, aligning with environmental conservation and poverty alleviation objectives. The EU's circular economy strategy advocates reducing, reusing, and recycling agricultural by-products to prolong resource life and develop new products. Based on the general review conducted in this chapter and recent studies, nutrient-rich by-products such as carrot roots, peels, leaves, and pomace can be utilized in various applications, including as food or animal feed, compost, fertilizer, natural dye, cosmeceuticals, pet food, crafting and arts, biofilms, additives across different sectors. These include uses in drugs, ceramics, and food products to provide therapeutic effects, enrich nutritional values, extend shelf life, create new formulations for innovative applications, and serve as raw materials for biodiesel production. Technological advances facilitate their conversion into valuable products, benefitting society, the economy, and the environment.

References

Agcam E, Akyıldız A, Balasubramaniam VM (2017) Optimization of anthocyanins extraction from black carrot pomace with thermosonication. Food Chem 237:461–470

Ahmad T et al (2019) Phytochemicals in *Daucus carota* and their health benefits-review article. Foods 8(9):1–22. https://doi.org/10.3390/FOODS8090424

Altiparmak D (2023) Utilisation of fermented residual carrot powders in tarhana production. Graduate School of Natural and Applied Sciences, Nigde Omer Halisdemir University

Ambroizak Z (1998) Health benefit and practical aspects of high fiber diets. Am J Clin Nutr 595:1242–1247

Arikan EB et al (2020) Production of bio-based pigments from food processing industry by-products (apple, pomegranate, black carrot, red beet pulps) using Aspergillus carbonarius. J Fungi 6(4):1–18. https://doi.org/10.3390/jof6040240

Arslan Bayrakçı H (2020) The usage of carrot and chickpea flour in production of highly nutritious gluten free pasta. Necmettin Erbakan University

Bao B, Chang KC (1994) Carrot pulp chemical composition, colour and water-holding capacity as affected by blanching. J Food Sci 59:1159–1161

Bas-Bellver C et al (2023) Physicochemical, technological and functional properties of upcycled vegetable waste ingredients as affected by processing and storage. Plant Foods Hum Nutr 78(4):710–719. https://doi.org/10.1007/s11130-023-01114-1

Çakır DN (2022) Determination of bioethanol production capacity of carrot pomaces. Graduate School of Natural and Applied Sciences, Ankara University. Available at: www.aging-us.com

Chantaro P, Devahastin S, Chiewchan N (2008) Production of antioxidant high dietary fiber powder from carrot peels. Let Food Sci Tech 41(10):1987–1994

Chen BH, Tang YC (1998) Processing and stability of carotenoid powder from carrot pulp waste. J Agric Food Chem 46:2312–2318

Chiboub W et al (2019) Valorization of the green waste from two varieties of fennel and carrot cultivated in Tunisia by identification of the phytochemical profile and evaluation of the antimicrobial activities of their essentials oils. Chem Biodivers 16(1). https://doi.org/10.1002/cbdv.201800546

Ching LS, Mohamed S (2001) Alpha-tocopherol content of 62 edible tropical plants. J Agric Food Chem 49:3101–3105

Clementz A, Torresi PA, Molli JS, Cardell D, Mammarella E, Yori JC (2019) Novel method for valorization of by-products from carrot discards. LWT Food Sci. Technol 100:374–380

da Silva Dias J (2014) Nutritional and health benefits of carrots and their seed extracts. Food Nutr Sci 5:2147–2156

Eliopoulos C et al (2022) Reintegration of food industry by-products: potential applications. Foods 11(22). https://doi.org/10.3390/foods11223743

Encalada AMI, Pérez CD, Flores SK, Rossetti L, Fissore EN, Rojas AM (2019) Antioxidant pectin enriched fractions obtained from discarded carrots (*Daucus carota* L.) by ultrasound-enzyme assisted extraction. Food Chem 289:453–460

Encalada AMI, Perez CD, Rossetti L, Rojas AM, Fissore EN (2020) Carrot pectin enriched fraction as a functional additive : Antioxidant and gelling effects in a model spreadable chia oil-in-water emulsion. Food Hydrocoll 108:106037

Ereshova N (2021) Investigation of the use of lambed plant (carrot, celery, potato, garlic, onion, turnip, sgar beet, Jerusalem artichoke, ginger) ashes in the glaze (1200 C). Anadolu University

Esposto BS et al (2022) TPP-chitosomes as potential encapsulation system to protect carotenoid-rich extract obtained from carrot by-product: A comparison with liposomes and chitosomes. Food Chem 397(August). https://doi.org/10.1016/j.foodchem.2022.133857

FAO (2021) Food and Agriculture Organization, Unep Food Waste Index Report 2021

FAO (2024) Food and Agriculture Organization, Carrot Production Report 2022

Ferreira MSL et al (2015) Formulation and characterization of functional foods based on fruit and vegetable residue flour. J Food Sci Technol 52(2):822–830. https://doi.org/10.1007/s13197-013-1061-4

Ferreira R et al (2021) Evaluation of fatty acids profile as a useful tool towards valorization of by-products of agri-food industry. Foods 10(11). https://doi.org/10.3390/foods10112867

Freepik (2023) Carrot, www.freepik.com. Available at: www.freepik.com.

Garcia ML et al (2002) Utilization of cereal and fruit fibres in low fat dry fermented sausages. Meat Sci 60(3):227–236

Gayas B, Shukla RN, Khan BM (2012) Physico-chemical and sensory characteristics of carrot pomace powder enriched defatted soyflour fortified biscuits. Int J Sci Res Publ 2(8):1–5

Giller K et al (2022) In vitro ruminal fermentation, methane production and nutrient degradability as affected by fruit and vegetable pomaces in differing concentrations. J Anim Physiol Anim Nutr 106(5):957–967. https://doi.org/10.1111/jpn.13656

Gonçalves EM, Pinheiro J, Abreu M, C.L. (2010) Carrot (*Daucus carota* L.) peroxidase inactivation, phenolic content and physical changes kinetics due to blanching. J Food Eng 97:574–581

Guerra L, Romanini D, Lopez S, Castelli V, Clementz A (2023) Upcycling of carrot discards into prebiotics (fructooligosaccharides) as high value food ingredients. Food Bioprod Process 138:172–180

Gull A, Prasad K, Kumar P (2015) Effect of millet flours and carrot pomace on cooking qualities, color and texture of developed pasta. LWT Food Sci Technol 63(1):470–474. https://doi.org/10.1016/j.lwt.2015.03.008

Gülsen HE (2014) Investigation of biogas production conditions from solid waste of apple and black carrot juice concentrate. Mersin University

Hanoglu A (2022) Investigation of the usage of carob, orange and carrot pulps in Turkish delight production. Ataturk University

Hernández-Ortega M et al (2013) Microwave dried carrot pomace as a source of fiber and carotenoids. Food Nutr Sci 04(10):1037–1046. https://doi.org/10.4236/fns.2013.410135

Horbowicz M, Kosson R, Grzesiuk A, Bski HD (2008) Anthocyanins of fruits and vegetables—their occurrence analysis and role in human nutrition. Veg Crops Res Bull 68:5–22

Hussein AMS et al (2015) Utilization of some fruits and vegetables by-products to produce high dietary fiber jam. Food Sci Qual Manag 37(2007):39–45. Available at: www.iiste.org

Jagtap SA, Adsule RN, Chavan UD (2000) Utilization of carrot pomace for preparation of toffees. Beverage Food World 27:30–32

Jayesree N, Kay P, Priyangaa A, Prasad N, Nagasundara R, Turki MSA, Charis MG, Wei C (2021) Chemosphere Valorisation of carrot peel waste by water-induced hydrocolloidal complexation for extraction of carotene and pectin. Chemosphere 272:129919

Kaisangsri N, Kowalski RJ, Wijesekara I, Kerdchoechuen O, Laohakunjit N, Ganjyal GM (2016) Carrot pomace enhances the expansion and nutritional quality of corn starch extrudates. LWT Food Sci Technol 68:391–399

Kamiloglu S et al (2017) Anti-inflammatory potential of black carrot (*Daucus carota* L.) polyphenols in a co-culture model of intestinal Caco-2 and endothelial EA.hy926 cells. Mol Nutr Food Res 61(2):1–11. https://doi.org/10.1002/mnfr.201600455

Kandemir AF (2022) Determination of digestibility and metabolizable energy values of carrot pulp silage by in-vitro gas method. Department of Animal Science, Eskişehir Osmangazi University

Kim SO (2023) Review of food upcycling in South Korea: regulation, limitation, and prospects. Food Sci Biotechnol 32(1):1–10. https://doi.org/10.1007/s10068-022-01184-y. Springer Nature Singapore

Kırbaş Z, Kumcuoglu S, Tavman S (2019) Effects of apple, orange and carrot pomace powders on gluten-free batter rheology and cake properties. J Food Sci Technol 56(2):914–926. https://doi.org/10.1007/s13197-018-03554-z

Kohajdová Z, Karovičová J, Jurasová M (2012) Influence of carrot pomace powder on the rheological characteristics of wheat flour dough and on wheat rolls quality. Acta Sci Pol Technol Aliment 11(4):381–387

Kumar N, Kumar K (2011) Development of carrot pomace and wheat flour based cookies. J Pure Appl Sci Technol 1(1):5–11

Kumar K, Kumar N (2012) Development of vitamin and dietary fibre enriched carrot pomace and wheat flour based buns. J Pure Appl Sci Technol 2(1):107–115

Kumari S, Grewal RB (2007) Nutritional evaluation and utilization of carrot pomace powder for preparation of high fiber biscuits. J Food Sci Technol 44(1):56–58

Lila MA (2004) Anthocyanins and human health: an in vitro investigative approach. J Biomed Biotechnol 5:306–313

López-Linares JC et al (2023) Carrot discard as a promising feedstock to produce 2,3-butanediol by fermentation with P. polymyxa DSM 365. Bioengineering 10(8). https://doi.org/10.3390/bioengineering10080937

Łusiak P et al (2024) Evaluation of physical parameters and spectral characterization of the quality of soaps containing by-products from the food industry. Sci Rep 1–10. https://doi.org/10.1038/s41598-024-54794-3. Nature

Mishra P, Bhatt D (2018) Evaluation of nutritional and physico-chemical characteristics of dietary fibre enriched with carrot pomace pasta. Int J Pharm Sci Res 9:4487–4491

Moure A, Cruz JM, Franco D, Dominguez JM, Sineiro J, Dominguez H, Nunez MJ, Carlos Parojo JC (2001) Natural antioxidants from residual sources. Food Chem 72:145–171

Mustafa A, Trevino LM, Turner C (2012) Pressurized hot ethanol extraction of carotenoids from carrot by-products. Molecules 17(2):1809–1818. https://doi.org/10.3390/molecules17021809

Nawirska A, Kwasniewska M (2005) Dietary fiber fractions from fruit and vegetable processing waste. Food Chem 91:221–225

Nguyen VT, Scarlett CJ (2016) Mass proportion, bioactive compounds and antioxidant capacity of carrot peel as affected by various solvents. Technologies 4(4):36

Nocolle C, Cardinault N, Aprikian O, Busserolles J, Grolier P, Rock E, Demigne C, Mazur A, Scalbert A, Amouroux P, Remesy C (2003) Effect of carrot intake on cholesterol metabolism and antioxidant status in cholesterol fed rats. Eur J Nutr 42:254–261

Onodera R et al (1997) Preparation of silages consisting of barley shochu distillery byproduct and pulps (juice residues) of Mandarin oranges and carrots. Bull Fac Agric Miyazaki Univ 43(2):145–150

Oonincx DGAB et al (2015) Feed conversion, survival and development, and composition of four insect species on diets composed of food by-products. PLoS One 10(12):1–20. https://doi.org/10.1371/journal.pone.0144601

Özcan MM, Chalchat JC (2007) Chemical composition of carrot seeds (*Daucus carota* L.) cultivated in Turkey: Characterization of the seed oil and essential oil. Grasas Aceites 58:359–365

Perotto G, Simonutti R, Ceseracciu L, Mauri M, Besghini D, Athanassiou A (2020) Water-induced plasticization in vegetable-based bioplastic films: A structural and thermo-mechanical study. Polymer 200:122598

Pérez P et al (2023) Effect of ultrasound and high hydrostatic pressure processing on quality and bioactive compounds during the shelf life of a broccoli and carrot by-products beverage. Foods 12(20). https://doi.org/10.3390/foods12203808

Pistol GC, Pertea AM, Taranu I (2024) The use of fruit and vegetable by-products as enhancers of health status of piglets after weaning: the role of bioactive compounds from apple and carrot industrial wastes. Vet Sci 11(1):1–21. https://doi.org/10.3390/vetsci11010015

Ricci A et al (2019) Vegetable by-product lacto-fermentation as a new source of antimicrobial compounds. Microorganisms 7(12). https://doi.org/10.3390/microorganisms7120607

Salari S, Ferreira J, Lima A, Sousa I (2024) Effects of particle size on physicochemical and nutritional properties and antioxidant activity of apple and carrot pomaces. Foods 13:710

Schieber A (2017) Side streams of plant food processing as a source of valuable compounds: selected examples. Annu Rev Food Sci Technol 8:97–112

Schieber A, Stintzing FC, Carle R (2001) By-products of plant food processing as a source of functional compounds—recent developments. Trends Food Sci Technol 12:401–405

Šeregelj V et al (2021) New concept of fortified yogurt formulation with encapsulated carrot waste extract. LWT 138:110732. https://doi.org/10.1016/j.lwt.2020.110732

Sharma HK, Kumar N (2017) Utilization of carrot pomace. In: Food processing by-products and their utilization. Wiley, Hoboken, pp 207–229. https://doi.org/10.1002/9781118432921.ch10

Sharma KD, Karki S, Thakur NS, Attri S (2012) Chemical composition, functional properties and processing of carrot—a review. J Food Sci Technol 49(1):22–32. https://doi.org/10.1007/s13197-011-0310-7

Shyamala Jr BN, Jamuna P (2010) Nutritional content and antioxidant properties of pulp waste from *Daucus carota* and beta vulgaris. Malays J Nutr 16(3):397–408

Singh B, Panesar PS, Nanda V (2006) Utilization of carrot pomace for the preparation of a value added product. World J Dairy Food Sci 1(1):22–27

Singh BK, Koley TK, Maurya A, Singh PM, Singh B (2018) Phytochemical and antioxidative potential of orange, red, yellow, rainbow and black coloured tropical carrots (*Daucus carota* subsp . *sativus* Schubl . & Martens). Physiol Mol Biol Plants 24:899–907

Siyakus S (2013) Ethanol production from carrot pomace. Ankara University Graduate School of Natural and Applied Sciences Department of Biology

Smigielski KB, Majewska M, Kunicka-Styczyñska A, Gruska R (2014) The effect of ultrasound-assisted maceration on the bioactivity, chemical composition and yield of essential oil from waste carrot seeds (*Daucus carota*). J Essent Oil Bear Plants 17(6):1075–1086

Solyom K et al (2020) Effect of temperature and moisture contents on dielectric properties at 2.45 GHz of fruit and vegetable processing by-products. RSC Adv 10(28):16783–16790. https://doi.org/10.1039/c9ra10639a

Staniszewska M et al (2005) Essential oils of wild and cultivated carrots—the chemical composition and antimicrobial activity. J Essent Oil Res 17(5):579–583

Stoll T, Schweiggert U, Schieber A, Carle R (2003a) Application of hydrolyzed carrot pomace as a functional food ingredient to beverages. J Food Agric Environ 1:88–92

Stoll T, Schweiggert U, Schieber A, Carle R (2003b) Process for the recovery of a carotene-rich functional food ingredient from carrot pomace by enzymatic liquefaction. Innov Food Sci Emerg 4:415–423

Szymanska-Chargot M et al (2017) Isolation and characterization of cellulose from different fruit and vegetable pomaces. Polymers 9(10). https://doi.org/10.3390/polym9100495

Tang YC, Chen BH (2000) Pigment change of freeze-dried carotenoid powder during storage. Food Chem 69(1):11–17

Tanska M, Zadernowsk R, Konopka I (2007) The quality of wheat bread supplemented with dried carrot pomace. Pol J Nat Sci 22:126–136

Tasbasi K (2022) The effect of the addition of carrot puree on physical, chemical and sensory properties of probiotic ice cream and vitality of bifidobacterium BB-12. Atatürk University

Turkoglu T (2017) The isolation of plant based proteins from carrot and potato including their wastes and their utilization in the preparation of a sample food. Ege University

Turksoy S et al (2011) Effect of black carrot (*Daucus carota* L. ssp. *sativus* var. atrorubens Alef.) fiber addition on the composition and quality characteristics of cookies. J Food Agric Environ 9(3–4):57–60

Upadhyay A, Sharma HK, Sarkar BC (2008) Characterization and dehydration kinetics of carrot pomace. Agric Eng Int 10:1–9

Vaz AA et al (2022) Physicochemical properties and bioaccessibility of phenolic compounds of dietary fibre concentrates from vegetable by-products. Foods 11(17). https://doi.org/10.3390/foods11172578

Vodnar DC, Calinoiu LF, Dulf FV, Stefanescu BE, Crisan G, Socaciu C (2017) Identification of the bioactive compounds and antioxidant, antimutagenic and antimicrobial activities of thermally processed agro-industrial waste. Food Chem 231:131–140

Wan YJ, Hong T, Shi HF, Yin JY, Koev T, Nie SP, Gilbert RG, Xie MY (2021) Probiotic fermentation modifies the structures of pectic polysaccharides from carrot pulp. Carbohydr Polym 251:117116

Yadav S et al (2018) Effect of wheat bran and dried carrot pomace addition on quality characteristics of chicken sausage. Asian Australas J Anim Sci 31(5):729

Yıldırım E (2022) Investigation on producing sugar-free dier fruit sheets (pestil) from off-grade apple and black carrots must and pomace added with powdered lemon peels and pomegranate and grape fruit seeds. Karamanoğlu Mehmetbey University

Ying D et al (2021) Nutrient-dense shelf-stable vegetable powders and extruded snacks made from carrots and broccoli. Foods 10(10). https://doi.org/10.3390/foods10102298

Zhang D, Hamauzu Y (2004) Phenolic compounds and their antioxidant properties in different tissues of carrots (*Daucus carota* L.). J Agric Food Environ 2:95–100

Bioactive Phytochemicals from Celeriac (*Apium graveolens* var. *rapaceum*) By-products

Samy Mohamed Galal and Nashwa Fathy Sayed Morsy

Contents

Abstract Celeriac (*Apium graveolens* L. var. *rapaceum*) is a versatile vegetable produced worldwide. It is known as root celery or turnip-rooted celery. It is a popular vegetable in Europe and North Africa. The edible part of the plant is its root. It has a unique flavor and texture when cooked. Celeriac is an essential source of natural phytochemicals such as flavonoids, phenolic acids, vitamins, dietary fibers, antioxidants, and essential oils. These bioactive compounds have gained attention due to their potential medicinal properties. Therefore, celeriac and its by-products can be used as value-added products in developing functional foods, nutraceuticals, or natural health products with potential health-promoting effects. The by-products can be utilized in non-health-related applications, including biodegradable packaging materials, animal feed ingredients, and compost for gardening.

S. M. Galal · N. F. S. Morsy (✉)
Food Science Department, Faculty of Agriculture, Cairo University, Giza, Egypt
e-mail: asadgalal@agr.cu.edu.eg; nanafsm@agr.cu.edu.eg

M. F. Ramadan (ed.), *Bioactive Phytochemicals in By-products from Leaf, Stem, Root and Tuber Vegetables*, https://doi.org/10.1007/978-3-031-80700-8_11

Keywords Celeriac · Celery root · By-products · Valorization · Bioactive components

Abbreviations

DPPH	2,2-diphenyl picrylhydrazyl
FRAP	Ferric-reducing antioxidant power
GC-MS	Gas chromatography-mass spectrometry
HPLC	High-performance liquid chromatography
TEAC	Trolox equivalent antioxidant capacity
TPC	Total phenolic content

1 Introduction

Celery is a member of the Apiaceae or Umbelliferae family. It exhibits a distinct flavor. It is one of the annual or perennial plants that grow throughout Europe and the tropical and subtropical regions of Africa and Asia (Gauri et al. 2015). There are three main subspecies of *A. graveolens*: leaf celery (*Apium graveolens* L. subsp. *secalinum*), stalk celery (*Apium graveolens* L. subsp. *dulce*), and root celery, also known as celeriac (*Apium graveolens* L. subsp. *rapaceum*). Celeriac is characterized by enlarged hypocotyl (Quiros 1993). Stalk celery and celeriac are consumed often as vegetables globally, whereas leaf celery is commonly cultivated and consumed in East Asian countries (Turner et al. 2021). It is known as "superfood" (Che et al. 2021). Celeriac has fine-grained white flesh and tastes much like a stalk of celery. Raw celeriac is a brittle, fractural plant with mild and specific flavor (Ilic et al. 2021). It is a perishable material with a pleasant aroma and prone to discoloration, which may influence the sensory quality and reduce consumer acceptance (Nurkhoeriyati et al. 2021). This vegetable's distinctive and characteristic taste and smell result from aldehydes, esters, terpenes, low molecular sugars, organic acids, and flavonoids, especially phthalides such as sedanolide and 3-*n*-butylphthalide, in the celeriac composition.

Celeriac can be eaten raw in salads, as cooked vegetables, in soups, or in a processed form such as celery salt made from a combination of salt and celeriac extract (Husband et al. 2011) (Fig. 1). The European Union produced 490,080 tons of celeriac in 2019 (Eurostat 2020).

2 Chemical Composition and Bioactive Compounds of Extracts from Celeriac By-products

Krešić et al. (2004) reported that one hundred grams of raw celeriac is composed of 12.24 g dry matter, 2.7 g total sugars, 0.68 g reducing sugars, 1.92 g non-reducing sugars, 6 mg ascorbic acid, 73.94 mg Ca, 3.84 mg Na, 308.33 mg K, and 52.01 mg

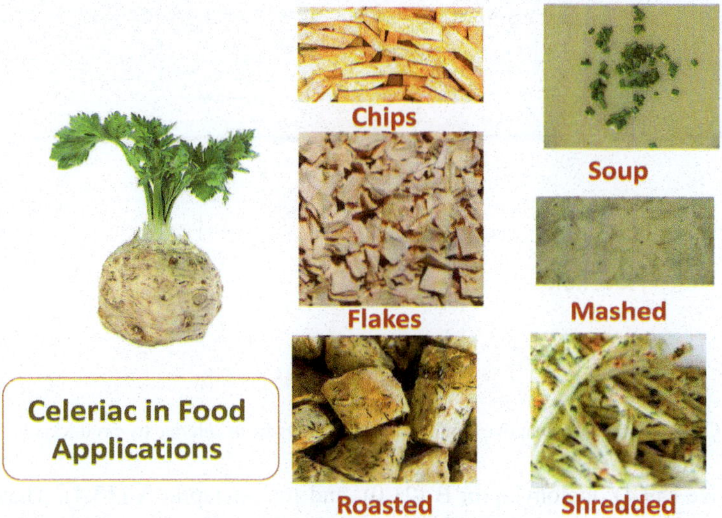

Fig. 1 Food applications of celeriac

P. Meanwhile, one hundred grams of edible portion of celeriac was found to contain 12 g dry matter, 9.2 g carbohydrate, 1.5 g protein, 0.3 g fat, 1 g ash, 300 mg K, 115 mg P, 100 mg Na, 43 mg Ca, 20 mg Mg, 0.7 mg Fe, 0.33 mg Zn, 0.158 mg Mn, 0.07 mg Cu, and 0.7 μg Se (Kosson et al. 2014; USDA 2019). Leclerc et al. (1991) found that the levels of vitamins B1 (thiamine), B2 (riboflavin), B6 (pyridoxine), niacin, and pantothenic acid in celeriac (DM, 9.29%) were 36, 47, 96, 629, and 269 μg/100 g fresh weight, respectively.

Celeriac is a good source of total dietary fiber, trace minerals, and phenolic compounds (Minarovičová et al. 2018). Celery root grown in the North Caucasus is a source of biologically active compounds: essential oil (2.04%), flavonoids in terms of rutin (1.44%), lipophilic substances (4.05%), carotenoids (3.03%), free organic acids (2.38%), tannins (8.06%), water-soluble polysaccharides (8.33%), pectin substances (3.45%), hemicellulose A (1.54%), and hemicellulose B (1.47%) (Sarmanovna 2019) (Fig. 2).

Nikolic et al. (2011) grouped Serbian celery root samples into three sizes: 55, 130, and 450 g. They found that the celery root ethanolic extract's total phenolic content (TPC) ranged from 265.4 to 368.5 μmol of chlorogenic acid/g dry extract. The highest TPC was recorded for the smallest root-size samples. High-Performance Liquid Chromatography (HPLC) analysis of the extracts indicated that the main components were genkwanin glucosides (30.0–35.8%), kaempferol glucosides (17.1–24.2%), and quercetin glucosides (2.23–17.8%).

Kaiser et al. (2013) reported that the TPC of the aqueous methanol extract of peeled celeriac was 123.4 mg GAE/100 g DM. HPLC-MS analysis showed that the major phenolic compounds (mg/100 g DM) of celeriac were apiin (180.7), diosmetin-apiosylglucoside/chrysoeriol-apiosylglucoside B (74.7), caffeic acid

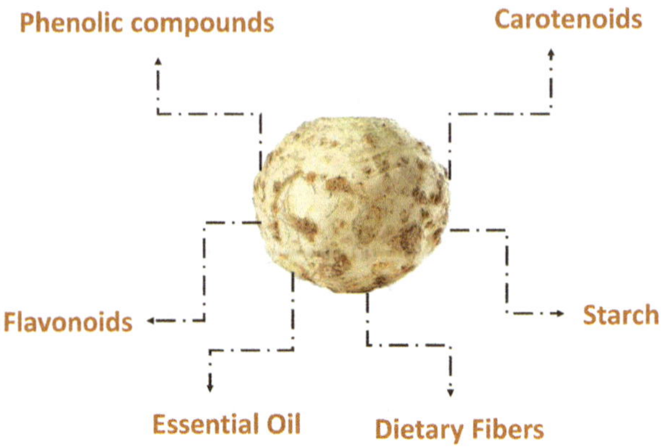

Fig. 2 Composition and bioactive compounds of extracts from celeriac by-products

derivative (59.4), malonylapiin B (21.0), and malonylapiin A (13.4). They found that steam-blanching peeled celeriac at 100 °C for 3 min caused a significant ($p < 0.05$) increase in malonylapiin A and malonylapiin B and a significant ($p < 0.05$) decrease in apiin.

Radziejewska-Kubzdela et al. (2014) studied the effect of washing (4 °C, 2 min) or soaking (4 °C, 10 min) in the tap water of shredded peeled celeriac roots grown in Poland on TPC during 12 days of storage at 4 °C. The shredded celeriac was packaged in plastic bags in air or a modified atmosphere containing 2/10/88 kPa O_2/CO_2/N_2. Storage of pre-treated samples (washed or soaked) for 1 day did not significantly ($p > 0.05$) affect the flavonoid content but decreased TPC by about 25%, regardless of the gas composition in the bag. The highest TPC (0.191 mg chlorogenic acid equivalent/g product) after 12 days of storage was noticed in non-pre-treated shredded celeriac packaged in the modified atmosphere.

Christiaens et al. (2015) evaluated celeriac steam peels as a source of pectin with unique functional properties. The cell wall material of the waste celeriac peels was isolated as alcohol-insoluble residue (AIR) using ethanol and acetone, followed by the extraction of water-soluble pectin (WSP) from boiling water. The total pectin content was 159 ± 5 mg/g AIR dry matter, and the galacturonic acid (GalA) content was 114 ± 2 mg/g AIR dry matter, while the GalA content in WSP reached 111 ± 2 mg/g. The isolated pectin is considered high-methoxylated pectin (Degree of methylation, DM > 50%) since the DM of WSP was 71 ± 3 and the DM of AIR was 55 ± 2.

Priecina et al. (2018) investigated the effect of steaming and drying techniques (convective and microwave-vacuum (MV)) on the biologically active components of peeled sliced celeriac roots grown in Latvia. Steam-blanching of samples before drying was performed at 95 ± 1 °C for 1.5 and 3 min. Drying in the convective dryer was carried out with controlled air circulation speed (1.2 ± 0.1 m/s) at 45 ± 1 °C for 12 h. Drying with an MV dryer was conducted for fresh, steamed for 1.5 min and 3 min samples at 35 ± 1 °C. Drying processes were carried out till moisture content was <10%. The major organic acids in the peeled celeriac roots were succinic acid (7597.84 mg/100 g DW), citric acid (5465.71 mg/100 g DW), malonic acid (3363.54 mg/100 g DW), and oxalic acid (806.75 mg/100 g DW). Carotenoid, phenolic, and flavonoid contents in the fresh celeriac samples were 5.77 ± 0.74 mg β-carotene equivalent/100 g DW, 466.63 ± 15.24 mg GAE/100 g DW and 348.28 ± 7.10 mg catechin equivalent/100 g DW, respectively. HPLC-DAD was used to determine individual phenolic compounds. Catechin (17.89 ± 1.78 mg/100 g DW), gallic acid (1.55 ± 0.15 mg/100 g DW), and sinapic acid (1.11 ± 0.11 mg/100 g DW) represented the major phenolic compounds in the fresh celeriac samples.

Carotenoid content decreased after convective drying by 89.4%, while after the MV drying process and using steam-blanching, carotenoids decreased by 72.4, 87.5, and 90.1%. These decreases in total carotenoids could be attributed to the oxidation process during drying and steam-blanching processes.

Flavonoid content decreased by 66.3% after convective drying, by 61.1% after MV drying, and by 64.1 and 64.5% after steam-blanching, followed by MV drying compared to fresh one. On the other hand, TPC in the samples increased significantly ($p < 0.05$) after processing compared with fresh celery roots. Their content increased by 82.9% using convective drying, while MV processing increased by 56.7, 56.5, and 18.7%, respectively.

Kulig et al. (2019) found that the HPLC analysis of the ethanolic extract of dried grounded root of *Apium graveolens* var. *rapaceum* was characterized by chlorogenic acid (123.8 μg/g), rutin (37.1 μg/g), trans-ferulic acid (28.8 μg/g), gallic acid (17.6 μg/g), and apigenin (7.5 μg/g).

Furanocoumarins are toxic chemicals produced in Apiaceae plants and some other plant families to provide antimicrobial and insecticidal activity. They have promising therapeutic prospects (Ahmed et al. 2020). The total furanocoumarins (psoralen, bergapten, xanthotoxin, and isopimpinellin) content ranged from 0.97 to 23.47 mg/kg fresh celeriac (Schulzová et al. 2012).

Radziejewska-Kubzdela et al. (2014) found that furanocoumarin content decreased by ~ 40% in the minimally processed (washed or soaked) celeriac after the first day of storage in air or modified atmosphere plastic bags.

Sanahuja et al. (2021) found that the TPC of the celery root samples grown in Spain ranged from 14.6 to 33.2 mg GAE/100 g FW. Nicetin et al. (2022) reported that the TPC of an aqueous extract of fresh celery root (*Apium graveolens* L. subsp. *rapaceum*), grown in Serbia, was 3.009 ± 0.001 mg/ml. HPLC-DAD of the extracts indicated that apigenin was the most dominant flavonoid (700.32 mg/kg), followed by kaempferol (600.26 mg/kg), chrysin (198.81 mg/kg), ρ-coumaric acid (81.54 mg/kg) luteolin (69.54 mg/kg), and caffeic acid (55.46 mg/kg).

Macleod and Ames (1989) extracted the volatile constituents of peeled celeriac cubes for 3 h with distilled H_2O using the Likens and Nickerson apparatus and 2-methylbutane before vacuum concentration. Gas Chromatography–Mass Spectrometry (GC-MS) indicated that these constituents consisted of monoterpene hydrocarbons (24.8%) and phthalides (37.4%). Major components were limonene (14.9%), 3-butyl-4,5-dihydrophthatide (sedanenolide) (13.4%) followed by trans-3-Butyl-3 α,4,5,6-tetrahydrophthalide (i.e., trans-sedanolide) (7.3%), cis-3-butyl-3 α,4,5,6-tetrahydrophthalide (i.e., cis-sedanolide) (5.3%), β-pinene (4.4%), α-terpinene (3.0%), and γ-terpinene (1.6%). They assessed the isolates sensorially after chromatographic separation using the gas chromatography-odor port assessment (GC-OPA) technique. They described the odor of fresh celeriac extract as strongly celery-like and possessing slight buttery and caramel overtones. Sanahuja et al. (2021) analyzed the volatile components in the crushed celery root samples by headspace solid-phase microextraction-GC-MS. The main volatile compounds in the samples were limonene (28.6%), γ-terpinene (23.9%), followed by β-cymene (13.5%), caryophyllene (9.9%), and β-pinene (6.7%). They attributed celery root's characteristic citrus and herbal aroma to terpenoids, sesquiterpenes, and monoterpenes.

Kesbiç (2023) extracted essential oil from dried celeriac leaves by hydrodistillation using a Clevenger apparatus. GC-MS analysis indicated that the identified volatile constituents were phthalide (3-isobutylidene) (49.42%), fenipentol (28.45%), and D-limonene (6.54%).

3 Biological and Functional Properties of Extracts and Bioactive Compounds from By-products

Celeriac is a rich source of biologically active phthalides. They play roles in the central nervous system and cardiac performance, aiding in anti-thrombotic modulation and protecting against cerebral ischemia and high blood pressure (Lin et al. 2005).

The antioxidant activity of celeriac is due to the presence of polyphenols, which are responsible, i.e., for free radicals scavenging, the ability to absorb single oxygen or transition metal chelation (Luiz et al. 2002; Jaworska et al. 2022).

Mühlbauer et al. (2003) reported that feeding rats a semi-purified diet mixed with air-dried powdered celeriac at a dosage of 1 g/day/rat, corresponding to 76.8 g/kg diet, suppressed significantly bone resorption and reduced the risk of osteoporosis.

Nikolic et al. (2011) found that the 2,2-diphenyl picrylhydrazyl (DPPH·) radical scavenging activity (EC_{50} value) of the phenolic extracts of Serbian celery root ranged between 2.41 and 3.14 mg/ml. The lowest EC_{50} value (the highest antioxidant activity) was recorded for the phenolic extracts of higher-sized roots (~450 g). Kaiser et al. (2013) found that the antioxidant activity of celeriac methanol extract, measured by ferric-reducing antioxidant power (FRAP) and Trolox equivalent antioxidant capacity (TEAC) assays, was 0.5 (mmol Trolox equivalent/kg DM) and 3.7 (mmol Trolox equivalent/kg DM), respectively. The DPPH antiradical activity of Latvian celeriac extract was 125.71 ± 0.59 μmol Trolox equivalent/100 g DW (Priecịna et al. 2018). Aboody (2021) extracted dried celery root with 70% ethanol using a cold maceration. The root extract exhibited high antioxidant activity against DPPH radicals (IC_{50} = 39.71 ± 0.035 μg/ml) compared to the IC_{50} value of ascorbic acid, 79.89 ± 7.89 μg/ml. The investigated extract demonstrated antimicrobial activity (zone of inhibition, mm) against gram-positive bacteria (*Bacillus aerogenes*, *B. coagulans*, *B. megaterium*, *B. subtilis*, *Lactobacillus leichmannii*, and *Staphylococcus aureus*) and gram-negative bacteria (*Kleibsella pneumoniae*, *Pseudomonas aeruginosa*, *Salmonella typhi*, and *Shigella* strains) extended between 6 and 16 mm. The extract possessed antifungal activity (8–11 mm) against fungal strains (*Aspergillus niger*, *A. flavus*, *Candida albicans*, *Cryptococcus neoformans*, *Trichophyton rubrum*).

Jaworska et al. (2022) extracted juice from celeriac and evaluated its quality and antioxidant activity. The celeriac plant, grown in Poland, was washed, peeled, and pulped before incubation with pectinase (10 mg/100 g of pulp) at pH 4.0 and 25 °C for 30 and 60 min. The juice was pressed out of the pulp using a hydraulic press. Control samples were prepared from unpeeled celeriac. The total soluble solids (TSS) of celeriac juices ranged from 7.2 to 8.3 °Brix. HPLC analysis indicated that celeriac juices' fructose, glucose, and sucrose levels were 2.8–5.7 g/l, 14.9–24.8 g/l, and 18.8–22.0 g/l, respectively. The total polyphenols of the juices ranged from 93 to 117 mg GAE/l. Twenty phenolic compounds were identified in the celeriac juices using ultra-performance liquid chromatography coupled with photodiode array-mass spectrometry detectors. The DPPH radical scavenging activity in celeriac juices ranged from 587 to 1108 mol Trolox/l. The juices' ferric-reducing antioxidant power (FRAP) ranged from 123 to 193 mol Trolox/l. The enzyme treatment enhanced the extraction of bioactive components. The juices enzymatically produced from unpeeled celeriac pulp were characterized by higher TSS, sugar levels, and antioxidant activity against DPPH radicals.

Kesbiç (2023) found that the celeriac leaf essential oil (CEO) concentration required to inhibit 50% of DPPH radicals (IC_{50}) was 30.52 ppm, close to that of vitamin C (29.93 ppm), indicating that CEO has good antioxidant activity.

Kolarovic et al. (2009) investigated the effect of celery root juice on the activities of xanthine oxidase (XOD), catalase (CAT), peroxidase (Px), glutathione peroxidase (GSHPx), and reduced glutathione (GSH) in blood hemolysate and liver homogenate of male rats (250–300 g). Rats were fed the standard laboratory diet *ad libitum*. Animals from the control group were drinking only water *ad libitum*. Celery juice (5%) was prepared by grounding celery root in a blender with distilled water. Animals from the juice-tested group drank only celery root juice instead of water for 14 days. Activities of XOD, CAT, Px, GSHPx, and GSH in liver homogenate were increased ($p < 0.05$), while activities of CAT and Px in blood hemolysate were decreased significantly ($p < 0.05$) by exclusively drinking celery root juice. Results indicated the antioxidant and hepatoprotective effects of celery root extract.

According to Bauermeister et al. (2009), one of central Europe's most common food allergies is an allergy to celery root. There are three known allergens in celeriac. Api g 1 is the major celeriac allergen, while Api g 4 is a minor allergen. Api g 5 belongs to flavoproteins. Following European Directive 2007/68/EC (which amends Directive 2000/13/EC), prepackaged foods in the EU must list celeriac on their labels. Husband et al. (2011) studied the impact of thermal and high-pressure (HP) processing on the immunoreactivity of the allergens in celeriac tissue. The allergenicity of celeriac and its allergen Api g 1 were both successfully decreased by the HP and heat processing combination.

Ranjbar et al. (2019) evaluated the anti-inflammatory effect of celery root ethanol dried extract at 100, 200, and 400 mg/kg doses in mice. The total flavonoid content of the extract was 0.0625 mg quercetin/g extract. They induced inflammation in the mice by subcutaneous injection of 0.02 ml of formaldehyde 2.5% after 30 min of the investigated extract's intraperitoneal injection (i.p.). Dexamethasone (15 mg/kg, i.p.) and normal saline were consumed in the positive and negative control groups. Caliper was used to measure the changes in paw thickness (mm). The extract showed maximum effectiveness at a concentration of 200 mg/kg. Lee et al. (2021) investigated the effect of celeriac extract at 1, 10, and 100 μg/ml on the production of pro-inflammatory cytokines such as Tumor necrosis factor α (TNF-α), Interleukin 6 (IL-6), Interleukin-1 β (IL-1β), and nitric oxide (NO) by lipopolysaccharide (LPS)-stimulated RAW 264.7 cells. Celeriac extract inhibited the production of TNF-α, IL-6, and NO without cytotoxicity.

Ahmed et al. (2022) investigated the anti-proliferative properties of ethanolic extract of Egyptian-grown celery root and its silver nanoparticles (AgNPs) against human cancer cell lines, including liver (HepG-2), colon (Caco-2), and breast (MCF-7) cancer cell lines. The root ethanolic extract suppressed the development of HepG-2, Caco-2, and MCF-7 tumor cells with IC_{50} values of 11.58, 7.13, and 6.58 μg/ml, respectively. In the same order, the AgNPs exhibited a greater cytotoxic effect (IC_{50} = 1.37, 1.37, and 2.36 μg/ml) on the examined cell lines.

4 Food and Non-food Applications of Extracts and Bioactive Compounds from By-products

4.1 Food Applications

Stoop and Pharr (1992) extracted mannitol dehydrogenase from celeriac root by fractionation with $(NH_4)_2\ SO_4$, followed by chromatography on a Fractogel DEAE column before concentration with $(NH_4)_2\ SO_4$. They used the extracted enzyme to catalyze the NAD-dependent oxidation of mannitol to mannose instead of fructose or glucose. HPLC identified the produced mannose.

Kaiser et al. (2013) prepared pasty celeriac by blanching peeled celeriac slices at 100 °C for 5, 7, and 10 min before mincing. The color characteristics of pasty celeriac were measured using a CR-300 chroma meter. During mincing, the unheated control sample turned brown due to peroxidase (POD) and polyphenol oxidase (PPO) activities. Thermal heating of celeriac slices before mincing decreased significantly ($p > 0.05$) POD and PPO activities caused an increase in the L^* value and gained lightness in the pasty celeriac products.

Minarovičová et al. (2018) studied the effects of wheat flour replacement at different levels (5, 7.5, 10, and 20%) by celery root powder (CRP) on rheological parameters and quality properties of enriched pasta. The CRP was prepared by washing, peeling, and slicing celery root before drying the by-product from juice extraction at room temperature. The chemical composition of CRP as a dietary fiber powder showed that it contained 91.61% dry matter, 7.72% protein, 1.15% fat, 5.77% ash, 62.19% total dietary fiber, 14.11% soluble dietary fiber, and 48.08% insoluble dietary fiber. Incorporating CRP increased water absorption, prolonged dough stability and development time, and reduced optimal cooking time. Moreover, cooked pasta with CRP was characterized by a brown color and a pleasant, acceptable vegetable flavor and taste.

Ivanova et al. (2020) improved the prepared yogurt's physicochemical and sensory characteristics and nutritional value by adding celery pectin as soluble dietary fiber. The fresh celery tubers (*Apium graveolens* var. *rapaceum*), Netherlands origin, were washed, peeled, chopped, and dried at 40 °C. Pectin was obtained with ammonium oxalate. Yogurt was prepared with standardized milk samples at 3.6% fat content. The extracted pectin was incorporated into the milk at 0.1, 0.2, and 0.3% concentrations. Sucrose was added at a level of 6% before pasteurization. The yield of the isolated pectin was 29%, with moisture and ash contents of 5.5% and 3.2%, respectively. The extracted pectin was characterized as high-methoxylated pectin with a degree of esterification of 75.8%, acetylation of 1.8%, and anhydrouronic acid content of 65.9%. The effect of the incorporation level of high-methoxylated pectin on the syneresis of yogurt was examined. Increasing the pectin level to 0.2% reduced significantly ($p < 0.05$) the quantity of separated serum from 25% (control sample) to 6.5%. The highest sensory scores were recorded for yogurt prepared with 0.2% celery pectin.

Kesbiç (2023) studied the effect of celeriac leaf essential oil (CEO) on the oxidative stability of fish oil during thermal oxidation. The fish oil samples enriched with CEO at 0.1, 0.5, 1, and 3% were subjected to thermal oxidation at 70 ± 0.5 °C for 24 h with continuous ventilation. Progress of oxidation was assessed with peroxide value determination. Results indicated that enriched fish oil with CEO at the investigated levels increased significantly ($p < 0.05$) the oxidative stability of fish oil.

4.2 Non-food Applications

Kowalska et al. (2020) evaluated the effects of soil amendment with dried celeriac leaves on the pathogenic bacteria *Burkholderia cepacia* (the most harmful soil-borne pathogen of onions). The leaves were crushed and added to the infested soil at 0.5, 1.0, and 2.0% (w/v) in the pot experiments. Adding the dried celeriac leaves at 0.5% or higher eliminated the pathogenic bacteria after 2 months of incubation. They attributed this effect to the biologically active compounds released from the decomposition of celeriac leaves, such as flavonoids and furanocoumarins.

Güroy et al. (2023) studied the effects of celeriac extracts on growth performance, immune response, and disease resistance against vibriosis caused by *Vibrio anguillarum* in European seabass (*Dicentrarchus labrax*). The dried leaves of celeriac, of Turkish origin, were extracted with 40% methanol before lyophilization. A commercial European seabass diet was formulated with celeriac extract at 0, 0.01, 0.05, and 0.1 g/100 g instead of wheat milling quality. European seabass fish were fed *ad libitum* with the experimental diets three times a day for 1 month. Feeding a seabass fish diet enriched with 0.05% celeriac extract increased the specific growth rate. At the end of the feeding experiment, 11 fish were injected intraperitoneally with the gram-negative marine bacterium *V. anguillarum*, and mortality was recorded during the following 8 days. The survival rate in the fish group fed the diet with 0.1% celeriac extract was greater than in the control group.

5 Valorization of By-products for Non-health Purposes

Kowalska et al. (2020) reported that the bioactive substances released from the decomposing celeriac leaves may support the beneficial microorganisms required to colonize the soil niche rapidly. Razavi et al. (2023) extracted celeriac essential oil with microwave hydrodistillation and used it as a green ionophore to prepare a polyvinyl chloride (PVC) membrane sensor to detect and determine Co^{+2} ion. The membrane that was produced with a PVC/acetophenone as a plasticizer/celeriac ionophore/sodium tetra-phenyl borate ratio of 33.0%/66.3%/2.0%/1.7% exhibited the best response characteristics to measure cobalt ion concentrations with 0.03×10^{-4} M detection limit.

References

Aboody MSA (2021) Cytotoxic, antioxidant, and antimicrobial activities of celery (*Apium graveolens* L.). Bioinformation 17(1):147–156. https://doi.org/10.6026/97320630017147

Ahmed S, Khan H, Aschner M, Mirzae H, Küpeli Akkol E, Capasso R (2020) Anticancer potential of furanocoumarins: Mechanistic and therapeutic aspects. Int J Mol Sci 21(16):5622. https://doi.org/10.3390/ijms21165622

Ahmed SST, Fahim JR, Youssif KA, Amin MN, Abdel-Aziz HMH, Khadra IA, Rateb ME, Abdelmohsen UR, Hamed ANE (2022) Comparative study of the chemical composition and anti-proliferative activities of the aerial parts and roots of *Apium graveolens* L. (celery) and their biogenic nanoparticles. S Afr J Bot 151:34–45. https://doi.org/10.1016/j.sajb.2021.11.002

Bauermeister K, Ballmer-Weber BK, Bublin M, Fritsche P, Hanschmann KMO, Hoffmann-Sommergruber K, Lidholm J, Oberhuber C, Randow S, Holzhauser T, Vieths S (2009) Assessment of component-resolved in vitro diagnosis of celeriac allergy. J Allergy Clin Immunol 124(6):1273–1281. https://doi.org/10.1016/j.jaci.2009.07.033

Che DN, Shin JY, Kim HR, Cho BO, Kang HJ, Oh H, Kim Y-S, Jang SI (2021) Citric acid and enzyme-assisted modification of flavonoids from celery (*Apium graveolens*) extract and their anti-inflammatory activity in HMC-1.2 cells. J Food Biochem 45(7):e13774. https://doi.org/10.1111/jfbc.13774

Christiaens S, Uwibambe D, Uyttebroek M, van Droogenbroeck B, van Loey AM, Hendrickx ME (2015) Pectin characterisation in vegetable waste streams: A starting point for waste valori-

sation in the food industry. LWT Food Sci Technol 61(2):275–282. https://doi.org/10.1016/j.lwt.2014.12.054
Eurostat (2020) Crop production in EU standard humidity. Eurostat, Luxembourg
Gauri M, Ali SJ, Khan MS (2015) A review of *Apium graveolens* (Karafs) with special reference to Unani medicine. Int Arch Integr Med 2(1):131–136
Güroy D, Güroy B, Bilen S, Kenanoğlu ON, Şahin İ, Terzi E, Karadal O, Mantoğlu S (2023) Effect of dietary celery (*Apium graveolens*) on the growth performance, immune responses, and bacterial resistance against *Vibrio anguillarum* of European seabass (*Dicentrarchus labrax*). Fish Physiol Biochem 49(1):75–95. https://doi.org/10.1007/s10695-022-01158-1
Husband FA, Aldick T, van der Plancken I, Grauwet T, Hendrickx ME, Skypala I, Mackie AR (2011) High-pressure treatment reduces the immunoreactivity of the major allergens in apple and celeriac. Mol Nutr Food Res 55(7):1087–1095. https://doi.org/10.1002/mnfr.201000566
Ilic J, Tomasevic I, Djekic I (2021) Influence of boiling, steaming, and sous-vide on oral processing parameters of celeriac (*Apium graveolens var. rapaceum*). Int J Gastron Food Sci 23:100308. https://doi.org/10.1016/j.ijgfs.2021.100308
Ivanova M, Petkova N, Todorova M, Dobreva V, Vlaseva R, Denev P, Hadjikinov D, Bouvard V (2020) Influence of citrus and celery pectins on physicochemical and sensory characteristics of fermented dairy products. Sci Study Res Chem Chem Eng Biotechnol Food Ind 21(4):533–545
Jaworska G, Szarek N, Hanus P (2022) Effect of celeriac pulp maceration by *Rhizopus sp.* pectinase on juice quality. Molecules 27(23):8610. https://doi.org/10.3390/molecules27238610
Kaiser A, Hartmann KI, Kammerer DR, Carle R (2013) Evaluation of the effects of thermal treatments on color, polyphenol stability, enzyme activities and antioxidant capacities of innovative pasty celeriac (*Apium graveolens* L. var. *rapaceum* (Mill.) DC.) products. Eur Food Res Technol 237:353–365. https://doi.org/10.1007/s00217-013-1998-6
Kesbiç O (2023) Protective effect of celeriac (*Apium graveolens*) leaf essential oil on temperature and oxygen-induced fish oil oxidation. Kastamonu Univ J Eng Sci 9(1):10–16. https://doi.org/10.55385/kastamonujes.1298104
Kolarovic J, Popovic M, Mikov M, Mitic R, Gvozdenovic L (2009) Protective effects of celery juice in treatments with doxorubicin. Molecules 14(4):1627–1638. https://doi.org/10.3390/molecules14041627
Kosson R, Anyszka Z, Grzegorzewska M, Golian J, Kohut M, Badełek E (2014) Postharvest quality of celeriac [*Apium graveolens* L. var. *rapaceum* (Mill.) Gaud.] depending on weed control methods. Prog Plant Protect 54(1):77–83. https://doi.org/10.14199/ppp-2014-014
Kowalska B, Szczech M, Smolińska U, Maciorowski R (2020) Application of waste material from celeriac as a control strategy for *Burkholderia cepacia* in soil. Crop Protect 130:105052. https://doi.org/10.1016/j.cropro.2019.105052
Krešić G, Lelas V, Šimundić B (2004) Effects of processing on nutritional composition and quality evaluation of candied celeriac. Sadhana 29:1–12
Kulig D, Matysiak M, Baldovska S, Štefániková J, Maruniaková N, Mňahončáková E, Árvay J, Galbavý D, Kolesárová A (2019) Screening of polyphenolic compounds from traditional medicinal herbs. J Microbiol Biotechnol Food Sci 9:487–491. https://doi.org/10.15414/jmbfs.2019.9
Leclerc J, Miller ML, Joliet E, Rocquelin G (1991) Vitamin and mineral contents of carrot and celeriac grown under mineral or organic fertilization. Biol Agric Hortic 7(4):339–348. https://doi.org/10.1080/01448765.1991.9754564
Lee J-H, Jeong H-J, Park J-S (2021) Effect of celeriac extract on the LPS-induced production of pro-inflammatory cytokines by RAW 264.7 cells. J Korea Converg Soc 12(2):295–300. https://doi.org/10.15207/JKCS.2021.12.2.295
Lin G, Chan SS, Chung H, Li S (2005) Chemistry and biological activities of naturally occurring phthalides. Stud Nat Prod Chem 32:611–669. https://doi.org/10.1016/S1572-5995(05)80065-1
Luiz M, Biasutti A, Garcia NA (2002) Micellar effect on the scavenging of singlet molecular oxygen by hydroxybenzenes. Redox Rep 7(1):23–28. https://doi.org/10.1179/135100002125000145
MacLeod G, Ames JM (1989) Volatile components of celery and celeriac. Phytochemistry 28(7):1817–1824. https://doi.org/10.1016/S0031-9422(00)97866-X

Minarovičová L, Lauková M, Kohajdová Z, Karovičová J, Dobrovická D, Kuchtová V (2018) Qualitative properties of pasta enriched with celery root and sugar beet by-products. Czech J Food Sci 36(1):66–72. https://doi.org/10.17221/242/2017-CJFS

Mühlbauer RC, Lozano A, Reinli A, Wetli H (2003) Various selected vegetables, fruits, mushrooms and red wine residue inhibit bone resorption in rats. J Nutr 133(11):3592–3597. https://doi.org/10.1093/jn/133.11.3592

Nicetin M, Pezo L, Pergal M, Lončar B, Filipovic V, Knežević V, Demir H, Filipovic J, Manojlović D (2022) Celery root phenols content, antioxidant capacities and their correlations after osmotic dehydration in molasses. Foods 11:1945. https://doi.org/10.3390/foods11131945

Nikolic N, Cvetkovic D, Todorovic ZB (2011) Characterization of content, composition, and antioxidant capacity of phenolic compounds in celery roots. Ital J Food Sci 23(2):214–219

Nurkhoeriyati T, Kulig B, Sturm B, Hensel O (2021) The Effect of pre-drying treatment and drying conditions on quality and energy consumption of hot air-dried celeriac slices: Optimisation. Foods 10(8):1758. https://doi.org/10.3390/foods10081758

Priecina L, Karklina D, Kince T (2018) The impact of steam-blanching and dehydration on phenolic, organic acid composition, and total carotenoids in celery roots. Innov Food Sci Emerg Technol 49:192–201. https://doi.org/10.1016/j.ifset.2018.01.008

Quiros CF (1993) Celery, *Apium graveolens* L. In: Kalloo G, Bergh BO (eds) Genetic improvement of vegetable crops. Pergamon, Oxford, pp 523–534

Radziejewska-Kubzdela E, Czapski J, Czaczyk K, Biegańska-Marecik R (2014) The effect of pretreatment and modified atmosphere packaging on contents of phenolic compounds and sensory and microbiological quality of shredded celeriac. J Sci Food Agric 94(6):1140–1148. https://doi.org/10.1002/jsfa.6381

Ranjbar AM, Vahidi A, Rezvani ME, Ramezani V, Boroumand M, Jahani Y (2019) Evaluation of anti-nociceptive and anti-inflammatory activities of *Apium graveolens* L. roots extract in mice. Res J Pharmacogn 6(3):69–75. https://doi.org/10.22127/rjp.2019.89467

Razavi R, Beitollahi H, Ahmadi SA, Salajegheh M (2023) Celeriac as ionphore for green Co^{+2} selective PVC membrane electrode by experimental and theoretical investigation. J Indian Chem Soc 100(10):101090. https://doi.org/10.1016/j.jics.2023.101090

Sanahuja AB, Landete MP, Martínez MID, Moya MSP, García AV (2021) Optimization of volatile compounds extraction from industrial celery (*Apium graveolens*) by-products by using response surface methodology and study of their potential as antioxidant sources. Foods 10(11):2664. https://doi.org/10.3390/foods10112664

Sarmanovna TZ (2019) Phytochemical study of odorous celery root (*Apium graveolens* L.) grown in the North Caucasus. Pharmacogn J 11(3):527–530. https://doi.org/10.5530/pj.2019.11.84

Schulzová V, Babička L, Hajšlová J (2012) Furanocoumarins in celeriac from different farming systems: a 3-year study. J Sci Food Agric 92(14):2849–2854. https://doi.org/10.1002/jsfa.5629

Stoop JM, Pharr DM (1992) Partial purification and characterization of mannitol: mannose 1-oxidoreductase from celeriac (*Apium graveolens* var. *rapaceum*) roots. Arch Biochem Biophys 298(2):612–619. https://doi.org/10.1016/0003-9861(92)90456-7

Turner L, Lignou S, Gawthrop F, Wagstaff C (2021) Investigating the factors that influence the aroma profile of *Apium graveolens*: A review. Food Chem 345:128673. https://doi.org/10.1016/j.foodchem.2020.128673

U.S. Department of Agriculture, Agricultural Research Service (USDA, ARS) (2019) Food Data Central. https://fdc.nal.usda.gov/fdc-app.html#/food-details/170400/nutrients

Bioactive Phytochemicals from Sweet Potato By-products

Ying Ping Chang and Kwan Kit Woo

Contents

Abstract In 2021, global sweet potato production hit 88.9 million metric tons, with China leading the pack, contributing 55% of the world's total. Sweet potatoes are a highly adaptable crop, capable of thriving in marginal growing conditions with a wide production geography, short production cycles, and high yield potential. It is the most efficient staple food crop, with 14–26 metric tons per hectare yield and higher nutrient density than cereals. However, the cultivation and processing of sweet potatoes generate significant organic waste in the form of solid and liquid by-

Y. P. Chang (✉)
Department of Agricultural and Food Science, Faculty of Science, Universiti Tunku Abdul Rahman, Jalan Universiti, Bandar Barat, Kampar, Perak, Malaysia
e-mail: changyp@utar.edu.my

K. K. Woo
Lee Kong Chian Faculty of Engineering and Science, Universiti Tunku Abdul Rahman, Bandar Sungai Long, Kajang, Malaysia
e-mail: wookk@utar.edu.my

M. F. Ramadan (ed.), *Bioactive Phytochemicals in By-products from Leaf, Stem, Root and Tuber Vegetables*, https://doi.org/10.1007/978-3-031-80700-8_12

products, including peelings, trimmings, leaves, stems, and stalks. These by-products contain macronutrients such as carbohydrates, proteins, and bioactive compounds or phytochemicals such as phenolics, anthocyanins, and carotenoids that can be extracted or used in various downstream processes and products. This chapter reviews the valuable compounds found in sweet potato peels, leaves, stems, and stalks and methods for extracting bioactive compounds and other valuable ingredients. It also discusses the functional properties of the extract and bioactive compounds and their applications in food and non-food products and explores opportunities for their utilization for non-health-related purposes. Finally, the chapter provides research gaps for further investigation in the valorization of sweet potato by-products.

Keywords Bioactive · Leaves · Peels · Phytochemicals · Sweet potato

1 Sweet Potato Production and Utilization

Sweet potato, an ancient crop, has long been recognized as an essential dietary energy source. Alongside cereals (such as rice, wheat, and corn), legumes (including soybeans, dry beans, and peas), and other tubers (such as cassava and potato), sweet potato is among the most significant food crops. It thrives in tropical and subtropical climates, even in nitrogen-poor soil, and can reach full maturity in as little as 4 months with minimal crop supervision (de Albuquerque et al. 2019). Compared to cereals, sweet potato is the most efficient staple food crop, yielding 14–26 metric tons per hectare while offering higher nutrient density (Truong et al. 2018). The edible storage root of sweet potato can sprout new plants, and this herbaceous perennial can also be propagated through stem cuttings.

The sweet potato plant comprises three primary parts: leaves, which conduct photosynthesis; stems and stalks, which transport energy to the roots and minerals and water from the root to the rest of the plant; and a root system, which absorbs nutrients and water from the soil, provides anchorage for the plant, and acts as an energy storage site by developing fleshy storage roots (Fig. 1). The plant's vines grow horizontally on the ground and stretch up to five meters long. There are three types of roots: storage, fibrous, and pencil, each serving distinct purposes and having varied development paths (Huaman 1992). The storage roots range from spindle-shaped to almost spherical, in length from a few centimeters to greater than 30 cm, and in weight from 0.1 kg to several kilograms. The skin and flesh contain carotenoid and anthocyanin pigments, which, when combined, determine a spectrum of colors ranging from whitish to yellow, orange, red, and purple in different sweet potato cultivars.

China is the leading producer of sweet potatoes, accounting for 55% of the world's total production, with global production reaching 88.9 million metric tons in 2021 (Statista 2024). Approximately 53% of the total sweet potato supply is utilized as food or seed, 40% as feed, and 7% is wasted (Akoetey et al. 2017).

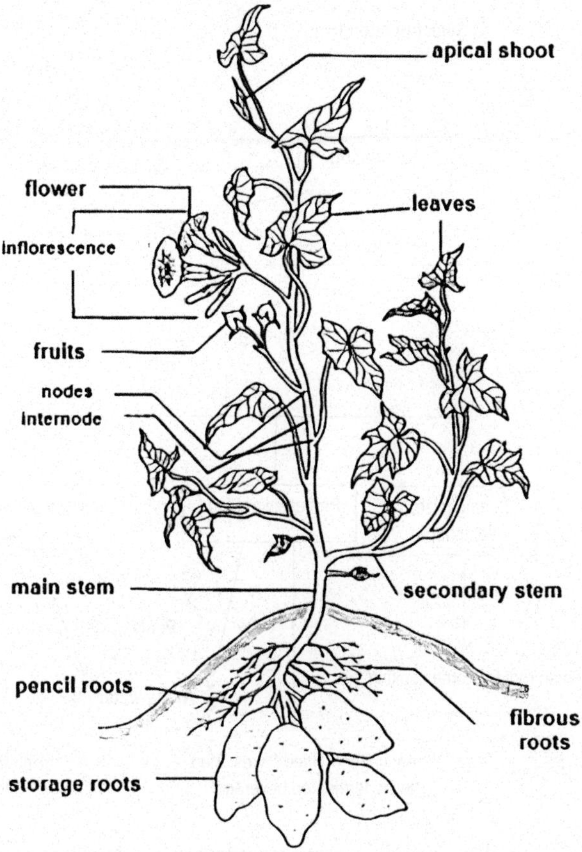

Fig. 1 Sweet potato plant parts (Huaman 1992) (retrieved from https://www.sweetpotatoknowledge.org/wp-content/uploads/2016/01/Systematic-Botany-and-Morphology-of-the-Sweetpotato-Plant.pdf)

Sweet potato roots are highly nutritious, making them highly sought after. They are versatile in taste, texture, and flesh color, ranging from white, cream, yellow, and orange to purple. They can be used in various recipes for household and culinary purposes, and they can also be processed into various primary and value-added products. Figure 2 illustrates the common wastes generated from different streams of sweet potato processing. The commercial processing methods for sweet potato products can differ based on location and market. In certain provinces of China, approximately 15–20% of sweet potato production is transformed into starch, which is then utilized in the textile industry or used as a feedstock to manufacture glucose syrups, noodles, and organic acids. Another processing technique is producing dried sweet potato chips, which can be ground into flour to be used in other food products. Countries such as the United States, with a robust food processing infrastructure, utilize canning and freezing as major forms of processing. (Akoetey et al. 2017).

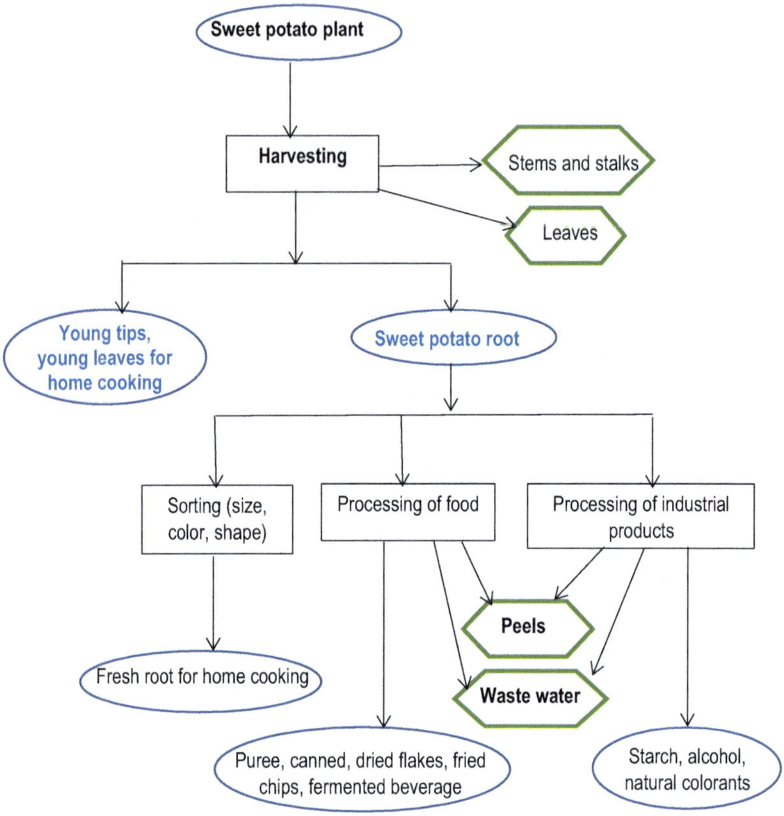

Fig. 2 Water and by-products generated (green hexagonal frame) during sweet potato harvesting and processing

Sweet potatoes generate by-products that vary based on the desired product and processing method during processing. Removal of vines, leaves, and roots that do not meet size grading requirements or are damaged during harvesting generates waste at the farm level. The food industry typically processes sweet potatoes into primary products, such as sweet potato flour or sweet potato purees, which can then be further processed into value-added products that are shelf-stable for year-round availability (Vithu et al. 2019). Ma (2019) reported that the demand for snack foods such as sweet potato dates, French fries, preserved sweet potato, and chips increased while the production of starch and other processed products from sweet potatoes was comparatively stable. Roots of all sizes and shapes can be utilized during primary processing, so the entire harvested crop is used, including the 30–40% off-grade from the fresh root markets. By-products generated during processing take the form of peels, trimmings, chunks of storage roots (sweet potato residue), and nutrient-rich wastewater. Approximately 6 cubic meters of water per ton of sweet potato roots are produced during the initial stages of starch production, and sweet potato proteins can be recovered from this wastewater for further utilization (Mu et al. 2017a).

The proportion of sweet potato plant mass can be categorized as follows: leaves constitute approximately 6–10%, stems contribute around 3–5%, while edible roots comprise most of the plant's mass at 80–90%. It is worth noting that the peels of sweet potatoes, in addition to being edible, are also a valuable source of nutrition and makeup around 5–10% of the root's total weight (Picha and Janick 2009). Sweet potato by-products, similar to the root itself, contain a variety of bioactive phytochemicals such as carotenoids, phenolic compounds, and terpenoids (Akoetey et al. 2017). These phytochemicals are known for their anticancer, anti-inflammatory, and antioxidant properties, making them highly beneficial for human health.

Given the potential health benefits of these bioactive phytochemicals, it would be worthwhile for scientists and entrepreneurs to explore their extraction and application. This chapter provides a comprehensive overview of the valuable compounds in sweet potato peels, leaves, stems, and stalks, as well as the methods used to extract them and their functional properties. Additionally, the chapter highlights the food and non-food applications of the extract and bioactive compounds derived from sweet potato by-products while exploring their potential for non-health-related purposes. Finally, recommendations are made for optimal extraction techniques and further utilization of sweet potato by-products.

2 Chemical Composition and Bioactive Compounds of Extracts from By-products

The valorization of sweet potato food processing by-products depends on their composition. The leaves, stems, and peel of sweet potatoes exhibit variations in composition due to factors such as cultivars, soil type, and environmental conditions. The macronutrient and selected micronutrient composition of these sweet potato by-products have been extensively studied and are presented in Tables 1 and 2. These values have been normalized based on dry matter and assuming that insoluble dietary fiber is equivalent to crude fiber.

Table 1 Nutritional composition of sweet potato by-products

Content (%)	Leaf	Stem and stalk	Dried peel	Dried residue
Ash	10.13–14.57	4.04–8.47	5.50–5.58	1.48–2.26
Crude protein	24.37–29.46	5.21– 8.93	4.32–5.78	1.69–4.10
Crude fat	2.56–6.82	1.79–3.25	1.08–3.61	0.19–1.43
Crude fiber	45.70–46.05	41.53–63.8	12.62–18.59	15.26–31.48
Reference	Ishida et al. (2000)	Ishida et al. (2000)	Ojokoh (2005); Agubosi et al. (2022)	Ju et al. (2017); Vannini et al. (2021)

Table 2 Micronutrient composition of sweet potato by-products

Content	Leaf	Stem and stalk	Dried peel	Dried residue
Calcium	1152.3–1449.6	745.2–1459.5	31.3–68.6	190.7–405.82
Potassium	2364.2–4953.4	1187.5–3521.5	11.44–38.35	196.39–531.29
Phosphorus	243.7–527.1	75.0–262.5	23.1–24.2	ND
Magnesium	612.4–708.6	169.7–558.6	5.77–10.37	80.2–139.77
Sodium	14.0–29.2	11.3–32.1	1.72–31.27	50.01–142.06
Copper	3.34–3.64	1.74–2.71	0.26–1.50	0.127–0.209
Zinc	3.95–6.86	1.36–4.19	0.79–1.11	0.314–0.533
Iron[a]	36.69–42.09	14.72–25.41	3.49–28.66	781.82–1817.69
Vitamin C	536.42–486.05	79.75–160.71	61.47–113.24	ND
Vitamin E	9.21–21.78	1.16–1.88	1.61–2.10	ND
Thiamin (B1)[a]	350.99–992.2	225.96–648.65	20.55–104.82	ND
Riboflavin (B2)[a]	1642.38–1968.99	225.96–558.56	1.36–2.27	ND
Carotene[a]	1807.95–3100.78	589.29–1785.28	2.10–1.61	ND
Reference	Ishida et al. (2000)	Ishida et al. (2000)	Agubosi et al. (2022); Marconato et al. (2020); Ojokoh (2005)	Ju et al. (2017)

ND not determined

[a]Values are expressed in μg/100 g while other values are expressed in mg/100 g

The leaves of sweet potatoes were found to contain relatively higher levels of ash (10.1–14.5%), protein (24.3–29.4%), and crude fiber (45.7–46.0%). In contrast, the stem and stalk contained higher levels of crude fiber (41.5–63.8%), lower protein content (5.21–8.93%), and lower ash content (4.04–8.47%) (Ishida et al. 2000). The peel, on the other hand, exhibited ash content (5.50–5.58%), crude protein (4.32–5.78%), and crude fat content (1.08–3.61%) similar to the stem and stalk but possessed relatively lower crude fiber content (12.6–18.5%) (Agubosi et al. 2022; Ojokoh 2005).

Both the peel and residue contain carbohydrates in the form of starch, sugar, and soluble fiber, such as pectin and hemicellulose, rather than cellulose, which is the primary component of crude fiber (Ju et al. 2017; Marconato et al. 2020; Vannini et al. 2021). The dried residue obtained after sweet potato starch extraction contained a significant amount of starch (47.83–65.49%) and total dietary fiber (21.31–27.27%) (Ju et al. 2017; Vannini et al. 2021). The low amount of protein (1.69–4.10%) in the dried residue is due to the highly soluble nature of sweet potato proteins, which dissolve and drain in the wastewater during sweet potato starch processing (Mu et al. 2017a). Interestingly, sweet potato protein is a unique, high-quality plant protein with a biological value of 72, higher than that of soybean (64) and peanut (59) proteins (Mu et al. 2017a). The high-quality protein and significant levels of total dietary fiber in different sweet potato by-products provide opportunities for valorization, which we will discuss in more detail.

Potassium and calcium are the predominant macrominerals found in all sweet potato by-products, among the inorganic matter. Calcium concentrations ranged from 1152.3–1449.6 mg/100 g in the leaves, 745.2–1459.5 mg/100 g in the stem and stalk, 190.7–405.82 mg/100 g in dried peels, and 190.7–405.8 mg/100 g in the dried residue. Meanwhile, potassium concentrations ranged from 2.36 to 4.95 mg/g in the leaves, 1.18–3.52 mg/g in the stem and stalk, 196.3–531.2 mg/100 g in dried residue, and 11.4–38.3 mg/100 g in dried peel. Sweet potato leaves are a great source of phosphorus, magnesium, and bioactive compounds like polyphenols and anthocyanins. They also contain vitamin C, B1 and B2, and carotene, a precursor to vitamin A. Research by Sun et al. (2014) confirms that sweet potato leaves from most cultivars provide important minerals like protein, potassium, phosphorus, calcium, magnesium, iron, and copper. In many Asian and African countries, sweet potato leaves are consumed as green vegetables due to their nutritional value and availability in sweet potato plantations.

To maintain good health, it is essential to have a balanced diet that includes a variety of macronutrients and micronutrients. For example, potassium is vital in maintaining fluid and electrolyte balance in body cells. If there is insufficient intake of potassium from the diet, it can lead to hypokalemia, which can cause life-threatening conditions in severe cases. Anti-nutrients such as phytic acid, cyanide, tannins, and oxalate are present in sweet potato leaves but at low levels and can be reduced through conventional cooking methods like boiling or blanching (Almazan 1995). Thus, consuming sweet potato leaves as part of a varied and balanced diet is unlikely to pose significant health risks related to anti-nutritional factors. Sweet potato leaves are a highly nutritious food source, and their consumption as part of a balanced diet can provide numerous health benefits. Thus, several publications have extensively reviewed the edible and health-promoting values of sweet potato leaves, including Alam (2021), Nguyen et al. (2021), and Sun et al. (2014).

Sweet potato by-products have gained increasing interest from researchers and the food industry due to their nutrient and micronutrient-rich content and their valuable bioactive phytochemicals such as carotenoids, phenolic compounds, and terpenoids. Among these phytochemicals, various health-promoting properties have been identified, making them a promising area of research. In 2020, Makori and colleagues evaluated four selected cultivars of sweet potato to determine the content of total phenolic (TPC), total flavonoid (TFC), and total anthocyanin (TAC), as well as the individual composition of these phytochemicals and amino acids, in the leaves, stalk, stem, peel, and flesh. The highest level of phytochemicals was found in the leaves, but the content varied depending on different factors such as location, cultivar, environmental conditions, and extraction method.

Colorimetric analysis is a common analytical method used to quantify TPC, TFC, and TAC. This method uses a suitable chromogen and a standard curve constructed using commercially available phenolic compounds like gallic and chlorogenic acids. TPC is typically expressed in milligrams of gallic acid equivalent

(GAE) or chlorogenic acid equivalent (CAE) per 100 g of dried weight (DW) or fresh weight (FW) or based on the weight of the extract. TFC is expressed in milligrams of quercetin equivalent (QE) or catechin equivalent (CE) per 100 g of DW or FW. Similarly, TAC is expressed in milligrams of cyanidin-3-glucoside equivalent (CGE) per 100 g of DW or FW or milligrams per 100 g of DW or FW.

Table 3 outlines the variety of total phenolic content (TPC), total flavonoid content (TFC), and total antioxidant capacity (TAC) that has been documented in sweet potato leaves from different countries using common cultivars. The methods for extraction and storage conditions used in various studies have slight variations. Typically, leaf samples that were either air-dried (Agbo et al. 2020; Ghamsemzadeh et al. 2022; Wayne et al. 2011), freeze-dried (Abong' et al. 2020; Makori et al. 2020; Sun et al. 2014), or frozen (Jia et al. 2022; Gama et al. 2023) were employed before batch solvent extraction. The solid-to-liquid ratio used in extraction varies, ranging from 1:1 (Gama et al. 2023), 1:3 (Wayne et al. 2011), 1:10 (Agbo et al. 2020), 1:20 (Sun et al. 2014; Makori et al. 2020), 1:25 (Jia et al. 2022), 1:50 (Ghamsemzadeh et al. 2022) to 3:200 (Abong' et al. 2020). The extraction temperature ranges from a chilled temperature of 4 °C to a warm temperature of 50 °C using alcoholic extraction. Ethanol 70% and methanol 80% are commonly used for extraction, while acidified alcohol is occasionally applied to extract anthocyanins specifically bound in the leaves (Jia et al. 2022; Gama et al. 2023). Some researchers repeated the extraction with fresh solvent for a short duration to improve the extraction efficiency, while others continuously shake or stirred the sample in the solvent for a specific duration. The extract was then analyzed directly or concentrated, followed by freeze-drying or freezing before further analysis. Optimizing extraction conditions is required to obtain the maximum yield and high bioactivity of phytochemicals from sweet potato leaves, providing sufficient crude extract for further in vitro and in vivo studies.

Makori et al. (2020) investigated the phytochemical composition of different edible parts of sweet potato, focusing on identifying individual phenolic compounds, flavonoids, anthocyanins, and carotenoids. Chromatographic analysis of sweet potato leaves extract revealed the presence of individual phenolic compounds such as 5-O-caffeoylquinic acid, 3-O-caffeoylquinic acid, 4-O-caffeoylquinic acid, caffeic acid, 4,5-di-O-caffeoylquinic acid, 3,5-di-O-caffeoylquinic acid, 3,4-di-O-caffeoylquinic acid, and 3,4,5-tri-O-caffeoylquinic acid. The type of flavonoid compounds that were identified in sweet potato leave extracts include myricetin, isoquercitrin, astragalin, quercitrin, quercetin, apigenin, and kaempferol. Most of these phenolic compounds were also found in extracts from sweet potatoes' stems, stalks, and peels. However, the phenolic contents of the stalk, stem, and peel were about half of those found in the leaves, and the total flavonoid contents were 20–40-fold lower than the leaves.

Sweet potato leaves have been found to contain anthocyanins, which are colored flavonoids that consist of red, purple, or blue pigmentation in plants, in addition to flavonoids and phenolic compounds. The major anthocyanidins present in sweet potato leaf extract include cyanidin, pelargonidin, and peonidin, which were extracted using acidified alcoholic extraction in studies conducted by Jia et al.

Table 3 Total phenolic content (TPC) and total flavonoid content (TFC) of sweet potato leaves

Country of origin	Extraction method	Total phenolic content	Total flavonoid content	Total anthocyanin content	Reference
China	Freeze-dried solid-to-liquid ratio: 1:20 Solvent: 70% ethanol Temperature: 50 °C Duration: 30 min Other assistance: ultrasonic treatment Improve efficiency and shelf-life extension: (i) Total 3 times extraction (ii) Rotary evaporated (iii) Freeze-dry	2.73–12.46 mg CAE/100 g DW	–	–	Sun et al. (2014)
		6050–12,080 mg CAE/100 g DW	100–4210 mg QE/100 g DW	790–1010 mg CGE/100 g DW	Makori et al. (2020)
	Frozen solid-to-liquid ratio: 1:25 Solvent: 95% alcohol: 1.5 N HCl (85:15) Temperature: 4 °C Duration: Overnight Other assistance: hexane extraction to remove polar lipids and interfering compounds, followed by ultrafiltration	48.43–125.76 GAE/100 g FW		3.22–109.60 mg/100 g FW	Jia et al. (2022)
Taiwan	Air-dried solid-to-liquid ratio: 1:3 Solvent: water Temperature: 100 °C Duration: 20 min Shelf-life extension: freeze-dry	28.1–130 mg GAE/g sample	18.0–72.7 mg QE/g sample		Wayne et al. (2011)
Ivory Coast, West Africa	Air-dried solid-to-liquid ratio 1:10 Solvent: 80% methanol (v/v) Temperature: room temperature Duration: 24 h Other assistance: Shaken with an orbital shaker incubator Shelf-life extension: oven-dried at 40 °C for 24 h	1285.83 mg GAE/g DM	660.0 mg QE/g DM		Agbo et al. (2020)

(continued)

Table 3 (continued)

Country of origin	Extraction method	Total phenolic content	Total flavonoid content	Total anthocyanin content	Reference
Kenya, Africa	Freeze-dried solid to liquid: 3:200 Solvent: 80% methanol Temperature: 25 °C Duration: 12 h Other assistance: Shaken on a mechanical shaker	4495.93–6801.09 mg GAE/100 g DW	4097.22–6941.51 mg CE/100 g DW		Abong' et al. (2020)
Malaysia	Freeze-dried solid to liquid: 1:50 Solvent: 80% methanol Temperature: room temperature Duration: 1 h Other assistance: swirl using an orbital shaker Shelf-life extension: frozen at –20 °C	4.47–8.11 mg GAE/g DW	1.87–3.95 mg QE/g DW		Ghamsemzadeh et al. (2022)
Brazil	Frozen leaves Solvent: aqueous methanol	55.3–266.0 mg GAE/100 g FW			Gama et al. (2023)
	Frozen leaves solid to liquid: 1 mg in 1 mL Solvent: 1:1 water/methanol solution containing 2 N HCl Temperature: room temperature Duration: 3 h Improve efficiency: 1. Continuous stirring 2. Filtration and rotary evaporation at 30 °C			29.9–75.8 mg CGE/100 g FW	

CAE chlorogenic acid equivalent, *QE* quercetin equivalent, *CGE* cynidin-3-glucoside equivalent

(2022) and Gama et al. (2023). Sweet potato leaf extract was found to have a total anthocyanin content that was approximately 40–100-fold higher than that of the peel and the root, as Makori et al. (2020) reported. In another study by Su et al. (2019), a novel anthocyanin, peonidin 3-caffeoyl-p-coumaryl sophoroside-5-glucoside, was identified among fourteen anthocyanins present in the leaves extract of purple-fleshed sweet potatoes. Anthocyanins in sweet potato roots and leaves contribute to a broad spectrum of bioactivities.

Carotenoids are pigments that give plants and microorganisms yellow, orange, and red colors. Sweet potato is an excellent source of carotenoids, especially beta-carotene, found in all parts of the plant. Abong' et al. (2020) observed that the concentration of β-carotene in sweet potato leaves of orange-fleshed varieties was significantly higher (ranging from 16.43 to 34.47 mg/100 g DW) than in the root. The carotenoids present in the leaves extract of sweet potato include β-xanthin, 13 cis β-carotene, all-trans β-carotene, and β-9 cis β-carotene. However, Hsu and Chen (2022) reported similar levels of total carotenoids in sweet potato peel and root, ranging from 271.80 to 450.88 μg/g DW. They successfully isolated and identified 11 carotenoids (all-trans-violaxanthin, cis-lutein, all-trans-lutein, all-trans-zeaxanthin, all-trans-Canthaxanthin (internal standard), 15 or 15′-cis-β-cryptoxanthin, all-trans-β-cryptoxanthin, 15 or 15′-cis-β-carotene, 13 or 13′-cis-β-carotene, all-trans-β-carotene, and all-trans-α-carotene) from the peel extract of sweet potato (TNG 66 cultivar) in Taiwan. The extraction of carotenoids from sweet potato by-products is a process that involves several challenges due to the nature of the matrices being processed. Carotenoids, more soluble in non-polar solvents such as hexane and toluene, require the removal of saponifiable matter from the extract through methanolic alkaline. The extraction step must be repeated to improve efficiency, while precautions must be taken to prevent degradation of the carotenoids, which are light-sensitive.

Recent studies have shown that bound terpenes, such as linalool and α-terpineol, are present in the peel and root extract of Garnet, Jewel, Japanese Beauregard, and New Orleans cultivars (Rodriguez et al. 2023). The peel contained linalool or α-terpineol in approximately twice the quantity in the root. Terpenes are secondary metabolites that contribute to the aroma of plant material. However, the high volatility of terpenes poses a challenge during their extraction and analysis.

The utilization of sweet potato by-products to extract phytochemicals has yielded compounds with antioxidative properties that have been evaluated through various measures, including 2,2-diphenyl-1-picrylhydrazyl (DPPH·) radical scavenging capacity, 2,2′-Azino-bis(3-Ethylbenzothiazoline-6-Sulfonic Acid) (ABTS) radical scavenging capacity, ferric reducing antioxidant power (FRAP), and inhibitory action against lipid oxidation. The antioxidative capacity of these compounds has been linked to anti-hypercholesterolemic, anti-obesity, and hepatoprotective effects, as they can regulate enzymes involved in fatty acid, cholesterol, and triglyceride metabolism in the liver. Recent studies have found that sweet potato leaves exhibit the highest antioxidative capacity among all parts of the plant, owing to high levels of bioactive compounds such as total phenolic and anthocyanin.

With the increasing demand for low-cost, functional ingredients that offer great health benefits, it is recommended that antioxidants entrapped in dietary fiber, known as antioxidant dietary fiber, be sourced from agroindustrial by-products. This helps to close the cycle in line with the principles of the circular economy and provides a rich source of antioxidant dietary fiber that can be obtained from different parts of sweet potato by-products generated through food processing.

Extensive research has been conducted on the dietary fiber content of sweet potatoes. Notably, the leaves (45.7–45.9 g/100 g DW) and stems and stalks (41.6–63.6 g/100 g DW) contain high levels of dietary fiber. The leaves contain a relatively higher content of soluble dietary fiber (5.77–6.83 g/100 g DW) compared to the stems and stalks (3.3–4.23 g/100 g DW), whereas the stems and stalks contain a relatively higher content of insoluble dietary fiber (38.3–59.6 g/100 g DW) compared to the leaves (39.1–39.9 g/100 g DW). Furthermore, dietary fiber extracted from sweet potato peel exhibits prebiotic activity through an in vitro fermentation model, whereby the gut microbiota could utilize the extracted dietary fiber. Specifically, the ultrasound-extracted compound enhances the relative abundance of short-chain fatty acid (SCFA)-producing bacteria while reducing the abundance of harmful bacteria, ammonia, and pH from 6.89 to 4.63 (Cao et al. 2022).

Pectin, a soluble dietary fiber, is also present in sweet potatoes. The recovery of pectin from by-products of food processing involves the pretreatment of the raw material to remove the small molecular weight carbohydrates and pigments, followed by pectin extraction, purification, and drying (Mu et al. 2017b). For instance, alkaline extraction (0.25 M NaOH) from sweet potato peel yielded 16.7% pectin with low methyl esterification (Abang Zaidel et al. 2017). Meanwhile, Joanne Toy et al. (2022) extracted 1.028% (w/w) dry basis of high-methoxyl (degree of esterification, DE: 61.6%) pectin from Margarita sweet potato leaves. Similarly, Hamidon et al. (2020) employed a citric acid solution to extract a high-methoxyl (DE: 58.5%) from sweet potato peels and yielded 65.2% pectin under optimized conditions (76 °C for 64 min at pH 1.2). Pectin is renowned for its various health-promoting activities, such as anti-hyperglycemia, hypolipidemic, and anti-hypercholesterolemic. It delays sugar and fat absorption in the digestive tract and affects bile acid metabolism (Dongowski and Lorenz 2004). Additionally, sweet potato pectin gel formation properties and emulsification capacity enable its use in various food applications (Mu et al. 2017b).

The protein content of different parts of sweet potatoes was compared, and it was found that sweet potato leaves have the highest protein content (Table 1). The amino acid composition of sweet potato leaves is of good quality, with lysine being the only or second limiting amino acid. This suggests that sweet potato leaves could be a valuable protein resource (Suárez et al. 2020; Tang et al. 2021). Moreover, sweet potato leaves may have high antioxidative capacities, which could be attributed to antioxidative enzymes such as superoxide dismutase (SOD), peroxidase (POX), ascorbate peroxidase (APX), and catalase (CAT). The high content of minerals such as sodium and potassium in sweet potato leaves often acts as a cofactor for these enzymes. Antioxidant enzyme activities are related to plant abiotic stress responses, such as salinity and drought (Ahmad et al. 2010). However, information on antioxidant enzymes in sweet potatoes and their expression under abiotic stress is limited.

Mondal et al. (2023) studied antioxidative enzyme activities in sweet potato leaves of BARI SP4 and BARI SP8 cultivars in response to salinity stress. They found that SOD activities varied between 0.075 and 0.15 U/g FW, POX ranged from 4 to 5.4 U/g, CAT activities varied between 1 and 4 U/g, while APX ranged from 100 to 350 U/mg FW. They concluded that the response of antioxidant enzymes in sweet potato leaves was not solely related to salt stress but was also modified or overlaid by factors such as leaf position, leaf age, duration of stress, and genotype. Castillo Leon et al. (2002) purified a peroxidase from sweet potato peel with a molecular mass of 37 kDa, an isoelectric point of 3.5, and a specific activity of 4900 U/mg protein. This sweet potato peroxidase exhibited substrate specificity on ferulic acid, which is not shown by other plant peroxidases. Therefore, further research is warranted on antioxidative enzymes from sweet potato leaves and peels to obtain information on their possible purification schemes, stability, and applications for potential sweet potato by-product valorization opportunities.

Sporamin is a sweet potato storage protein extensively studied due to its multiple biological functions in the plant, including insect-resistance capability, inhibitory activity, and reactive oxygen species scavenging activity. According to Senthilkumar and Yeh (2012), this protein plays a significant role in facilitating the growth and development of sweet potatoes. Typically, sporamin is expressed in response to wounding and other abiotic stresses and is found primarily in the root and peel. Recent research has shown that sporamin possesses various health-promoting activities, including antidiabetic and antioxidant properties, owing to its pepsin and trypsin inhibitory activity. Specifically, the sporamin from the outer peel and the white-skinned Calopo cultivar has been found to resist digestive enzyme activities (Maloney et al. 2012). Additionally, sporamin exhibits significant emulsifying capacity due to the intermolecular disulfide linkage between sporamin A, sporamin B, and aggregations formed at the oil–water interface (Guo and Mu 2011).

Mu et al. (2014) proposed a foam separation method to recover highly soluble protein from sweet potato starch processed water that yielded a maximum recovery of 84.1% and an enrichment ratio of 1.3 under optimized conditions of influent protein concentration, influent volume, foaming time, air flow rate, and pH. Mu et al. (2017a) attempted to isolate protein from sweet potato flesh using the salting-out principle, starting with extraction using a 1:2 (solid to 1% citric acid) ratio, followed by ammonium sulfate precipitation. They also used the isoelectric precipitation principle, starting with extraction using a 1:2 (solid to 1% sodium bisulfite solution) ratio, followed by adjusting the pH to 4 (near the protein's isoelectric point) to obtain sweet potato proteins.

However, research on isolating protein from starch residue collected from sweet potato processing is scarce and warrants further investigation. The common protein extraction method from a solid matrix involves adding dilute saline or alkaline solution to increase protein solubility, followed by homogenization, facilitating protein molecules' diffusion from the solid matrix to the aqueous extract. A concentration step after extraction may be required to improve the efficiency of protein isolation through salting-out or isoelectric precipitation.

3 Biological and Functional Properties of Extracts and Bioactive Compounds from By-products

The sweet potato plant is known for its diverse pharmacological properties, including antioxidant, anticancer, anti-inflammatory, and aphrodisiac effects. The types of phytochemical compounds present in each plant part differ, and the pharmacological activity and mechanism of action depend on factors such as the type of phytochemical compound, plant variety, extraction solvent, and plant part. A literature review on sweet potato by-products reveals that sweet potato leaves are rich in bioactive compounds such as polyphenolics, carotenoids, proteins, and soluble dietary fiber. At the same time, the peel has a higher level of phytochemicals compared to the root.

Numerous studies have focused on using extracts or isolated phytochemicals from sweet potato leaves and peels *in vitro* or *in vivo* animal models. For example, Cao et al. (2022) tested four types of dietary fiber isolated enzymatically after water extraction assisted by different means using an *in vitro* fermentation model. They concluded that sweet potato peel dietary fiber could improve the health of the intestinal microecology by increasing beneficial bacteria and decreasing harmful bacteria upon administration. Similarly, Bernabeu et al. (2024) found that an aqueous extract obtained through pulsed electric fields extraction on sweet potato peels, especially of the purple variety, could increase the growth of specific strains of the probiotic *Lactobacillus* genus. Incorporating the pulsed electric field-assisted aqueous extract of sweet potato peel also promoted the production of short-chain fatty acids after a 48-h fermentation with *L. gasseri*, showing typical characteristics of potential prebiotics. These findings suggest that sweet potato by-products have immense health-promoting properties and can be used as a possible source of prebiotics or probiotics, depending on the extraction method used.

Anthocyanins found in leaves have demonstrated significant potential as an anticancer agent, as evidenced by their ability to effectively trigger apoptosis in the MCF-7 breast cancer cell line, HCT-116, as well as in HeLa cell lines (Vishnu et al. 2019). Additionally, sweet potato peels are a promising source of anticancer bioactive compounds. Asadi et al. (2017) have reported that the administration of diets containing 10% purple sweet potato flesh, 10% purple sweet potato peel, or 0.12% anthocyanin-rich extract for 18 weeks resulted in reduced adenoma numbers in APCMIN mice, effectively inhibiting tumor cell proliferation. Furthermore, an isolated 16-amino-acid peptide known as *Ipomoea batatas* anticancer peptide (IbACP) extracted from sweet potato leaves demonstrated inhibitory effects on the human pancreatic cancer Panc-1 cell line (Vincent Chang et al. 2013). However, the exact mechanism of the anticancer effects is yet to be fully elucidated. Further research is needed to identify key bioactive compounds or combinations with cancer-prevention potential from sweet potato by-products.

Lyophilized leaf powder derived from three different sweet potato cultivars demonstrated significant efficacy in suppressing foodborne pathogens, such as *E.coli* 0157:H7, *Bacillus cereus* and *Staphylococcus aureus*, as per a study conducted by Islam in 2008. Further studies on bioactive compound extraction from sweet potato by-products using different solvents revealed that the antibacterial ability of the extract was impacted by the type of extraction solvents used. Naz et al. (2017) showed that the 95% methanolic extract of the red variety's peel exhibited antibacterial activity against multi-drug resistant *S. aureus*, while the 95% ethanolic extract of the white variety's peel was effective against *B. subtilis* and gram-positive *Pasteurella multocida*. In another study, Kusuma et al. (2017) evaluated a 70% ethanolic extract of sweet potato leaves and observed a dose-dependent growth inhibition zone through an agar well-diffusion assay against *Shigella dysenteriae*. The extract contained flavonoids, tannins, steroids, and polyphenolic compounds. However, Pochapski et al. (2011) found that the 70% ethanolic sweet potato leaf extract had minimal effect on *Streptococcus* mutants, *Streptococcus mitis*, *S. aureus*, and *Candida albicans* growth. The differences in the antibacterial effect of ethanolic sweet potato leaf extract could be attributed to variations in cultivars, harvesting time, defoliation, and sweet potato growing conditions.

Research has found that ingesting sweet potato leaves with a high-fat diet has an anti-obesity effect. An experimental animal group was fed a high-fat diet supplemented with freeze-dried leaves powder. The group showed less weight gain and adipose tissue weight gain than those not given sweet potato leaves. Kurata et al. (2017) found that rats fed sweet potato leaves had lower plasma triglyceride and total cholesterol levels than those fed a high-fat diet alone. The rats also had significantly lower total hepatic cholesterol levels. The anti-obesity effect may be due to decreased preadipocyte 3T3-L1 cell proliferation, leading to lower lipid accumulation in differentiated 3T3-L cells. According to Lee et al. (2015), the inhibition of lipogenesis may be due to the down-regulation of the transcription factors, C/EBPα and SREBP-1, during 3T3-L1 adipocyte differentiation. Mu et al. (2017c) reported that Wistar rats fed on dietary fiber extracted from sweet potato residue exhibited a lower body weight than those in the control and high-fat diet groups, suggesting that sweet potato dietary fiber may help prevent obesity.

Studies have demonstrated that the extract of purple sweet potato leaves has a potential antidiabetic effect. This is because it positively affects glucose uptake in 3T3-L1 adipocyte cells. Lee et al. (2015) reported that this effect is achieved through enhanced expression of Glut4 and regulation of the PI3K/AKT pathway. Similarly, Shih et al. (2020) found that feeding high doses of air-dried white sweet potato leaves to STZ-induced diabetic mice significantly lowered blood glucose and triglyceride levels. This also activated the insulin-signaling pathway, as per their 2020 report. *In vitro* α amylase and α glucosidase inhibition assays have been conducted to test for the potential antidiabetic properties of plant extract. Elevated postprandial blood glucose levels are related to carbohydrate digestibility, which is why limiting the activity of carbohydrate digestive enzymes such as α-amylase and α-glucosidase in the intestinal tract is one of the strategies to reduce postprandial hyperglycemia, as suggested by Gong et al. (2020).

Luo et al. (2021) isolated protocatechualdehyde, ethyl caffeate and quercetin from sweet potato leaves and found that ethyl caffeate was the most potent inhibitor of both α-glucosidase and α-amylase. Its inhibitory effect on α-glucosidase and α-amylase was 6.77 times lower and 13.1 times lower than that of acarbose. Yustiantara et al. (2021) conducted a feeding trial using ethanolic extract of sweet potato leaves on STZ-induced rats, and the results showed significant antihyperglycemic activity due to flavonoids acting as antioxidants. There was a linear correlation between flavonoid antioxidant agent levels and antihyperglycemic activity. Additionally, the anthocyanin compound in the ethanolic extract helped regulate blood sugar levels by inhibiting excess sugar absorption and protecting pancreatic beta cells (Yustiantara et al. 2021).

A few studies examined sweet potato leaf extract's potential direct or indirect effects on cardiovascular disease prevention. Atherosclerosis, for instance, can arise due to low-density lipoprotein (LDL) oxidation (Wang et al. 2016). Additionally, angiogenesis, which involves the growth of new blood vessels from existing ones, can lead to diseases like atherosclerosis through extracellular matrix degradation, endothelial cell migration, and proliferation. Methanolic extract of sweet potato leaves can inhibit LDL oxidation in both human subjects and *in vitro*, owing to the antioxidant properties of the leaf polyphenolic compounds (Nagai et al. 2011). Furthermore, freeze-dried sweet potato tip powder has an inhibitory effect on Angiotensin-Converting Enzyme (ACE) activity *in vitro*, suggesting that atherosclerosis could be delayed by administering the powder (Cui et al. 2011). Sweet potato leaf extract has hypolipidemic and anti-atherosclerotic properties in male Wistar rats with hypercholesterolemia (Ntchapda et al. 2021). Treatment with the extract resulted in reduced aortic thickness and surface area in the animal model. In contrast, hypercholesterolemic animals that did not receive treatment developed more conspicuous atherosclerotic plaques. These findings justify using sweet potato leaves in traditional medicine for their anti-atherosclerogenic effects.

The cholesterol metabolism is closely linked to liver health, as evidenced by the findings of various studies. In a survey conducted by Mu et al. (2017c), Wistar rats fed a high-fat diet exhibited a significant increase in total cholesterol levels from 36 mg/mL (control group) to 206.99 mg/mL, resulting in a large area of stem cell fatty degeneration in the liver tissue. The presence of white fat particles indicated moderate cellular degeneration. However, the administration of middle doses of sweet potato dietary fiber significantly reduced the total cholesterol levels and decreased the degree of liver degeneration and the amount of fat particles. In contrast, the high-dose sweet potato dietary fiber group showed only minimal signs of liver degeneration and few fat particles. The study concluded that 10% sweet potato dietary fiber could prevent the fatty liver induced by a high-fat diet.

Furthermore, as reported by Mahfudh et al. (2021), sweet potato leaf extract exhibited similar hepatoprotective activity. The researchers fed ethanolic leaf extract to a 2-month-old male Wistar rat. This resulted in a significant decrease in cholesterol and triglyceride levels compared to the negative control group. Additionally, a reduction in serum glutamic pyruvic transaminase (SGPT) and serum glutamic oxaloacetic transaminase (SGOT) was observed, indicating the hepatoprotective effect of sweet potato leaf extract. Purple sweet potato leaf extract exerted similar effects on male C57BL/6 mice, as reported by Kang and Lee (2021). Blood alcohol concentration and SGPT levels significantly reduced compared to the levels observed in the untreated group, indicating the potential effectiveness of sweet potato leaf extract in protecting the liver from alcohol-induced hepatotoxicity and fatty liver.

The ethanolic extract of sweet potato peels is known for its high antioxidative capacity, which has been found to play a significant role in wound healing. In a recent study conducted by Silva-Correa et al. (2021), a 1% gel formulation containing acidified ethanolic extract of purple sweet potato peels was found to be effective in healing wounds in male BALB/c mice aged 3½ months with induced injury. The researchers proposed that the wound-healing mechanism of the peel extract could be attributed to phytochemicals such as anthocyanins and phenolic compounds, which possess antioxidant, antimicrobial, and anti-inflammatory properties. These findings suggest that the ethanolic extract of sweet potato peels may have potential therapeutic applications in the management of wounds (Table 4).

Sweet potato leaves, stems, stalks, and peels are excellent sources of nutrients and phytochemicals that offer several potential health benefits. Sweet potato leaves and peels, mainly, are rich in nutrients and functional compounds such as complex carbohydrates, protein, soluble and insoluble dietary fiber, micronutrients (i.e., vitamins and minerals), antioxidants, and other bioactive compounds. The antioxidants and other bioactive compounds present in sweet potato leaves can help reduce the risks associated with disease by promoting more favorable antioxidant status, free radical-scavenging capacities, and thwarting processes involved in disease pathogenesis.

Bioactive compounds in sweet potato leaves are potent anti-obesity, hepatoprotective, cardioprotective, antidiabetic, and antimicrobial agents in disease prevention. *In vitro* and *in vivo* studies using extracts or bioactive compounds of sweet potato by-products have demonstrated promising disease prevention effects, mainly through the synergistic interaction of phytochemicals in preventing oxidative stress, enhancing immune responses, and reducing lipid peroxidation.

Table 4 Health-promoting properties of extract and bioactive compound of sweet potato by-products

Biological properties	Sweet potato parts	Extract	Bioactive compound	References
Prebiotic activity • Increasing the abundance of beneficial bacteria (e.g., *Bifidobacterium*, *Faecalibacterium*, and *Prevotella*) and reducing the abundance of harmful bacteria (e.g., *Proteobacteria*, *Romboutsia*, and *Dorea*) promote the production of short-chain fatty acids	Peel		Dietary fiber	Cao et al. (2022)
• Improved the growth kinetics of probiotic bacteria, *Lactobacillus* members and also the production of short-chain fatty acids (SCFA)	Peel	Pulse electric field-assisted aqueous extract		Bernabeu et al. (2024)
Anticancer 1. Induce apoptosis in MCF-7 breast cancer cell lines	Leaf, peel		Acylated anthocyanin: peonidin and cyanidin derivatives	Vishnu et al. (2019)
2. Inhibition of tumor cell proliferation → reduced adenoma number in APCMIN mice	Peel		Anthocyanins	Asadi et al. (2017)
• Able to dose-dependently inhibit Panc-1, a pancreatic cancer line, cell proliferation. • Induce and promote apoptosis	Leaf		16-amino acid peptide	Vincent Chang et al. (2013)
Antibacterial activity 1. Multi-drug resistant *S. aureus* 2. *Bacillus subtilis* 3. Gram-positive *Pasteurella multocida*	Peel	Methanolic extract of the red variety and ethanolic extract of the white variety	May be due to flavonoids, tannin, steroid, and polyphenolic compounds	Naz et al. (2017)
Inhibit the growth of gram-positive and gram-negative foodborne pathogens: *Escherichia coli* 0157:H7, *Bacillus cereus,* and *Staphylococcus aureus*	Leaves	Freeze-dried leave powder		Islam (2008)

(continued)

Table 4 (continued)

Biological properties	Sweet potato parts	Extract	Bioactive compound	References
show a growth inhibition zone (agar well diffusion assay) in a dose-dependent manner against *Shigella dysenteriae* ATCC13313	Leaves	Ethanolic extract		Kusuma et al. (2017)
Anti-atherosclerotic/ cardioprotective activities • Hypercholesterolemic rats fed 500 and 600 mg/kg extract showed reduced aorta thickness and surface area	Leaf	Aqueous extract	May be due to flavonoid compounds	Ntchapda et al. (2021)
Antidiabetic activities • Inhibitor of α-glucosidase and • Inhibitor of α-amylase activity	Leaf		Protocatechualdehyde, ethyl caffeate, and quercetin	Luo et al. (2021)
• Antihyperglycemic in STZ-induced diabetic rat	Leaf	Ethanolic extract	May be due to flavonoid or anthocyanin compounds	Yustiantara et al. (2021)
Anti-obesity effect • The rat group that ingested a 5% freeze-dried powder of sweet potato leaves showed a significant decrease in weight gain andadipose tissue weight without reducing feed intake	Leaf	Freeze-dried leaf powder	This may be due to polyphenols and dietary fiber	Kurata et al. (2017)
• The weight gain of rats fed by sweet potato dietary fiber was significantly slower than that of the high-fat diet group, and their weights were lower than those of the control group, especially after day 28	Residue		Dietary fiber	Mu et al. (2017c)
Hepatoprotective activity • Administration of the leaf extract to the treatment animal reduced serum glutamic pyruvic transaminase (SGPT) and serum glutamic oxaloacetic transaminase (SGOT) activity, protecting the liver from oxidative damage	Leaf	Ethanolic extract	May be due to quercetin	Mahfudh et al. (2021)

(continued)

Table 4 (continued)

Biological properties	Sweet potato parts	Extract	Bioactive compound	References
• High-dose sweet potato dietary fiber protects the liver from cellular damage and prevents fatty liver caused by a high-fat diet in Wistar rats	Residue		Dietary fiber	Mu et al. (2017c)
Wound-healing effect • Wound-induced mice treated with the peel ethanolic extract showed a higher percentage of wound closure, also evidenced in histopathological changes	Peel	Acidified ethanolic extract	This may be due to the combined effect of phenolic compounds and anthocyanins.	Silva-Correa et al. (2021)

4 Food and Non-food Applications of Extracts and Bioactive Compounds from Sweet Potato By-products

Application of sweet potato by-products' bioactive compounds spans both food and non-food applications, driven by their rich content in various compounds, including (poly)phenols, terpenoids, tannins, saponins, glycosides, alkaloids, and phytosterols. These compounds are primarily present in the peel and flesh of sweet potatoes, with different colors of the root yielding diverse types of (poly)phenols and carotenoids (Alam et al. 2016; Kurata et al. 2019; Luo et al. 2021; Sun et al. 2018; Teow et al. 2007; Wang et al. 2016). For example, carotenoids, phenolic acids, and flavonoids were recovered from orange and yellow flash varieties, while anthocyanins and phenolic acids were from the purple flash (Alam et al. 2016; Kurata et al. 2019). Research conducted with various varieties of sweet potatoes in Uganda revealed that unpeel sweet potatoes possessed higher bioactive compounds (Amagloh et al. 2022), indicating potential values in the peels.

Bioactive compounds have been broadly studied due to their physiological benefits, antioxidant properties, anti-inflammatory effects, immunity enhancement, blood sugar reduction, and cholesterol reduction. Consequently, they have found applications as food supplements, functional foods, natural preservatives, and natural colorants.

Initially explored in the nutraceutical and functional food sectors, sweet potato by-products have gained attention for their antioxidant, antibacterial, antidiabetic, and antifungal properties. That promotes health benefits or disease prevention (Pattnaik et al. 2021). Notably, compounds such as 4-O-caffeoylquinic acid,

1,3-di-O-caffeoylquinic acid, and 3,5-di-O-caffeoylquinic acid exhibited active antioxidant activities and were isolated from methanolic and hydromethanolic extracts of sweet potato *(Ipomoea batatas* [L.] *Lam*) tuber flour (Jung et al. 2011). The extracts foresee an excellent source of food ingredients with bioactive compounds. Furthermore, sweet potato bioactive compounds have demonstrated therapeutic properties, including hepatoprotection, antibacterial, antihistamine (Meira et al. 2012), inhibiting HIV replication (Mahmood et al. 1993), hypoglycemic (Okudaira et al. 2005), radical-scavenging (Islam et al. 2003), and antimutagenic (Yoshimoto et al. 2002) activities.

These bioactive compounds can be used as ingredients to enhance the nutritional properties of products such as functional food, snacks, desserts, beverages, and supplements. For diversified usage, sweet potato by-products can be processed into various forms such as flour, powder, puree, extracts, or fiber. Sweet potato flour can be used as a gluten-free alternative bakery product such as breads, Meng et al. (2022) found that flour and dough's properties were improved by replacing wheat flour with sweet potato flour. The texture and sensory quality of the bread were better. Therefore, it is suggested that cakes, breads, cookies, and snacks may use sweet potato flour as an alternative to improve antioxidant, polyphenol, and dietary fiber content. Meat products such as sausages, meat balls, and petties that require flour as binding agents could consider sweet potato flour as the alternative to complement the gluten allergy populations.

The dehydrated form of sweet potato maybe a good source of natural food coloring agents, flavoring agents, or nutritional supplements in food and beverage products. Sweet potatoes in various varieties (yellow, orange, and purple) may produce sweet potato powder with diverse color tones to be used as a natural colorant in processed food such as confectionaries, beverages, and dairy products. Natural foods are susceptible to color change and off-flavor due to harvesting procedures, processing conditions, storage methods, etc. Therefore, synthetic colorants and flavoring agents are commonly added to improve the appearance and flavor of processed food products. Due to recent health issues and consumer awareness, natural food additives as synthetic alternatives have drawn much attention. Sweet potato by-product powder may be introduced as a substitute. The bioactive compound in sweet potato by-product powders is expected to benefit the nutritional value of processed food. At the same time, the antioxidant and antimicrobial activity of sweet potatoes by-products is expected to improve the shelf-life of the food products.

Food products such as soups, sauces, dips, spreads, and baked goods can include sweet potato puree from its by-products to add flavor, texture, and nutritional value. The extracts from sweet potato by-products, which are rich in bioactive compounds, can be applied as natural food additives, preservatives, or functional ingredients in various food and beverage products. Furthermore, the bioactive compounds extracted from sweet potatoes are Generally Recognized as Safe (GRAS), exempting them from premarket approval requirements. Therefore, using these bioactive compounds as food ingredients or additives is exempt from the premarket approval requirements. These compounds can substitute synthetic counterparts in food supplements, functional foods, natural food preservatives, and natural food colorants, aligning with current consumer preferences for clean labeling and natural ingredients. However, challenges remain regarding the stability and availability of sweet potato by-products. More research is required to optimize product yield and stability, scale up production strategies, and establish optimum dosages for health benefits.

Additionally, bioactive compounds, polyphenol compounds, and anti-inflammatory, antimicrobial, and anticancer properties have been reported in sweet potatoes and hold promising applications in cosmetics, pharmaceuticals, and nutraceuticals. Liu et al. (2022) found that white-fleshed sweet potato extract (SPE) has anti-obesity effects on a high-fat diet (HFD)-induced obese mice model. The report proved that SPE may be consumed as a dietary supplement for weight control and treating hepatic steatosis.

A study on the biological activity of sweet potato tips and tubers (Hong et al. 2022) suggests sweet potato tips and tubers may improve eye disease whitening effects, which can serve as functional food and cosmetic material. Cosmetic products such as creams, lotions, and serums may incorporate sweet potato by-product extracts to achieve a whitening effect. Another study by Chan et al. (2010) concluded that anthocyanin water extracts from purple potatoes were effective against UV-B rays. The dosage of 0.61 mg in 100 g of cream prepared successfully absorbed 46% of the UV radiation. Anthocyanins are groups of antioxidants, and incorporating these compounds in cosmetics may help protect the skin from damage caused by free radicals and UV radiation.

To obtain the valuable compounds in sweet potato by-products, techniques such as microencapsulation and nanoencapsulation can help preserve the composition of these compounds and ensure their optimal benefits in end products. In terms of pharmaceutical, health care, and cosmetics products, homogeneity distribution of the ingredients is critical. Technologies such as supercritical carbon dioxide (CO_2) have been reported to be able to manipulate the particle sizes of the encapsulates. Therefore, drug loading in the products in various temperatures, pressure, and flow ratios can be managed (Ghasemi et al. 2022). Thus, bioactive compounds extracted from sweet potato by-products may utilize these methods to achieve optimum benefit in the end products.

Sweet potato by-products offer a versatile and sustainable source of bioactive compounds with extensive applications in food and non-food sectors, ranging from functional foods to skincare products and pharmaceuticals. Figure 3 summarizes the various potential applications of sweet potato by-products.

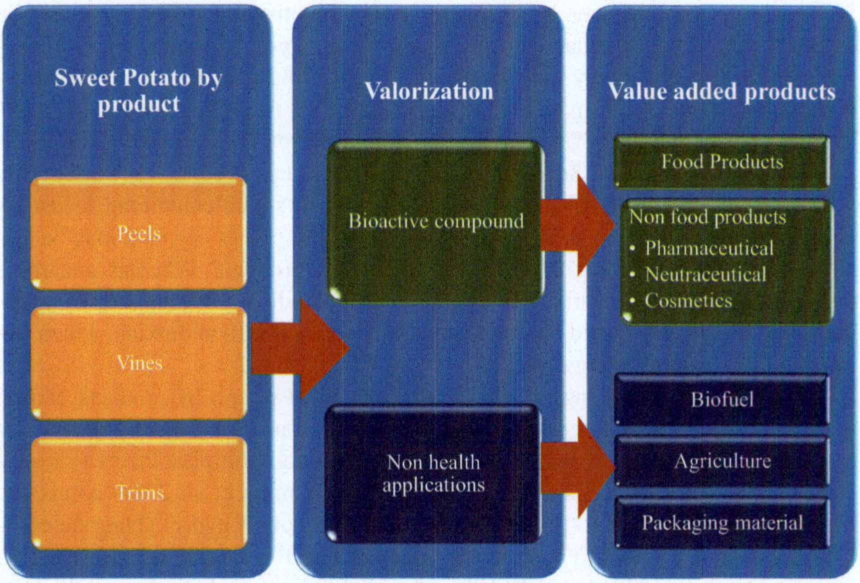

Fig. 3 Potential applications of sweet potato by-products

5 Valorization of By-products for Non-health Purposes

Valorization of sweet potato by-products can extend beyond food and pharmaceutical applications to include sectors such as agriculture and bioenergy. Vermicomposting, a straightforward and low-cost application of agricultural waste products, is one possibility. A study done by vermicomposting sweet potato waste revealed improved nutritional composition (C, N, P, and K) (Nedunchezhiyan et al. 2011). On the other hand, biofertilizers may be another alternative to sweet potato by-products for agricultural purposes. Biofertilizers produced from solid-state fermentation of sweet potato waste may offer a better alternative. Studies have shown promising results on the physical characteristics of watermelon, papaya, and banana waste when treated with biofertilizers (Lim and Matu 2015).

Additionally, anaerobic digestion of organic waste is expected to yield significant amounts of electrical energy and biofertilizer, reducing carbon dioxide emissions, landfill area usage, and leachate production (Yong et al. 2021). Their report revealed that anaerobic digestion of 50% organic waste is expected to yield 3941 MWh/day of electrical energy and 2500 t/day biofertilizer. These results will benefit the environment by reducing 2735 t/day of carbon dioxide (CO_2) emission, saving up to 1128 m^2/day of landfilling area and 481 m^3/day of leachate. Therefore, directing sweet potato by-products toward these applications is crucial for environmental benefits.

The conversion of agricultural waste to biofuel can be achieved through various strategies. Enzymatic hydrolysis of cellulose and hemicellulose followed by fermentation with yeast or bacteria to produce bioethanol is a well-understood industrial approach. Bioethanol is commonly made by yeast *Saccharomyces cerevisiae*, which can effectively convert glucose or other hexoses to ethanol. Besides, biobutanol production from anaerobic *Clostridium* has recently gained much interest due to its suitability for current engines, high energy density, and lower corrosiveness than bioethanol. Solventogenic *Clostridium* can produce industrial-scale solvents such as acetone, butanol, and ethanol from starch-containing feedstock (Karthick and Nanthagopal 2021). Therefore, a similar processing path may be feasible to produce biobutanol from sweet potato by-products. Whereas anaerobic digestion is another alternative in biogas production, utilizing sweet potato by-product as the feedstock to yield biogas such as methane and carbon dioxide, which can be used as renewable energy sources.

In addition to producing renewable energy, sweet potato by-products can be recovered as biodegradable packaging materials. The process involves the extraction of starch from sweet potato by-products, blending with plasticizers or cross-linking agents to form a film. A recent report on (Vannini et al. 2021) composite from sweet potato solid residues blended with poly(3-hydroxybutyrate-co-3-hydroxyvalerate) (PHBV) composite demonstrated thermal stability up to 260 °C. Composite blended with 10% wt. sweet potato residues showed 30 MPa tensile strength and 3.2% strain. These biocomposites, made from renewable resources, are believed to be biodegradable, aligning with circular economy principles and offering a potential pathway for the downstream recovery of sweet potato by-products.

6 Conclusions

Utilizing the residual materials of sweet potatoes to extract bioactive phytochemicals is a cost-effective approach that can enhance the value of various industries. The bioactive phytochemicals found in sweet potatoes possess health benefits that can promote overall well-being and prevent chronic illnesses. To explore the extraction, identification, and application of these compounds, additional research is necessary, which has the potential to lead to further innovation and the development of new products. Furthermore, more scientific evidence is required to examine how the bioactive compounds in sweet potato leaves interact synergistically and their influence on gene expression and subsequent disease pathogenesis. Due to the limited research on antioxidant enzymes in sweet potatoes and their expression under abiotic stress, further exploration of antioxidative enzymes extracted from sweet potato leaves and peels is needed to gather information on their potential purification schemes, stability, and applications for sweet potato by-product valorization opportunities.

Sweet potato by-products are a valuable source of antioxidant dietary fiber, containing significant amounts of dietary fiber and antioxidative compounds. Derivatization of antioxidant dietary fiber into prebiotic can be achieved by employing various thermo-mechanical or enzymatic treatments. These findings suggest that sweet potato by-products can be transformed into valuable ingredients or materials for food and non-food applications by adapting existing technology or developing innovative approaches.

References

Abang Zaidel DN, Hamidon NH, Mat Zahir N (2017) Extraction and characterization of pectin from sweet potato (*Ipomoea batatas*) peels using alkaline extraction method. Acta Hortic 1152:211–218. https://doi.org/10.17660/ActaHortic.2017.1152.29

Abong' GO, Muzhingi T, Okoth MW, Ng'ang'a F, Ochieng PE, Mbogo DM, Malavi D, Akhwale M, Ghimire S (2020) Phytochemicals in leaves and roots of selected Kenyan orange fleshed sweet potato (OFSP) varieties. Int J Food Sci 2020:3567972. https://doi.org/10.1155/2020/3567972

Agbo EA, Kouassi K, Gouekou DA, Méité S, Gbogouri AG, Brou K (2020) Antioxidant activities in sweet potatoes leaves steamed with spices. J Food Res 9:41–49. https://doi.org/10.5539/jfr.v9n4p41

Agubosi OCP, Imudia FD, Alagbe JO (2022) Evaluation of the nutritive value of air-dried and sun-dried sweet potato (*Ipomoea batatas*) peels. Eur J Life Saf Stability 14:43–51. Retrieved from http://ejlss.indexedresearch.org/index.php/ejlss/article/view/452

Ahmad P, Jaleel CA, Salem MA, Nabi G, Sharma S (2010) Roles of enzymatic and nonenzymatic antioxidants in plants during abiotic stress. Crit Rev Biotechnol 30:161–175. https://doi.org/10.3109/07388550903524243

Akoetey W, Britain MM, Morawicki RO (2017) Potential use of byproducts from cultivation and processing of sweet potatoes. Ciência Rural 47(05):e20160610. https://doi.org/10.1590/0103-8478cr20160610

Alam MK (2021) A comprehensive review of sweet potato (*Ipomoea batatas* [L.] Lam): Revisiting the associated health benefits. Trends Food Sci Technol 115:512–529. https://doi.org/10.1016/j.tifs.2021.07.001

Alam M, Rana Z, Islam C (2016) Comparison of the proximate composition, total carotenoids and total polyphenol content of nine orange-fleshed sweet potato varieties grown in Bangladesh. Foods 5:64. https://doi.org/10.3390/foods5030064

Almazan AM (1995) Antinutritional factors in sweetpotato greens. J Food Compos Anal 8:363–368. https://doi.org/10.1006/jfca.1995.1031

Amagloh FC, Kaaya AN, Yaya B, Chelangat DM, Katungisa A, Amagloh FK, Tumuhimbise GA (2022) Bioactive compounds and antioxidant activities in peeled and unpeeled sweet-potato roots of different varieties and clones in Uganda. Future Foods 6:100183. https://doi.org/10.1016/j.fufo.2022.100183

Asadi K, Ferguson LR, Philpott M, Karunasinghe N (2017) Cancer-preventive properties of an anthocyanin-enriched sweet potato in the APCMIN mouse model. J Cancer Prev 22(3):135–146. https://doi.org/10.15430/JCP.2017.22.3.135

Bernabeu M, Salgado-Ramos M, Barba FJ, Collado C, Castagnini JM (2024) Impact of sweet potato peels extracts obtained by pulsed electric fields on the growth of probiotic strains from *Lactobacillus* genus. Innov Food Sci Emerg Technol 92:103590. https://doi.org/10.1016/j.ifset.2024.103590

Cao Y, Tian B, Zhang ZG, Yang K, Cai M, Hu W, Guo Y, Xia Q, Wu W (2022) Positive effects of dietary fiber from sweet potato [*Ipomoea batatas* (L.) Lam.] peels by different extraction methods on human fecal microbiota *in vitro* fermentation. Front Nutr 9:986667. https://doi.org/10.3389/fnut.2022.986667

Castillo Leon J, Alpeeva IS, Chubar TA, Galaev IY, Csoregi E, Sakharov IY (2002) Purification and substrate specificity of peroxidase from sweet potato tubers. Plant Sci 163:1011–1019. https://doi.org/10.1016/S0168-9452(02)00275-3

Chan CF, Lien CY, Lai YC, Huang CL, Liao WC (2010) Influence of purple sweet potato extracts on the UV absorption properties of a cosmetic cream. J Cosmet Sci 61(5):333–341. PMID: 20939973

Cui L, Liu CQ, Li DJ, Song JF (2011) Effect of processing on taste quality and health-relevant functionality of sweet potato tips. Agric Sci China 10(3):456–462. https://doi.org/10.1016/S1671-2927(11)60025-4

de Albuquerque TMR, Sampaio KB, de Souza EL (2019) Sweet potato roots: Unrevealing an old food as a source of health promoting bioactive compounds—a review. Trends Food Sci Technol 85:277–286. https://doi.org/10.1016/j.tifs.2018.11.006

Dongowski G, Lorenz A (2004) Intestinal steroids in rats are influenced by the structural parameters of pectin. J Nutr Biochem 15(4):196–205. https://doi.org/10.1016/S0955-2863(03)00080-9

Gama ABND, Silva EAD, Andrade Junior VCD, Brito OG, Costa ALD, Cavasin PY, Carvalho EEN (2023) Potential use of sweet potato leaves for human consumption. Pesq Agrop Brasileira 58:e02991. https://doi.org/10.1590/S1678-3921.pab2023.v58.02991

Ghamsemzadeh A, Omidvar V, Jaafar HZE (2022) Polyphenolic content and their antioxidant activity in leaf extract of sweet potato (*Ipomoea batatas*). J Med Plants Res 6:2971–2976. https://doi.org/10.5897/JMPR011.1353

Ghasemi K, Tasnim S, Mahmud SP (2022) Nano/microencapsulation and slurries: A review of fundamentals, categories, fabrication, models and applications. Sustain Energy Technol Asses 52:102084. https://doi.org/10.1016/j.seta.2022.102084

Gong L, Feng D, Wang T, Ren Y, Liu Y, Wang J (2020) Inhibitors of α-amylase and α-glucosidase: Potential linkage for whole cereal foods on prevention of hyperglycemia. Food Sci Nutr 8:6320–6337. https://doi.org/10.1002/fsn3.1987

Guo Q, Mu TH (2011) Emulsifying properties of sweet potato protein: Effect of protein concentration and oil volume fraction. Food Hydrocoll 25:98–106. https://doi.org/10.1016/j.foodhyd.2010.05.011

Hamidon NH, Zaidel DNA, Jusoh YMM (2020) Optimization of pectin extraction from sweet potato peels using citric acid and its emulsifying properties. Recent Patents Food Nutr Agric 11:202–210

Hong CY, Jo YJ, Kim MY, Chung MN, Choi EK, Kim YB, Chung MN, Choi E-K, Kim Y-B, Lee J, Jeong HS (2022) Biological activities of sweet potato (*Ipomoea batatas* L.) tips and tubers. Food Sci Nutr 4041–4048. https://doi.org/10.1002/fsn3.2999

Hsu H-Y, Chen B-H (2022) A comparative study on inhibition of breast cancer cells and tumors in mice by carotenoid extract and nanoemulsion prepared from sweet potato (*Ipomoea batatas* L.) peel. Pharmaceutics 14:980. https://doi.org/10.3390/pharmaceutics14050980

Huaman Z (1992) Systemic botany and morphology of the sweet potato plant. Technical information bulletin 25. International Potato Centre, Lima, Peru. 22 p. Retrieved from https://www.sweetpotatoknowledge.org/wp-content/uploads/2016/01/Systematic-Botany-and-Morphology-of-the-Sweetpotato-Plant.pdf. Accessed 8 Apr 2024

Ishida H, Suzuno H, Sugiyama N, Innami S, Tadokoro T, Maekawa A (2000) Nutritive evaluation on chemical components of leaves, stalks and stems of sweet potatoes (*Ipomoea batatas* poir). Food Chem 68:359–367. https://doi.org/10.1016/s0308-8146(99)00206-x

Islam S (2008) Antimicrobial activities of *Ipomoea batatas* (L.) leaf. J Food Agric Environ 6(1):14–17. https://doi.org/10.1234/4.2008.1070

Islam S, Yoshimoto M, Ishiguro K, Yamakawa O (2003) Bioactive compounds in *Ipomoea batatas* leaves. ISHS Acta Hortic 2:693–699. https://doi.org/10.17660/ActaHortic.2003.628.88

Jia R, Tang C, Chen J, Zhang X, Wang Z (2022) Total phenolics and anthocyanins contents and antioxidant activity in four different aerial parts of leafy sweet potato (*Ipomoea batatas* L.). Molecules 27:3117. https://doi.org/10.3390/molecules27103117

Joanne Toy YH, See JW, Huang D (2022) Physicochemical and functional characterisation of pectin from margarita sweet potato leaves. Food Chem 385:132684

Ju D, Mu T, Sun H (2017) Sweet potato and potato residual flours as potential nutritional and healthy food material. J Integr Agric 16:2632–2645. https://doi.org/10.1016/S2095-3119(16)61601-5

Jung JK, Lee SU, Kozukue N, Levin CE, Fiedman M (2011) Distribution of phenolic compounds and antioxidative activities in parts of sweet potato (*Ipomoea batata* L.) plants and in home processed roots. J Food Compos Anal 24:29–37. https://doi.org/10.1016/j.jfca.2010.03.025

Kang H, Lee SG (2021) Protective effect of purple sweet potato leaf (*Ipomoea batatas* Linn Convolvulaceae) against alcohol-induced liver damage in mice. Trop J Pharm Res 20(2):301–308. https://doi.org/10.4314/tjpr.v20i2.12

Karthick C, Nanthagopal K (2021) A comprehensive review on ecological approaches of waste to wealth strategies for production of sustainable biobutanol and its suitability in automotive applications. Energy Convers Manag 239:114219. https://doi.org/10.1016/j.enconman.2021.114219

Kurata R, Kobayashi T, Ishii T, Niimi H, Niisaka S, Kubo M, Kishimoto M (2017) Influence of sweet potato (*Ipomoea batatas* L.) leaf consumption on rat lipid metabolism. Food Sci Technol Res 23(1):57–62. https://doi.org/10.3136/fstr.23.57

Kurata R, Sun HN, Oki T, Okuna S, Ishiguro K, Sugawara T (2019) Sweet potato chemistry, processing and nutrition. In: Mu TH, Singh J (eds) Sweet potato polyphenols. Elsevier, Amsterdam, The Netherlands, pp 177–222

Kusuma SAF, Wahyuni UT, Zuhrotun A (2017) Evaluation of antibacterial activity of Indonesian varieties sweet potato leaves extract from Cilembu against *Shigella dysenteriae* ATCC 13313. Asian J Pharm Clin Res 10(2):377–380

Lee SL, Chin TY, Tu SC, Wang YJ, Hsu YT, Kao MC, Wu YC (2015) Purple sweet potato leaf extract induces apoptosis and reduces inflammatory adipokine expression in 3t3-l1 differentiated adipocytes. Evid Based Complement Altern Med. https://doi.org/10.1155/2015/126302

Lim SF, Matu SU (2015) Utilization of agro-wastes to produce biofertilizer. Int J Energy Environ Eng 6:31–35. https://doi.org/10.1007/s40095-014-0147-8

Liu T, Wu F, Chen K, Pan B, Yin X, You Y, Song Z, Li D, Huang D (2022) Sweet potato extract alleviates high-fat-diet-induced obesity in C57BL/6J mice, but not by inhibiting pancreatic lipases. Front Nutr 9:1016020. https://doi.org/10.3389/fnut.2022.1016020

Luo D, Mu T, Sun H (2021) Profiling of phenolic acids and flavonoids in sweet potato (*Ipomoea batatas* L.) leaves and evaluation of their anti-oxidant and hypoglycemic activities. Food Biosci 39:100801. https://doi.org/10.1016/j.fbio.2020.100801

Ma DF (2019) Global market trends, challenges, and the future of the sweet potato processing industry. In: Mu TH, Singh J (eds) Sweet potato chemistry, processing and nutrition. Academic, London, UK, pp 381–392

Mahfudh N, Sulistyani N, Syakbani M, Dewi AC (2021) The antihyperlipidaemic and hepatoprotective effect of *Ipomoea batatas* L. leaves extract in high-fat diet rats. Int J Public Health 10(3):558–564

Mahmood M, Moore PS, Tommasi ND, Simone FD, Colman S, Hay AJ, Pizza C (1993) Inhibition of HIV infection by caffeoylquinic acid derivatives. Antivir Chem Chemother 4:236–240. https://doi.org/10.1177/095632029300400406

Makori SI, Mu T-H, Sun H-N (2020) Total polyphenol content, antioxidant activity, and individual phenolic composition of different edible parts of 4 sweet potato cultivars. Nat Prod Commun 15(7):1–12. https://doi.org/10.1177/1934578X20936931

Maloney KP, Truong V-D, Allen JC (2012) Chemical optimization of protein extraction from sweet potato (*Ipomoea batatas*) peel. J Food Sci 77:E307–E311. https://doi.org/10.1111/j.1750-3841.2012.02921.x

Marconato AM, Hartmann GL, Santos MMR, Amaral LA, Souza GHO, Santos EF, Novello D (2020) Sweet potato peel flour in hamburger: effect on physicochemical, technological and sensorial characteristics. Braz J Food Technol 23:e2019115. https://doi.org/10.1590/1981-6723.11519

Meira M, Pereira DE, David JM, David JP (2012) Review of the genus Ipomoea: traditional uses, chemistry and biological activities. Rev Bras 22:682–713. https://doi.org/10.1590/S0102-695X2012005000025

Meng H, Xu C, Wu M, Feng Y (2022) Effects of potato and sweet potato flour addition on properties of wheat flour and dough, and bread quality. Food Sci Nutr 10(3):689–697. https://doi.org/10.1002/fsn3.2693

Mondal S, Burgert S, Asch J, Rahaman EHMS, Asch F (2023) Salinity effects on the activities of ROS scavenging enzymes in leaves of two sweet potato clones. J Agron Crop Sci 209:84–853. https://doi.org/10.1111/jac.12657

Mu T-H, Liu Y, Zhang M, Sun H-N (2014) Protein recovery from sweet potato starch wastewater by foam separation. Sep Sci Technol 49:2255–2260

Mu T, Sun H, Zhang M, Wang C (2017a) Sweet potato proteins. In: Jones G, Chan K (eds) Sweet potato processing technology. Academic, London, pp 49–119

Mu T, Sun H, Zhang M, Wang C (2017b) Sweet potato pectin. In: Jones G, Chan K (eds) Sweet potato processing technology. Academic, London, pp 183–261

Mu T, Sun H, Zhang M, Wang C (2017c) Sweet potato dietary fiber. In: Jones G, Chan K (eds) Sweet potato processing technology. Academic, London, pp 121–181

Nagai M, Tani M, Kishimoto Y, Iizuka M, Saita E, Toyozaki M, Kamiya T, Ikeguchi M, Kondo K (2011) Sweet potato (*Ipomoea batatas* L.) leaves suppressed oxidation of low density lipoprotein (LDL) in vitro and in human subjects. J Clin Biochem Nutr 48(3):203–208

Naz S, Naqvi SAR, Khan ZA, Mansha A, Ahmad M, Zahoor AF, Hussain Z (2017) Antioxidant, antimicrobial and antiproliferative activities of peel and pulp extracts of red and white varieties of *Ipomoea batatas* (L) Lam. Trop J Pharm Res 16(9):2221–2229

Nedunchezhiyan M, Jata SK, Byju G, Veena SD (2011) Effect of tuber crop wastes/byproducts on nutritional and microbial composition of vermicomposts and duration of the vermicomposting process. J Bot 801703. https://doi.org/10.1155/2011/801703

Nguyen HC, Chen C-C, Lin K-H, Chao P-Y, Lin H-H, Huang M-Y (2021) Bioactive compounds, antioxidants, and health benefits of sweet potato leaves. Molecules 26:1820. https://doi.org/10.3390/molecules26071820

Ntchapda F, Tchatchouang FC, Miaffo D, Maidadi B, Vecchio L, Talla RE, Bonabe C, Etet PFS, Dimo T (2021) Hypolipidemic and anti-atherosclerogenic effects of aqueous extract of *Ipomoea batatas* leaves in diet-induced hypercholesterolemic rats. J Integr Med 19(3):243–250. https://doi.org/10.1016/j.joim.2021.02.002

Ojokoh AO (2005) Changes in nutrient and antinutrient contents of sweet potato (*Ipomea batatas*) peels subjected to solid substrate fermentation. Biosci Biotechnol Res Asia 3:17–20. Retrieved from https://www.biotech-asia.org/vol3no1/changes-in-nutrient-and-antinutrient-contents-of-sweet-potato-ipomea-batatas-peels-subjected-to-solid-substrate-fermentation/

Okudaira R, Kyanbu H, Ichiba A, Toyokawa T (2005) *Ipomoea* extracts with disaccharidase-inhibiting activities. Jpn J Sanitary Zool 5:213–221

Pattnaik M, Pandey P, Martin GO, Mishra HN, Ashokkumar M (2021) Innovative technologies for extraction and microencap-sulation of bioactives from plant-based food waste and their applications in functional food development. Foods 10(2):279. https://doi.org/10.3390/FOODS10020279

Picha DH, Janick J (2009) Sweet potato: Origins, production, and utilization. In: Janick J (ed) Horticultural review, vol 35. Wiley, pp 157–217

Pochapski MT, Fosquiera EC, Esmerino LA, Dos Santos EB, Farago PV, Santos FA, Groppo FC (2011) Phytochemical screening, antioxidant, and antimicrobial activities of the crude leaves extract from *Ipomoea batatas* (L.) Lam. Pharmacogn Mag 7(26):165–170

Rodriguez G, Prinyawiwatkul W, Aryana KJ, King JM, Xu Z (2023) Bound form terpenes in sweet potatoes and their distribution in flesh and peel of different cultivars. Int J Food Sci Technol 58(11):5773–5780. https://doi.org/10.1111/ijfs.16675

Senthilkumar R, Yeh K-W (2012) Multiple biological functions of sporamin related to stress tolerance in sweet potato (*Ipomoea batatas* Lam). Biotechnol Adv 30:1309–1317

Shih CK, Chen CM, Varga V, Shih LC, Chen PR, Lo SF, Shyur LF, Li SC (2020) White sweet potato ameliorates hyperglycemia and regenerates pancreatic islets in diabetic mice. Food Nutr Res 64. https://doi.org/10.29219/fnr.v64.3609

Silva-Correa CR, Ortiz-Noriega CM, Villarreal-La Torre VE, Calderón Peña AA, Aspajo-Villalaz CL, Guerrero-Espino LM, Sagástegui-Guarniz WA, González-Siccha AD, González-Blas MV, Cruzado-Razco JL, Del Rosario-Chávarri J (2021) Effect of a gel based on *Ipomoea batatas* (purple sweet potato) on dermal wound healing in mice. Pharm J 13(6):1720–1726

Statista (2024) Sweet potato production worldwide from 2010 to 2021 (in million metric tons). Retrieved from https://www.statista.com/statistics/812343/global-sweet-potato-production/. Accessed 8 Apr 2024

Su X, Griffin J, Xu J, Ouyang P, Zhao Z, Wang W (2019) Identification and quantification of anthocyanins in purple-fleshed sweet potato leaves. Heliyon 5:e01964. https://doi.org/10.1016/j.heliyon.2019.e01964

Suárez S, Mu T, Sun H, Añón MC (2020) Antioxidant activity, nutritional, and phenolic composition of sweet potato leaves as affected by harvesting period. Int J Food Prop 23(1):178–188. https://doi.org/10.1080/10942912.2020.1716796

Sun H, Mu T, Xi L, Zhang M, Chen J (2014) Sweet potato (*Ipomoea batatas* L.) leaves as nutritional and functional foods. Food Chem 156:380–389. https://doi.org/10.1016/j.foodchem.2014.01.079

Sun H, Zhang P, Zhu Y, Lou Q, He S (2018) Antioxidant and prebiotic activity of five peonidin-based anthocyanins extracted from purple sweet potato (*Ipomoea batatas* (L.) Lam.). Sci Rep 8:5018

Tang CC, Ameen A, Fang BP, Liao MH, Chen JY, Huang LF, Zou HD, Wang ZY (2021) Nutritional composition and health benefits of leaf-vegetable sweet potato in South China. J Food Compos Anal 96:103714. https://doi.org/10.1016/j.jfca.2020.103714

Teow CC, Troung V-D, McFeeters RF, Thompson RL, Pecota KV, Yencho GC (2007) Antioxidant activities, phenolic and β-carotene contents of sweet potato genotypes with varying flesh colours. Food Chem 103:829–838. https://doi.org/10.1016/j.foodchem.2006.09.033

Truong VD, Avula RY, Pecota KV, Yencho GC (2018) Sweet potato production, processing, and nutritional quality. In: Siddiq M, Uebersax MA (eds) Handbook of vegetables and vegetable processing. Wiley, Oxford, UK, pp 811–838

Vannini M, Marchese P, Sisti L, Saccani A, Mu T, Sun H, Celli A (2021) Integrated efforts for the valorization of sweet potato by-products within a circular economy concept: biocomposites for packaging applications close the loop. Polymers 13(7):1048. https://doi.org/10.3390/polym13071048

Vincent Chang H-S, Diane Yang H-A, Lin H-H, Pearce G, Ryan CA, Chen Y-C (2013) IbACP, a sixteen-amino-acid peptide isolated from *Ipomoea batatas* leaves, induces carcinoma cell apoptosis. Peptides 47:148–156

Vishnu VR, Renjith RS, Mukherjee A, Anil SR, Sreekumar J, Jyothi AN (2019) Comparative study on the chemical structure and *in vitro* antiproliferative activity of anthocyanins in purple root tubers and leaves of sweet potato (*Ipomoea batatas*). J Agric Food Chem 67(9):2467–2475

Vithu P, Sanjaya KD, Kalpana R (2019) Post-harvest processing and utilization of sweet potato: A review. Food Rev Int 35:726–762. https://doi.org/10.1080/87559129.2019.1600540

Wang S, Nie S, Zhu F (2016) Chemical constituents and health effects of sweet potato. Food Res Int 89:90–116. https://doi.org/10.1016/j.foodres.2016.08.032

Xu S, Bai Z, Jin B, Xiao R, Zhuang G (2014) Bioconversion of wastewater from sweet potato starch production to *Paenibacillus polymyxa* biofertilizer for tea plant. Sci Rep 4:4131. https://doi.org/10.1038/srep04131

Yong ZJ, Bashir MJ, Hassan MS (2021) Biogas and biofertilizer production from organic fraction municipal solid waste for sustainable circular economy and environmental protection in Malaysia. Sci Total Environ 776:145961. https://doi.org/10.1016/j.scitotenv.2021.145961

Yoshimoto M, Yahara S, Islam M, Islam MS, Yamakawa O (2002) Antimutagenicity of mono-, di-, and tricaffeoylquinic acid derivatives isolated from sweet potato (*Ipomoea batatas* L.) leaf. Biosci Biotechnol Biochem 66:2336–2341. https://doi.org/10.1271/bbb.66.2336

Yustiantara PS, Warditiani NK, Sari PMNA, Dewi NLKAA, Ramona Y, Jawi IM, Wirasuta IMAG (2021) Determination of TLC fingerprint biomarker of *Ipomoea batatas* (L.) Lam leaves extracted with ethanol and its potential as antihyperglycemic agent. Pharmacia 68(4):907–917. https://doi.org/10.3897/pharmacia.68.e71334

Wayne LC, Lai YC, Yuan MC, Hsu YL, Chan CF (2011) Antioxidative activity of water extract of sweet potato leaves in Taiwan. Food Chem 127:1224–1228. https://doi.org/10.1016/j.foodchem.2011.01.131

Bioactive Phytochemicals from Potato (*Solanum tuberosum*) By-products

Moncef Chouaibi

Contents

Abstract The potato is a tuber with high global demand and importance in the human diet. It is used especially by food industries to manufacture processed products (French fries, chips, and purees), and consequently, significant quantities of

M. Chouaibi (✉)
Higher School of Food Industries of Tunisia, University of Carthage, Elkhadra City, Tunisia

Department of Chemical and Food Engineering (DICIV), Università degli Studi di Salerno | UNISA, Fisciano, Italy
e-mail: moncef.chouaibi@esiat.ucar.tn

M. F. Ramadan (ed.), *Bioactive Phytochemicals in By-products from Leaf, Stem, Root and Tuber Vegetables*, https://doi.org/10.1007/978-3-031-80700-8_13

potato peel are generated as a by-product. Potato peels are rich in compounds like ascorbic acid, anthocyanins, glycoalkaloids, phenolic compounds, and flavonoids. These biomolecules are related to good health because they protect the body's cells from oxidative processes and have antioxidant, anti-inflammatory, antitumor, and anticancer properties. Therefore, new extraction methods were applied to recover these compounds and their application. The most extensive extraction of these compounds is maceration, ultrasound-assisted extraction (UAE), microwave-assisted extraction (MAE), and supercritical fluid extraction (SFE). This current work aims to describe the physicochemical composition of potato peels. Besides, the phytochemical composition showed that the peeled potatoes are rich in phenolic compounds, especially chlorogenic acid and anthocyanins. The most common application of peeled potatoes was in food industries, especially cereal products, which can also be employed in food systems to extend the shelf-life of food products.

Keywords Potatoes · Peels · Bioactive compounds · Phytochemical composition · Potential uses

1 Introduction

The potato (*Solanum tuberosum*) is the third most used crop globally, behind rice and wheat (International Potato Center 2013). Potatoes are versatile crops that can be cultivated in diverse environments. They are currently grown in 100 different countries. Potato crops can yield 9.2 M calories per acre, which is more than that of maize (7.5 M), rice (7.4 M), wheat (3 M), and soybean (2.8 M) (Ensminger et al. 1994). Potatoes are an excellent candidate for biofortification to address malnourishment in developing countries, feed a growing global population, and fit into a healthy part of the American diet.

Potatoes combine several health-promoting vitamins and nutrients. Notably, they are a key source of vitamin C. One serving of boiled potatoes (150 g) provides about 25% of the vitamin C, making potatoes a larger contributor to the world's vitamin C intake than lemons, oranges, and other seasonal fruits and vegetables. They also contain high levels of phenolic compounds, carotenoids, chlorogenic acid, and vitamins B6 and E, which slow the aging process and prevent inflammation, chronic disease, cardiovascular disease, and cancer. Interestingly, potassium amounts in potatoes exceed even those found in foods typically associated with essential minerals, such as bananas, avocadoes, and orange juice. They contain no fat or gluten. The protein amount is quite low.

Nonetheless, the quality is very high, approaching that of egg whites. Potatoes are an excellent source of carbohydrates and an essential source of energy. However, peeled potatoes have a lower caloric density than all other common side dishes, such as rice, pasta, or bread. This is due to their fiber, resistant starch, high water content, and special dry matter composition. Fiber and resistant starch also make potatoes exceptionally satiating, a key reflection in diets and weight loss procedures.

Potatoes (*Solanum tuberosum* L.) are starchy tuberous crops that represent the world's fourth most important crop. According to the "Food and Agricultural Organisation of the United Nations," 376 million metric tonnes of potatoes were produced globally in 2021. It is supposed that 8000 kilotons of potato peel waste might be generated in 2030, with related greenhouse gas emissions of 5 million tons of CO equivalent. Waste generated throughout the food supply chain, from potato farms to forks, contributes considerably to global warming. Apart from animal feed, technical processes in the potato processing industry generate waste organic residues with high amounts of bioactive compounds and carbohydrates, and thus extraction and bioconversion processes can produce high-value products. These include the isolation of functional ingredients for the formulation of nutraceuticals and pharma products, bioenergy-related products, enzymes, and fertilizers for the market, hence contributing to a more responsible production and consumption and, overall, circular economy.

The extraction of plant bioactive molecules has been largely explored and well-established by many authors as an efficient route for the reutilization of food processing waste, especially when considering the recent technological advances in molecular separations and identifications (Wijngaard et al. 2012). Peel potatoes can be a rich source of bioactive compounds due to their high contents of phenolic compounds with recognized health-promoting properties, such as antioxidant activity, that can also be employed in food systems to extend the shelf-life of food products (Albishi et al. 2013; Friedman et al. 2018).

2 Composition of Potato By-products

2.1 Water

The main physical and chemical properties of industrial potato peel are water and carbohydrates. Potato peel is mostly composed of water, representing almost 90%. However, the 10% of solids mainly contain organic matter (89.3%) (Table 1). Similar results were reported in the literature by Jimenez-Champi et al. (2023). As expected, carbon and oxygen are the main elements present in potato peels, accounting for 47.3 and 32.9%, respectively. The sulfur content was below the detection

Table 1 Chemical composition of raw potato peel (PP), g per 100 g

Composition (%)	Value
Water	83.3–85.1
Protein	1.2–2.3
Total lipids	0.1–0.4
Total carbohydrates	8.7–12.4
Starch	7.8

Adapted from Javed et al. (2019)

limit. Other authors have reported similar elemental analyses for potato peel samples (Jimenez-Champi et al. 2023).

2.2 Minerals

Minerals have many essential functions in the potato plant, such as structural, energetic, or osmotic, and they are components of various enzymes. The need for potato plants depends on many parameters, such as weather conditions, soil type, nutrient content in the soil, location, timing and the dose of fertilization, and the requirements of the variety. Maximum yield can be achieved through optimal and balanced nutrition (van Eck 2007). An increase in the mineral element supplied to the plant in fertilizers leads to the rise in the content of this element in the potato plant and tubers. However, the uptake of elements by plants is mutually influenced by the ratio of elements in the soil, as well as by the distribution of elements in the plant body (White et al. 2009).

The use of minerals in the human body from potatoes is relatively high because they contain high levels of vitamin C, which increases the utilization of minerals (especially iron and zinc). Bioavailability is also augmented because of low amounts of phytic acid and oxalates (which reduce mineral bioavailability in many other crops) (White et al. 2009). The mineral amount is larger in the skin and near the skin than in the flesh, as mentioned by Boškovic-Rakocevic et al. (2018) and Rybácek (1988).

Potassium is the most abundant mineral in potatoes, as it contains about 400 mg/100 g FW according to USDA, but it can vary between 150 and 1386 mg/100 g FW (Nassar et al. 2012). According to the National Institutes of Health, USA, the recommended daily intake of potassium is 4700 mg per capita, and 100 g of boiled potatoes contributes up to 16% of the recommended daily intake. Since most of the potassium is in the skin and near the skin, the best way to preserve as much potassium as possible in potatoes is to boil or bake them with the skin. Boiling diced or shredded potatoes reduces potassium levels by 50% and 75%, respectively (Bethke and Jansky 2008). Potassium helps control blood pressure, and it can reduce the risk of stroke.

Potatoes contain much more potassium than other side dishes. Potatoes (baked with skin) even contain more potassium (544 mg/100 g) than foods associated with high potassium levels such as bananas (358 mg/100 g), orange juice (200 mg/100 g), and avocado (485 mg/100 g) (Dietary Guidelines Advisory Committee 2015).

2.3 Protein

Potato protein content is 1.7–2.1 g/100 g FW (Vreugdenhil and Bradshaw 2007) and about 10 g/100 g DM (range of 8.7–14.2 g/100 g), which is quite low, but the amino acid composition has an exceptionally very high biological value (90–100%), which is similar to the whole egg standard (100%), and higher than cow's milk (84–88%), beef (83–92%), or beans (73%) (Vreugdenhil and Bradshaw 2007). Potato protein contains high levels of all nine essential amino acids (including lysine, which is usually low in plant sources such as rice or pasta) but low in sulfur amino acids (Camire et al. 2009; Woolfe 1987). The main proteins contained in potatoes are patatin (patatin can be an allergen for some people, but it is well reduced by cooking) and a group of inhibitors of proteases (Camire et al. 2009). The contribution of protein intake from potatoes can be higher than intake from legumes, which have a high protein content but are not eaten in such large quantities (Bárta et al. 2008). Potato protein content is mainly influenced by variety, growing conditions, and fertilization. Growing potatoes at higher altitudes increases protein content. Since proteins are 16% nitrogen, nitrogen fertilizers have a significant effect on protein content (Bártová et al. 2009).

Potato protein content is also influenced by the cooking method. Raw potatoes contain 2 g/100 g FW, boiled potatoes 1.8 g/100 g FW, and baked potatoes 2.2 g/100 g FW (Vreugdenhil and Bradshaw 2007). According to Pęksa et al. (2018), the protein content decreases during storage (6.4% after 3 months and 19% after 6 months). Potatoes have a similar protein content (2 g/100 g) as other vegetables and fruits, with the exception of legumes and beans, with 5–8 g/100 g of protein (Gebhardt and Thomas 2002).

2.4 Total Lipids

Potatoes are low in fat and low in calories compared to other common side dishes. Boiled or baked potatoes contain only 0.1 g of fat/100 g FW, which is less than boiled pasta (0.5 g/100 g FW) and boiled rice (1.95 g/100 g FW) (Priestley 2006). The fat content increases rapidly after frying, so fried potato products such as crisps and French fries contain 36 g/100 g and 19 g/100 g of fat, respectively (Gebhardt and Thomas 2002). Most of the lipids in potatoes are structural elements of biological membranes (phospholipids, glycols, and galactolipids) and neutral lipids (acylglycerols and free fatty acids) (Ramadan and Oraby 2016).

2.5 Starch

Potatoes usually contain 15–20% starch in fresh weight, and they can range from 11.0 to 30.4%, with an average of 18.8% (Singh and Kaur 2016). Approximately 70% of the dry matter is starch (Lu et al. 2011). The starch content is influenced by genotype and maturity, as late varieties usually contain more starch, and very early varieties are low in starch (van Eck 2007). Too high a dose of nitrogen fertilizer reduces the starch content (Hajšlová et al. 2005; Vokál 2013). Starch is composed of amylopectin and amylose in a relatively stable ratio of 3:1 (Akyildiz et al. 2015; Woolfe 1987). Amylopectin is highly branched, being formed of 2,000 to 200,000 glucose units. Its inner chains are formed of 20–24 glucose subunits. Dissolved amylopectin starch has a lower tendency of retrogradation (a partial recrystallization after cooking—a part of the staling process) during storage and cooling. Potato starch contains typical large oval spherical granules ranging in size from 5 to 100 μm (Liu et al. 2007). From a digestive point of view, starch is divided into rapidly digestible starch (RDS), slowly digestible starch (SDS), and resistant starch (RS). RDS is digested in the small intestine within 20 min, SDS is also digested in the small intestine within 20–120 min, and RS is not digested, and it reaches the large intestine, where microbes ferment it. RDS and SDS consist mainly of amylopectin and have a higher glycemic index, and RS consists primarily of amylose, which helps maintain gut health. The starch in raw potatoes is substantially resistant to amylolytic enzymes, so it acts like resistant starch (RS) (Camire et al. 2009). After cooking (at 70 °C), starch gelatinizes, making it more accessible to amylolytic enzymes. Amylopectin with a highly branched structure is easier to digest than linear amylose, so amylopectin has a higher response for a glycemic index of potato. Part of the starch (more amylose) is also resistant to digestion in the small intestine after cooking, and it behaves like fiber and has similar health benefits in the gut. After cooking and cooling, potatoes high in amylose-starches have greater retrogradation, which results in more crystalline structure and resistance to digestion (amylopectin is digested easily because the retrogradation is less common). Potatoes contain 68% RDS, 3% SDS, and 3.9% RS immediately after cooking, but 44% RDS, 23% SDS, and 7% RS after cooling cooked potatoes (Monro et al. 2009). Thus, cooked and cooled potatoes have a higher RS content than cooked hot potatoes, and therefore, they improve gut health, such as fiber, and have a lower glycemic effect. Raatz et al. (2016) found significant differences between cooking methods and RS content, with the highest RS found in cooled (boiled or baked) potatoes (4.3%). Reheated potatoes (3.5%) and baked potatoes had more RS (3.6%) than hot-boiled potatoes (2.4%) (Camire et al. 2009; Šimková et al. 2013).

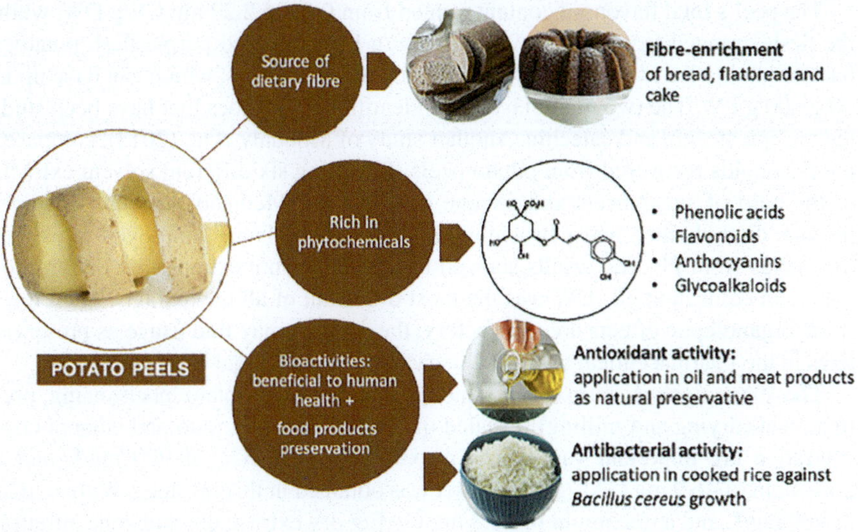

Fig. 1 Potato peels and their main interesting features for the food industries

3 Phytochemical Composition

Bioactive substances such as flavonoids, phenolic compounds, glycoalkaloids, and anthocyanins are abundant in potato peels. These substances are linked to good health because they have anti-inflammatory, antitumor, anticancer, and antioxidant traits that shield the body's cells from oxidative processes. Food processing operations produce large amounts of potato peel in the potato industry. Biomolecules with high added value are abundant in the leftover biomass from potato peels. Enzymatic techniques were used to extract the starch, minerals, and carbs from the potato skin (Fig. 1).

3.1 Phenolic Compounds

Chlorogenic acid (CGA) is the most prevalent phenolic acid in potatoes, and it is found in varying amounts in both the skin and the flesh. The chlorogenic acid content of fresh potato flesh and peel ranges from 30 to 900 mg/kg and 1000–4000 mg/kg, respectively. Other phenolic acids are present in trace amounts, ranging from 0 to 30 mg/kg. Four cultivars of purple flesh and two cultivars of yellow flesh were tested for antioxidant activity and total phenolic content (TPC). In contrast to the cultivars of yellow-fleshed potatoes, which displayed 2.96 g of gallic acid equivalent (GAE) per kg DM, the cultivars of purple-fleshed potatoes had an average of 4.68 g of GAE per kg DM.

The peel's total flavonoid content ranged from 0.06 to 2.29 mg CE/g DW, while the flesh's ranged from 0.51 to 9.50 mg CE/g DW. Red and purple flesh potatoes have around twice as many phenolics as white flesh potatoes, which can have up to 30 g/100 g FW. The two main flavonoids identified in potatoes that have been studied are epicatechin and catechin. Another study of Mohdaly et al. (2013), compared the flavonoids recovered from potato peels (PP) using six different solvent extracts to the pulp of sugar beets and sesame cake extract. Methanol, ethanol, acetone, hexane, diethyl ether, and petroleum ether were the solvents employed to extract flavonoids from PP. The results showed that the methanol extract with 0.96 mg of quercetin equivalent g/L DW was the most dominant of all the extracts. Therefore, if the organoleptic effects are satisfactory, the extracts may find usage as preservatives in the pharmaceutical and/or culinary industries (Khanal et al. 2024).

The extraction of antioxidants and phenolics from potato peels using pre-treatment (drying and milling) revealed that the liquid-solid ratio and ethanol concentration are important variables. An extract with an IC_{50} of 900 mg/L and a maximum of 46.9 mg GAE/g dry extract was obtained in this instance. With an IC_{50} of 900 mg/L and a maximum of 22.8 mg GAE/g dry extract, ethanol concentration was the sole relevant factor in the extraction of potato peel without pre-treatment. After three sequential extractions employing a multi-stage cross-flow with solvent, approximately 6.18 mg GAE/g PPt may be recovered. A concentration of 402.9 mg GAE/L was achieved in the third stage of a multi-stage cross-flow with feed (Khanal et al. 2024).

3.2 *Anthocyanins*

The amount of anthocyanidin in potatoes varies according to the peel's color. For instance, acylated glycosides of pelargonidin, which range from 20 to 200 mg/100 g FW, are the main component of those with red peels. Petunidin and pelargonidin acylated glucosides can reach 200–500/100 g in the purple types (Lachman and Hamouz 2005; Peña and Restrepo 2013). Petunidin, pelargonidin, peonidin, and malvidin are the results of identifying certain anthocyanidins in colored PPs (purple and red types). Malvidin was discovered in the HeiJingang potato, indicating that this anthocyanin is exclusively present in the purple types of potatoes, whereas Makori et al. (2022) found peonidin and pelargonidin in all PP samples. Similarly, only HeiJingang and Favorita potatoes contained pelargonidin, with values of 7.7 and 0.2 mg/100 g DW, respectively.

In their evaluation of the anthocyanidin content in several PPs, Bellumori et al. (2020) found that petunidin was the main anthocyanin, with values ranging from 22.1 to 4.3 mg/100 g DW. However, pelargonidin was only found in the Sangre de

Toro (12.4 mg/200 g DW) and Yawar Manto (6.2 mg/100 g DW) kinds of potatoes, but malvidin was not found in any of them. The phenolic components of ten colored potato peels were examined by Sampaio et al. in 2021. They were able to recognize up to six acylated anthocyanidin peaks in purple and red potato types (Rosemary, Red Cardinal, and Red Emmalie).

While peonidin was only found in the red potatoes, petunidin and malvidin were found in all PP samples. With values ranging from 37.8 to 107.6 mg/100 g DW, petunidin was the main anthocyanidin found in purple potato types. Malvidin was also found in all purple potatoes, with the exception of Shetland Black. Overall, the authors discovered that the peels of purple-skinned potatoes contained more anthocyanins than those of red-skinned potatoes, with a total anthocyanin concentration of 1.3 mg/100 g DW.

3.3 *Glycoalkaloids*

Due to their toxicity to microbes, viruses, and insects, glycoalkaloids, which are secondary metabolites, shield potato plants from these threats. The primary glycoalkaloids found in PPs are α-solanine and α-chaconine. Because of the danger of colic and other stomach issues, including vomiting, diarrhea, and gastroenteritis, their consumption should not exceed 3–5 mg/kg of body weight (Rodríguez-Martínez et al. 2021). Glycoalkaloids are phytochemicals that are extensively researched in therapeutic settings despite their toxicity since, depending on dosage and usage circumstances, they may have antitumor, anticancer, and anti-inflammatory effects (Benkeblia 2020).

At doses of 30 μM or less, α-solanine can promote apoptosis in human JEG-3 choriocarcinoma cells and prevent the proliferation of various cancer cells (Gu et al. 2021). According to Zhang et al. (2021), α-solanine has demonstrated anticancer activity against leukemic cell lines (THP-1 and MV4-11) in vitro. It has also been shown to induce apoptosis and inhibit the proliferation of malignant cells in bladder cancer (Dong et al. 2022), suppress SGC-7901 cells that promote gastric cancer (Li et al. 2022), and inhibit the growth of tumors and the proliferation of human colorectal cancer cells. α-chaconine has been investigated as a possible anti-inflammatory agent. According to Lee et al. (2015), pro-inflammatory cytokines (IL-1β, IL-6, and TNF-α) and several inflammatory mediators, including COX-2, decrease α-chaconine.

4 Potential Uses

4.1 *Enhancement of Bacteria "Strains from* Lactobacillus *Genus"*

A pulsed electric field (PEF) was used to help value sweet potato (Sp) peels to produce extracts that promote the growth of possible probiotic bacteria. White sweet potatoes (WSp, "O'Henry"), white with purple (WPSp or "Violeta Roja"), orange (OSp or "California"), and purple (PSp, "Pepita") were the four types that were examined. The PEF extracts from PSp had a much larger amount of flavonoids, phenolics, and carbohydrates (12.5 mg/mL, 4.34 mg GAE/g, and 1.73 mg GAE/g, respectively) than those from maceration extraction.

Quantitative nuclear magnetic resonance (qNMR) analysis was used to determine that the primary sugars extracted were sucrose, fructose, and glucose. The total concentration of the PSp-PEF crude extracts was 94 mg/g PSp, which represents a 92% improvement over the conventional extraction method. These extracts enhanced the kinetics of probiotic bacteria and Lactobacillus growth and short-chain fatty acid (SCFA) production. According to Bernabeu et al. (2024), this characteristic should be mentioned because they are regarded as useful platform chemicals that can be converted into high-added-value products, suggesting an additional method of valorization.

The need to find novel, sustainable food for consumption and to process it using cutting-edge technologies makes this effort particularly relevant. A successful, noteworthy conversion of leftover sweet potato (Sp) peels to possible prebiotics would be made possible by the suitability of PEF-assisted extraction to be scaled up at the industrial level. This would increase the recovery of these compounds by PEF technology in comparison to conventional extraction methods and the health benefits.

4.2 *Cadmium Elimination*

A highly hazardous element called cadmium finds its way into the environment through the wastewater of sectors like electronics. Because of its extreme toxicity, international organizations control the release of wastewater containing cadmium into the environment. Researchers have looked at surface adsorption as one of the cadmium removal techniques because of its many benefits, including easy operation, high removal efficiency, cost-effectiveness, and broad access to various adsorbents. Cadmium was extracted from wastewater using the porous, crystalline mesopore and magnetic BPP/MnFeO@ZIF-8 material. Using RSM-CCD, the impact of parameters was investigated. The cadmium adsorption data followed the PSO kinetic. In the BPP/MnFe2O @ZIF-8 Cd^{2+} adsorption, the heterogeneous surface played a more significant impact. The process of adsorption was endothermic and spontaneous. The produced adsorbent cleaned the effluent from the shipbuilding sector (Foroutan et al. 2023).

4.3 Wheat Bread

The creation of functional food is setting the standard for improving the health benefits of diets. One of the easiest and safest ways to deal with some of these problems is to take prebiotics. However, more research is still needed to determine which basic food systems best transmit their useful qualities. Therefore, by creating bread that is both functional and nutritionally balanced, this study seeks to address the issue mentioned above. Wheat, haricot bean flour, and orange flashing sweet potato peel (OFSPP) were combined to create composite flour. Using the necessary techniques, the quality parameters of the created bread, the functional, pasting, and physical characteristics of composite flours, as well as the β-carotene and proximate content, were ascertained and examined.

The β-carotene level (ND–731 μg/100 g) and dietary fiber content (2–8.4 g/100 g) of the composite flours have also increased. The designed flour's protein level rose from 11.25 to 17.4 g/100 g. The addition of OFSPP and haricot bean flour up to 10% and 20%, respectively, did not significantly impair the bread's sensory acceptability, according to the consumer-based sensorial evaluation. However, more research on the gut microbiomes of humans and animals is still required to support the results of this study (Admasu et al. 2024).

4.4 Cake and Cookies

Since cakes are typically eaten as baked goods worldwide, adding nutrients to them is a good method to improve people's nutritional status. When making cakes, potato peel flour (PPF) can help attain this goal. The effects of sun and cabinet drying on the physicochemical, antioxidant, and functional characteristics of PPF (dried potato peel flour) were compared. The PPF-incorporated cakes' physical, textural, and sensory qualities were assessed. The potato peel was blanched (at 90–95 °C for 2–3 min), dried (at 65 and 30 °C), and ground to create the PPF. Its physical and chemical characteristics were then examined. According to the results, cabinet-dried PPF has a larger percentage of protein (11.3%), lipids (7.23%), fiber (12.9%), and carbohydrates (55.4%) than sun-dried PPF. It also has a lower percentage of moisture (7.61%) and ash (9.63%). Additionally, the cabinet-dried PPF exhibited greater mineral and antioxidant activity than the sun-dried PPF. It was discovered that the cabinet-dried PPF had superior functional qualities to the sun-dried PPF in terms of bulk density, foaming capacity, emulsifying activity, swelling capacity, and water and oil absorption capacity.

Different proportions of cabinet-dried PPF to wheat flour (100: 0, 96: 4, 94: 6, and 92: 8) were used to make composite cakes. When PPF was added, the cakes' height, volume, and weight increased in comparison to the cakes made with 100% wheat flour as a control. The cakes enhanced with 4% PPF received the greatest score for the sensory qualities in the sensory evaluation. In order to create composite cakes with improved physical and organoleptic properties, it was recommended to incorporate 4% cabinet-dried PPF with wheat flour (WF) (Akter et al. 2023).

4.5 *Biofilm for Cheese Preservation*

On the basis of starch (S) and agro-industrial wastes like potato (P) and/or quince (Q) peels, edible and biodegradable films for food packaging were created. The swelling behavior, water contact angles, water vapor sorption capacity, and water vapor permeability results all demonstrated that the films were hydrophilic. Perhaps as a result of the presence of plasticizing sugars and pectin, quince peel films were the most flexible.

The edible film of the cheeses was assessed throughout ripening, along with their sensory qualities and consumer acceptance. The SPQ (starch (S), potato peels (P), and quince (Q) peels) formulation film was selected for the proof of concept with cured cheeses. Compared to the control and natamycin-coated cheeses, the formed film made the cheese harder. Additionally, the film did not alter the cheeses' sensory qualities, making it a viable film for use in this kind of food product (Coimbra et al. 2023).

4.6 *Bioethanol Production*

A starchy by-product produced in large quantities during the industrial processing of potatoes is potato peel waste (PPW). It can be utilized as a renewable and inexpensive feedstock to produce second-generation bioethanol. To speed up this process, two experimental designs and response surface evaluations were carried out to enable very high gravity fermentations (VHGF) using PPW as feedstock. Saccharomyces cerevisiae Ethanol Red®, a resilient and thermotolerant yeast strain, was used. The first one concentrated on optimizing the liquefaction and enzymatic hydrolysis phases, allowing for a yield of 80.4% at 72 h of fermentation and a maximum ethanol concentration of 116.5 g/L. According to Rodríguez-Martínez et al. (2023), the second one concentrated on optimizing the pre-saccharification and fermentation stages, which further boosted process productivity and resulted in a maximum ethanol concentration of 108.8 g/L and a yield of 75.1% after 54 h of fermentation.

Because of these results, an intensified pre-saccharification and simultaneous saccharification and fermentation (PSSF) process for producing ethanol from PPW was defined. This process involved short liquefaction and pre-saccharification times of 2 and 10 h, respectively, at an enzyme loading of 80 U/g PPW of Viscozyme and 5 UE/g PPW of SAN Super, as well as a higher fermentation temperature of 34 °C because a thermotolerant yeast was used. Overall, the described PSSF procedure enabled achieving a high ethanol concentration and yield (104 g/L and 71.9%, respectively) standing at high productivity with only 54 h of fermentation under these circumstances and using only PPW without any supplements.

4.7 Biohydrogen Production

The potato industry has struggled with the sustainable handling of potato peel waste. The viability of using potato peel waste for dark fermentation-based biohydrogen generation was examined as well as the impact of different inoculum enrichment techniques (acid, aeration, heat-shock, and base) on process efficiency. When applying several inoculum enrichment techniques, it was found that the hydrogen output varied greatly. Significant differences were also seen in the metabolic pathway and bacterial community structure among the various enriched cultures. For the fermenter inoculated with aeration-treated culture, the maximum abundance of *Clostridium sensu* stricto significantly contributed to the highest process efficiency. This study closes a gap in the best inoculum preparation technique for biohydrogen fermentation of potato peel waste and proposes a viable recycling strategy (Cao et al. 2022).

5 Conclusion

Considering the nutritional value and chemical composition of potato peels, along with the bioactivity and technological properties of their extracts, the sustainable valorization of potato processing by-products presents great interest for the food and pharmaceutical industries that could increase the overall added value and minimize the environmental impact of this food crop. These bioactive compounds are associated with good health because they protect the body's cells from oxidative processes and have antioxidant, anti-inflammatory, antitumor, and anticancer properties. Hence, new methods to recover these compounds and their application to food, cosmetic, and pharmaceutical matrices are being explored.

References

Admasu F, Fentie GE, Admassu H, Shin J-H (2024) Functionalization of wheat bread with prebiotic dietary insoluble fiber from orange-fleshed sweet potato peel and haricot bean flours. LWT Food Sci Technol 200:116182. https://doi.org/10.1016/j.lwt.2024.116182

Akter M, Anjum N, Roy F, Yasmin S, Sohany M, Mahomud MS (2023) Effect of drying methods on physicochemical, antioxidant and functional properties of potato peel flour and quality evaluation of potato peel composite cake. J Agric Food Res 11:100508. https://doi.org/10.1016/j.jafr.2023.100508

Akyildiz A, Polat S, Agçam E, Fenercioglu H (2015) Potato and potato-processing technology. In: Hui YH, Özgül Evranuz E (eds) Handbook of vegetable preservation and processing. CRC, Boca Raton, FL, pp 589–627

Albishi T, John JA, Al-Khalifa AS, Shahidi F (2013) Phenolic content and antioxidant activities of selected potato varieties and their processing by-products. J Funct Foods 5(2):590–600. https://doi.org/10.1016/j.jff.2012.11.019

Bárta J, Čepl J, Diviš J, Hamouz K, Jůzl M, Vacek J (2008) Brambory. In: Kvalita rostlinných produktů na prahu 3. tisíciletí. Výzkumný ústav pivovarský a sladařský ve spolupráci s komisí jakosti rostlinných produktů ČAZV, Praha

Bártová V, Veronika, Bárta J, Jan, Diviš, Jirí, Švajner, Josef, Peterka, Department of Applied Plant Biotechnology (2009) Crude protein content in tubers of starch processing potato cultivars in dependence on different agro-ecological conditions. J Cent Eur Agric 10(1):57–66

Bellumori M, Silva NAC, Vilca L, Andrenelli L, Cecchi L, Innocenti M, Balli D, Mulinacci N (2020) A study on the biodiversity of pigmented Andean potatoes: Nutritional profile and phenolic composition. Molecules 25(14):3169. https://doi.org/10.3390/molecules25143169

Benkeblia N (2020) Potato glycoalkaloids: Occurrence, biological activities and extraction for biovalorisation—a review. Int J Food Sci Technol 55(6):2305–2313. https://doi.org/10.1111/ijfs.14330

Bernabeu M, Salgado-Ramos M, Barba FJ, Collado MC, Castagnini JM (2024) Impact of sweet potato peels extracts obtained by pulsed electric fields on the growth of probiotic strains from *Lactobacillus* genus. Innov Food Sci Emerg Technol 92:103590. https://doi.org/10.1016/j.ifset.2024.103590

Bethke PC, Jansky SH (2008) The effects of boiling and leaching on the content of potassium and other minerals in potatoes. J Food Sci 73:H80–H85. https://doi.org/10.1111/j.1750-3841.2008.00782.x

Boškovic-Rakocevic L, Dinic Z, Dugalic G, Dugalic M, Mladenovic J, Đuric M (2018) Effect of different rates and methods of application of NPK-fertilizers on the quality of potato tubers. Acta Agriculturae Serbica 23:101–110. https://doi.org/10.5937/AASer1845101B

Camire ME, Kubow S, Donnelly DJ (2009) Potatoes and human health. Crit Rev Food Sci Nutr 49:823–840. https://doi.org/10.1080/10408390903041996

Cao J, Xu C, Zhou R, Duan G, Lin A, Yang X, You S, Zhou Y, Yang G (2022) Potato peel waste for fermentative biohydrogen production using different pretreated culture. Bioresour Technol 362:127866. https://doi.org/10.1016/j.biortech.2022.127866

Coimbra P, Marona B, Henriques MHF, Campos L, Gomes DMGS, Vitorino C et al (2023) Edible films based on potato and quince peels with potential for the preservation of cured cheese. Food Packag Shelf Life 40:101176. https://doi.org/10.1016/j.fpsl.2023.101176

Dietary Guidelines Advisory Committee (2015) Scientific report of the 2015 Dietary Guidelines Advisory Committee: Advisory Report to the Secretary of Health and Human Services and the Secretary of Agriculture. U.S. Department of Agriculture, Agricultural Research Service, Washington, DC

Dong Y, Hao L, Shi Z, Fang K, Yu H, Zang G, Fan T, Han C, Yang D-H (2022) Solasonine induces apoptosis and inhibits proliferation of bladder cancer cells by suppressing NRP1 expression. J Oncol 2022:1–15. https://doi.org/10.1155/2022/7261486

Ensminger A, Ensminger ME, Konlande JE (1994) Foods & nutrition encyclopedia. CRC, Boca Raton, FL

Foroutan R, Peighambardoust SJ, Ghojavand S, Farjadfard S, Ramavand B (2023) Cadmium elimination from wastewater using potato peel biochar modified by ZIF-8 and magnetic nanoparticle. Colloid Interface Sci Commun 55:100723. https://doi.org/10.1016/j.colcom.2023.100723

Friedman M, Huang V, Quiambao Q, Noritake S, Liu J, Kwon O et al (2018) Potato peels and their bioactive glycoalkaloids and phenolic compounds inhibit the growth of pathogenic trichomonads. J Agric Food Chem 66:7942–7947. https://doi.org/10.1021/acs.jafc.8b01726

Gebhardt SE, Thomas RG (2002) Nutritive value of foods. U.S. Department of Agriculture, Agricultural Research Service, Nutrient Data Laboratory, Beltsville, MD

Gu T, Yuan W, Li C, Chen Z, Wen Y, Zheng Q, Yang Q, Xiong X, Yuan A (2021) α-Solanine inhibits proliferation, invasion, and migration, and induces apoptosis in human choriocarcinoma JEG-3 cells in vitro and in vivo. Toxins 13(3):210. https://doi.org/10.3390/TOXINS13030210

Hajšlová J, Schulzová V, Slanina P, Janné K, Hellenäs KE, Andersson C (2005) Quality of organically and conventionally grown potatoes: Four-year study of micronutrients, metals, secondary metabolites, enzymic browning and organoleptic properties. Food Addit Contam 22:514–534. https://doi.org/10.1080/02652030500137827

International Potato Center (2013) Potato facts and figures. http://cipotato.org/potato/facts. Retrieved December 2014

Javed A, Ahmad A, Tahir A, Shabbir U, Nouman M, Hameed A (2019) Potato peel waste its nutraceutical, industrial and biotechnological applications. Aims Agric Food 4:3807–3823

Jimenez-Champi D, Romero-Orejon FL, Moran-Reyes A, Muñoz AM, Ramos-Escudero F (2023) Bioactive compounds in potato peels, extraction methods, and their applications in the food industry: a review. CyTA J Food 21(1):418–432. https://doi.org/10.1080/19476337.2023.2213746

Khanal S, Karimi K, Majumdar S et al (2024) Sustainable utilization and valorization of potato waste: state of the art, challenges, and perspectives. Biomass Convers Biorefin 14:23335–23360. https://doi.org/10.1007/s13399-023-04521-1

Lachman J, Hamouz K (2005) Red and purple coloured potatoes as a significant antioxidant source in human nutrition—a review. Plant Soil Environ 51(11):477–482. https://doi.org/10.17221/3620-pse

Lee KG, Lee SG, Lee HH, Lee HJ, Shin JS, Kim NJ, An HJ, Nam JH, Jang DS, Lee KT (2015) α-Chaconine isolated from a *Solanum tuberosum* L. cv Jayoung suppresses lipopolysaccharide-induced pro-inflammatory mediators via AP-1 inactivation in RAW 264.7 macrophages and protects mice from endotoxin shock. Chem Biol Interact 235:85–94. https://doi.org/10.1016/j.cbi.2015.04.015

Li T, Chen N, Chen Y, He B, Zhou Z (2022) Solasonine induces apoptosis of the SGC-7901 human gastric cancer cell line in vitro via the mitochondria-mediated pathway. J Cell Mol Med 26(12):3387–3395. https://doi.org/10.1111/jcmm

Liu Q, Tarn R, Lynch D, Skjodt NM (2007) Physicochemical properties of dry matter and starch from potatoes grown in Canada. Food Chem 105:897–907. https://doi.org/10.1016/j.foodchem.2007.04.034

Makori SI, Mu TH, Sun HN (2022) Profiling of polyphenols, flavonoids and anthocyanins in potato peel and flesh from four potato varieties. Potato Res 65(1):193–208. https://doi.org/10.1007/s11540-021-09516-x

Mohdaly AAA, Hassanien MFR, Mahmoud A, Sarhan MA, Smetanska I (2013) Phenolics extracted from potato, sugar beet, and sesame processing by-products. International Journal of Food Properties, 16(5):1148–1168. https://doi.org/10.1080/10942912.2011.578318

Monro J, Mishra S, Blandford E, Anderson J, Genet R (2009) Potato genotype differences in nutritionally distinct starch fractions after cooking, and cooking plus storing cool. J Food Compos Anal 22:539–545. https://doi.org/10.1016/j.jfca.2008.11.008

Nassar AMK, Sabally K, Kubow S, Leclerc YN, Donnelly DJ (2012) Some canadian-grown potato cultivars contribute to a substantial content of essential dietary minerals. J Agric Food Chem 60:4688–4696. https://doi.org/10.1021/jf204940t

Pęksa A, Miedzianka J, Nemś A (2018) Amino acid composition of flesh-coloured potatoes as affected by storage conditions. Food Chem 266:335–342. https://doi.org/10.1016/j.foodchem.2018.06.026

Peña CB, Restrepo LP (2013) Compuestos fenólicos y carotenoides en la papa : revisión. Actualización En Nutrición 14(1):25–32

Priestley H (2006) How to think like consumers … and win! In: Haase NU, Haverkort AJ (eds) Potato developments in a changing Europe. Wageningen Academic, Wageningen, pp 189–198

Raatz SK, Idso L, Johnson LK, Jackson MI, Combs GF (2016) Resistant starch analysis of commonly consumed potatoes: Content varies by cooking method and service temperature but not by variety. Food Chem 208:297–300. https://doi.org/10.1016/j.foodchem.2016.03.120

Ramadan MF, Oraby HF (2016) Fatty acids and bioactive lipids of potato cultivars: an overview. J Oleo Sci 65:459–470. https://doi.org/10.5650/jos.ess16015

Rodríguez-Martínez B, Gullón B, Yáñez R (2021) Identification and recovery of valuable bioactive compounds from potato peels: A comprehensive review. Antioxidants 10(10):1–18. https://doi.org/10.3390/antiox10101630

Rodríguez-Martínez B, Coelho E, Gullón B, Yáñez R, Domingues L (2023) Potato peels waste as a sustainable source for biotechnological production of biofuels: Process optimization. Waste Manag 155:320–328. https://doi.org/10.1016/j.wasman.2022.11.007

Rybácek V (1988) Brambory. SZN, Praha

Sampaio SL, Petropoulos SA, Dias MI, Pereira C, Calhelha RC, Fernandes Â, Leme CMM, Alexopoulos A, Santos-Buelga C, Ferreira ICFR, Barros L (2021) Phenolic composition and cell-based biological activities of ten coloured potato peels (Solanum tuberosum L.). Food Chem 363:130360. https://doi.org/10.1016/j.foodchem.2021.130360

Šimková D, Lachman J, Hamouz K, Vokál B (2013) Effect of cultivar, location and year on total starch, amylose, phosphorus content and starch grain size of high starch potato cultivars for food and industrial processing. Food Chem 141:3872–3880

Singh J, Kaur L (2016) Advances in potato chemistry and technology. Academic, San Diego, CA

van Eck HJ (2007) Genetics of morphological and tuber trait. In: Vreugdenhil D, Bradshaw J, Gebhardt C, Govers F, MacKerron DKL, Taylor MA, Ross HA (eds) Potato biology and biotechnology: advances and perspectives. Elsevier, Oxford, pp 91–115

Vokál B (2013) Brambory: šlechtení, pestování, užití, ekonomika. Profi, Praha

Vreugdenhil D, Bradshaw J (eds) (2007) Potato biology and biotechnology: advances and perspectives.1st edn. Elsevier, Oxford

White PJ, Bradshaw JE, Finlay M, Dale B, Ramsay G, Hammond JP, Broadley MR (2009) Relationships between yield and mineral concentrations in potato tubers. HortScience 44:6–11. https://doi.org/10.21273/HORTSCI.44.1.6

Wijngaard H, Ballay M, Brunton N (2012) The optimisation of extraction of antioxidants from potato peel by pressurised liquids. Food Chem 133(4):1123–1130. https://doi.org/10.1016/j.foodchem.2011.01.136

Woolfe JA (1987) The potato in the human diet.1st edn. Cambridge University Press, Cambridge

Zhang H, Tian F, Jiang P, Qian S, Dai X, Ma B, Wang M, Dai H, Sha X, Yang Z, Zhu X, Sun X (2021) Solasonine Suppresses the proliferation of acute monocytic leukemia through the activation of the AMPK/FOXO3A axis. Front Oncol 10:10. https://doi.org/10.3389/fonc.2020.614067

Bioactive Phytochemicals in Taro (*Colocasia esculenta*) By-products

Rajan Sharma, Natascha Cheikhyoussef, Ahmad Cheikhyoussef, Masixole Makhaba, and Ahmed A. Hussein

Contents

Abstract Taro (*Colocasia esculenta* L.) is one of the underutilized roots and tuber crops that have the potential to be used as a future smart food to combat malnutri-

R. Sharma · M. Makhaba · A. A. Hussein (✉)
Chemistry Department, Cape Peninsula University of Technology, Cape Town, South Africa
e-mail: MOHAMMEDAm@cput.ac.za

N. Cheikhyoussef
Ministry of Higher Education, Technology, and Innovation (MHETI), Windhoek, Namibia

A. Cheikhyoussef
Science and Technology Division, Multidisciplinary Research Services, Centre for Research Services, University of Namibia, Windhoek, Namibia

M. F. Ramadan (ed.), *Bioactive Phytochemicals in By-products from Leaf, Stem, Root and Tuber Vegetables*, https://doi.org/10.1007/978-3-031-80700-8_14

tion and hunger in several parts of the world. Taro is one of the world's oldest domesticated crops and belongs to the Araceae family, which contains 106 and 2500 genera and species, respectively. Nigeria, China, Cameroon, Ghana, and Papua New Guinea are the World's top taro-producing countries. Taro leaves and stems are the main by-products with promising potential applications. The leaves are rich in proteins, dietary fiber, and micronutrients and low in calories. Several antinutrients (tannins, phytates, and oxalates) are present in these by-products, limiting their wide use in the same way for the taro corms. Alkaloids, flavonoids (catechin, kaempferol, orientin, rutin, quercetin, and vicenin), saponins, sterols, phenolic acids (caffeic, gallic, benzoic, chlorogenic, and ellagic acids), and lectins (tarin) are the main bioactive phytochemicals from taro's by-products with antiatherogenic, antimicrobial, antioxidant, antidiabetic, anticarcinogenic, antihypertensive, and insecticidal, hepatoprotective, immunoprotective, and neuroprotective properties. Taro by-products have several food applications in the production of gluten-free noodles, bread, and extruded snacks, as well as non-food applications in bioethanol and pharmaceuticals. Gastronomic applications for taro leaves include making traditional dishes and various value-added food products such as purees, sauces, stews, and soups with enhanced nutritional and sensory characteristics. This chapter presents the latest highlights on Taro's by-products, bioactives, potential applications, health benefits, and biological activities.

Keywords Taro · Antinutrients · By-products · Tarin · Food applications

Abbreviations

AGP	Arabinogalactan protein
ALP	Alkaline phosphatase
ALT	Alanine aminotransferase
AST	Aspartate aminotransferase
CAGR	Compound annual growth rate
COX-2	Cyclooxygenase-2
DM	Dry matter
FP	Fresh product
HFSD	High fat sucrose diet
iNOS	Inducible nitric oxide synthase
ITC	International Trade Centre
LDL	Low-density lipoprotein
PCE	Percentage control efficacy
TI	Trypsin inhibitors
WSC	Water soluble carbohydrates

1 Introduction and Economic Values of Each Discussed By-product

Taro (*Colocasia esculenta*), also known as cocoyam, dasheen, or talas, is the most cultivated species in the genus *Colocasia.* Regarding production by weight, it ranks as the fourth most cultivated tuber crop, following cassava, potato, and sweet potato. Regarding consumption, it is the second-most important stable root crop, right behind sweet potato (Legesse and Bekele 2021). Taro is rich in carbohydrates and is a viable alternative to wheat flour. This crop grows even in poor soil conditions, requiring minimal farm inputs. As a result, it is often grown by resource-limited farmers and is regarded as a good source of food security (Kapoor et al. 2022; Onsay et al. 2022).

The global taro market is expected to witness a steady compound annual growth rate (CAGR) of 3.5% from 2024 to 2029 (Mordor Intelligence 2024). Asia Pacific is currently the fastest-growing market for taro. In 2020, the United States was the leading taro importer, accounting for 35.7% of the global market, followed by Japan, Vietnam, Australia, and New Zealand. North America is the largest market for Taro, and it dominates globally due to the high consumer demand. According to the International Trade Centre (ITC), in 2021, Global import value was valued at USD 181,258 thousand, which increased by 21% compared to 2020. In 2021, the United States and Canada were among the major importing countries, and the United States is the largest importer of taro in the North American region. During the same year, a significant portion of global taro production, nearly 86.3%, came from major producing countries like China, Ecuador, Costa Rica, Fiji, and Indonesia. Among these countries, China had the highest production value of 1.9 million metric tons, followed by Fiji. While major African producers like Nigeria, Cameroon, and Ghana contribute significantly to global taro production, data on their international trade in this crop remains scarce. This lack of information could be attributed to inadequate documentation and domestic focus, where taro production in these African countries primarily serves domestic food security and nutritional needs, with less emphasis on international export.

2 Chemical Composition and Bioactive Compounds of Extracts from By-products

Bioactive phytochemicals found in taro (*Colocasia esculenta*) by-products have garnered significant attention due to their potential health benefits (Mitharwal et al. 2022; Temesgen and Retta 2015; Shah et al. 2022).

2.1 *Taro Powder*

The phytochemical components of taro powder and its processed foods have been investigated, revealing a reduction in phytochemical components upon processing, with significant decreases in phenols, flavonoids, and saponins (Alcantara et al. 2013). Despite these reductions, taro retains valuable bioactive compounds that can benefit its health.

2.2 *Taro Leaves*

Two main pharmacologically active phyto-constituents in taro are flavonoids and triterpenoids. Phytochemical analysis of the methanolic extract of fresh leaves of taro resulted in the isolation of eight flavonoids (Fig. 1): orientin (**1**), isoorientin (**2**), isovitexin (**3**), vicenin-2 (**4**), orientin-7-*O*-glucoside (**5**), isovitexin 4`-*O*-glucoside (**6**), glucosyl vitexin (**7**), and luteolin-7-*O*-glucoside (**8**) (Iwashina et al. 1999).

Other phytochemical compounds reported from Taro leaves include carotenoids such as lutein, zeaxanthin, phytoene, and carotene (Champagne et al. 2011), anthraquinones, caffeic acid, catechin, chlorogenic acid, coumaric acid, glucaric acid, scoparin, shaftoside, and sinapic acid (Dwivedi et al. 2016; Ehiobu and Ogu 2018). Taro leaves are considered a rich source of micronutrients such as copper (0.8 mg 100 g^{-1}), iron (10–11.7 mg 100 g^{-1}), potassium (0.2–1.8 g 100 g^{-1}), and zinc (0.8–4.2 mg 100 g^{-1}) (Gupta et al. 2019; Mitharwal et al. 2022).

A recent study by Kumar et al. (2023) reported the impact of household processing methods on the nutritional and anti-nutritional properties of Taro leaves. They noted that soaking Taro leaves for 8–12 h, cooking time from 30–60 min, and blanching from 1–3 min, followed by sun drying for the leaves led to a significant increase in crude fiber (25.7–29.65%), protein (4.33–15.6%) content as well as the calcium and iron contents were increased significantly (16.38% and 5.9%, respectively). Furthermore, the same treatment for the Taro leaves significantly decreased

1 $R_1 = R_2 = H$, $R_3 = C$-glucosyl
2 $R_1 = C$-glucosyl, $R_2 = R_3 = H$
5 $R_1 = H$, $R_2 =$ glucosyl, $R_3 = C$-glucosyl
8 $R_1 = R_3 = H$, $R_2 =$ glucosyl

3 $R_1 = C$-glucosyl, $R_2 = R_3 = H$
4 $R_1 = R_2 = C$-glucosyl, $R_3 = H$
6 $R_1 = C$-glucosyl, $R_2 = H$, $R_3 =$ glucosyl
7 $R_1 = R_3 = H$, $R_2 = C$-glucoglucosyl

Fig. 1 Chemical structures of the flavonoids isolated from the leaves of taro

the ash (20.34–28.22%), fat (5.7–31.4%), oxalic acid (27.07–35.32%), and tannic acid (96%) (Kumar et al. 2023). Oxalates decrease the palatability of taro, and they are responsible for taro's acridity (Soudy et al. 2010).

2.3 *Taro Roots*

Taro roots are rich in starch, consisting of 28% amylose (**8**) and 72% amylopectin (**9**), and anthocyanins (Fig. 2), including cyanidin 3-*O*-glucoside (**10**), pelargonidin 3-*O*-glucoside (**11**), and cyanidin 3-*O*-rhamnoside (**12**) (Cambie and Ferguson 2003). Oxalates are found in taro corms as calcium oxalate, a compound that can irritate the digestive system, which prevents them from being eaten raw or undercooked and can form insoluble crystals that form kidney stones and reduce calcium absorption (Ferdaus et al. 2023). In addition to oxalates, trypsin inhibitors (TI) can interfere with protein digestion and absorption (Ferdaus et al. 2023).

Taro, an underexploited tuber crop, contains a variety of bioactive compounds such as tarin, taro polysaccharides, α-amylase inhibitors, mono galactosyl diacylglycerols, di galactosyl diacylglycerols, polyphenols, and nonphenolic antioxidants (Pereira et al. 2018; Ferdaus et al. 2023). These compounds contribute to the anticancer, immunomodulatory, antioxidant, and anti-inflammatory properties associated with taro (Baro et al. 2023). Studies have shown that taro possesses antineoplastic

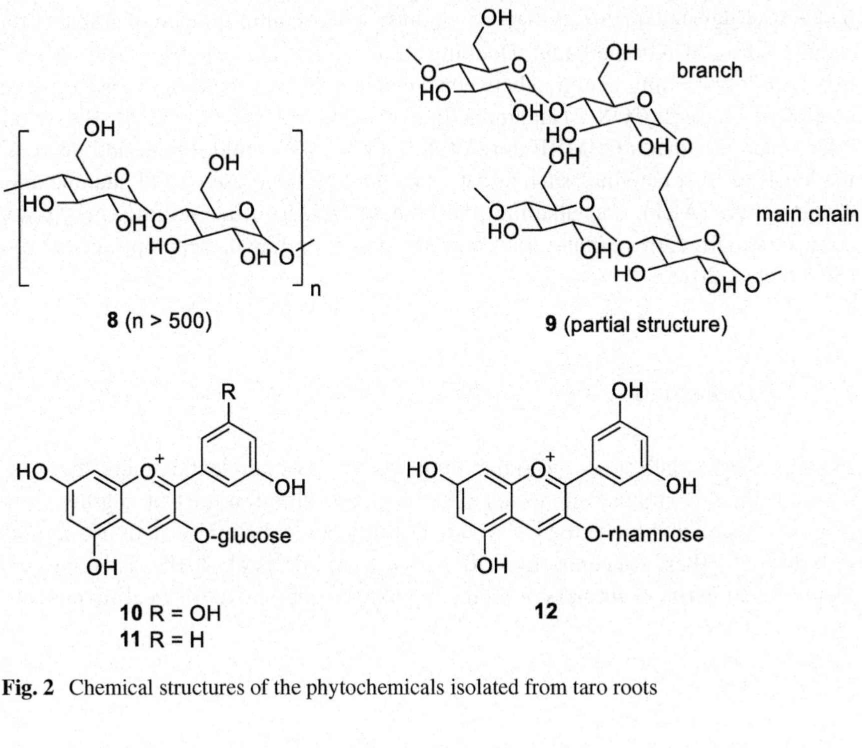

Fig. 2 Chemical structures of the phytochemicals isolated from taro roots

and protective agents, making it a promising candidate for further exploration in natural medicine.

3 Biological and Functional Properties of Extracts and Bioactive Compounds from By-products

The functional profile of Taro and its encapsulating properties has been reported by Zubair et al. (2023), and a very recent study summarized its bioactive compounds and their biological activities were presented by Wudali et al. (2024).

3.1 *Taro Roots*

The phytochemicals of taro roots contribute to the anticancer, immunomodulatory, antioxidant, and anti-inflammatory properties associated with taro. Studies have shown that taro possesses antineoplastic and protective agents, making it a promising candidate for further exploration in natural medicine (Pereira et al. 2018). Colocasia esculenta methanolic root extract (CEMRE) has shown anti-inflammatory potential in in vitro and in vivo models of inflammation (Baro et al. 2023). A dose of 400 mg/kg from CEMRE could significantly inhibit the paw inflammation, which was elevated due to carrageenan induction with inhibition, comparable to the standard drug of Indomethacin. Docking studies uncovered the promising interaction with low binding energies between Sinapic acid with both the target proteins Cyclooxygenase-2 (COX-2) and inducible nitric oxide synthase (iNOS) (Baro et al. 2023). The fact that a CEMRE dose of 400 mg/kg BW could inhibit and decrease the levels of liver enzymes such as aspartate aminotransferase (AST), alanine aminotransferase (ALT), and alkaline phosphatase (ALP) in the serum when it was compared to the carrageenan-induced group, thus revealing its hepatoprotective role (Saikia et al. 2018).

3.2 *Taro Leaves*

In an investigation to study the anthelmintic potential of taro leaf extracts, the aqueous and ethanol extracts were tested at varying concentrations (10–50 mg/mL). The results revealed that the extracts exhibited significant anthelmintic activity, particularly at the highest concentration (50 mg/mL) (Kubde et al. 2010). This suggests that taro leaf extracts might be a promising natural alternative for treating parasitic worm infections.

A study explored the free radical scavenging efficacy associated with taro whole-leaf juice. The results demonstrated that the juice effectively prevented the initiation of lipid peroxidation reactions triggered by free radicals generated by hepatotoxins. This finding suggests that taro whole-leaf juice possesses significant free radical scavenging efficacy, potentially protecting against oxidative stress and its associated health consequences (Patil and Ageely 2011).

Researchers tested the ethanol extract of taro leaves for its antidiabetic activity in rats with alloxan-induced diabetes. The results revealed that the extract, administered at 400 mg/kg, significantly lowered blood glucose levels in diabetic rats, suggesting its potential as a promising antihyperglycemic agent (Kumawat et al. 2010). Taro leaves have been shown to inhibit aldose reductase activity. This is significant because inhibiting aldose reductase is essential for managing diabetes and preventing its complications. In a study to examine the effect of taro leaf extract in inhibiting erythrocyte aldose reductase activity and increasing hemoglobin, it was found that there was a significant increase ($p < 0.05$) in the hemoglobin in the high-fat sucrose diet (HFSD) induced rats treated with 200 and 400 mg/KgBW of taro leaf extract when compared with the control group of rats (Sulistiani et al. 2020). However, there was no significant difference ($p > 0.05$) in erythrocyte aldose reductase levels with 200 mg/KgBW and 400 mg/KgBW doses.

A study investigated the antimicrobial activity of an aqueous extract from taro leaves against various bacterial and fungal strains. The tested bacteria included both Gram-positive (*Streptococcus mutans* and *Bacillus subtilis*) and Gram-negative (*Klebsiella pneumoniae*, *Pseudomonas fragi*, and *Escherichia coli*) species, while the fungal strains were *Aspergillus niger* and *Candida albicans* (Singh et al. 2011). The aqueous extract of taro leaves displayed the strongest activity against *S. mutans* among the selected microbial strains. This suggests that the extract from taro possesses promising antimicrobial properties against certain bacteria and fungi. These results offer encouraging evidence for the potential use of taro leaves aqueous extract as a natural infection treatment. In another study, the anti-fungal properties of the alcoholic and aqueous extract of taro leaves were studied against *Alternaria solani* and *A. ricini* (Mengane 2015). The alcoholic extract showed higher anti-fungal activity than the aqueous extract, with the alcoholic extract demonstrating 100% control efficacy (PCE) at 25% concentration.

A study investigated the potential of taro as a prebiotic, focusing on its ability to support the growth and adherence of beneficial *Lactobacillus* bacteria in the gut. Using an in vitro human digestion system to mimic the intestinal environment, researchers found that taro significantly boosted the growth of *L. acidophilus*, *L. paracasei*, and *L. plantarum* compared to inulin and glucose. Notably, *L. paracasei* exhibited strong self-aggregation and adherence to Caco-2 cells (Saxby et al. 2019). These findings highlight taro as a promising strain-specific prebiotic that could be used as a dietary aid to promote a healthy gut microbiome.

The anti-inflammatory property of an ethanolic extract from taro leaves was investigated in Wistar rats, whereby three models were employed: carrageenan-induced left hind paw edema, carrageenan-induced pleurisy, and cotton pellet-induced granuloma. The results demonstrated that the extract significantly reduced inflammation ($p < 0.05$) compared to both the standard treatment and untreated control groups (Biren et al. 2007). These results indicate that the ethanolic extract of taro leaves possesses promising anti-inflammatory potential.

Tarin has been identified as a potential immunomodulator and COX-inhibitor lectin found in Taro (Pereira et al. 2018). Taro starch is associated with health benefits in managing some health conditions, including chronic liver problems, gall bladder diseases, pancreatic disease, and ulcers (Emmanuel-Ikpeme et al. 2007). Other functional roles of crude fiber of taro by-product have been reported to facilitate the delivery of micro-components, digesting glucose, slowing the cycle of reabsorption of cholesterol, reducing intestinal transit time, low-density lipoprotein (LDL) cholesterol reduction, constipation prevention, raising food holding ability in the water, and blood glucose and insulin levels reduction (Temesgen and Retta 2015).

4 Food and Non-food Applications of Extracts and Bioactive Compounds from By-products

4.1 Food Applications

Taro is a tuber exceptionally rich in carbohydrates, comprising 73–80% on dry matter (DM) basis (Jane et al. 1992; Kaushal et al. 2015). This composition is primarily starch (77.9% DM) with a minimal amount of crude fiber (1.4% DM) (Onwueme and Charles 1994). Because of its high carbohydrate content, this tuber is a crucial staple food in developing countries across Africa, the West Indies, the Pacific, and Asia.

Apart from providing a good amount of carbohydrates, taro corms are a valuable source of nutrients. Larger servings can contribute significantly to dietary protein intake, especially when consumed more than once daily (Alcantara et al. 2013). Taro is also a good source of vitamins and minerals, including thiamine, riboflavin, iron, phosphorus, zinc, and vitamin B6. It is also an excellent source of vitamin C, niacin, potassium, copper, and manganese (Kaushal et al. 2015). Among minerals, taro contains potassium as the most abundant mineral (between 3.23–5.30 g/kg on the fresh product (FP) basis), followed by low contents of calcium (100–450 mg/kg FP) and magnesium (190–370 mg/kg FP) with sodium as the least abundant macroelement (0–3 mg/kg FP) (Wills et al. 1983). Taro contains anti-nutritional factors like oxalates, ranging from 2.05% to 4.21% of its dry matter (DM). These compounds are believed to contribute to the acrid taste of taro tubers, which can sometimes limit their consumption (Nip et al. 1994; Kaushal et al. 2015). Notably, taro offers more vitamin B-complex than whole milk, making it a nutritious addition to a balanced diet (Soudy et al. 2010). Gums (mucilage) (Njintang et al. 2014; Mijinyawa et al. 2018; Andrade et al. 2020) and special starch granules (small sizes) are two important macronutrients from Taro (Kaushal et al. 2015; Singla et al. 2020), which makes it very suitable to prepare various foods including flour (Kaur et al. 2013; Peluola-Adeyemi and Idowu 2014), extruded snacks (Rodríguez-Miranda et al. 2011), fermented taro (Muller et al. 2005), taro chips (Ikpeme-Emmanuel et al. 2009), paste (Lai et al. 1998; Njintang et al. 2008), starch (Antonio-Estrada et al. 2009), cookies (Nip et al. 1994; Waheed et al. 2010), noodles (Maga et al. 1993; Del Rosarlo et al. 1999), taro ice products (Lu et al. 2002), and weaning food (Ikpeme-Emmanuel et al. 2009).

Taro mucilage has unique properties to act as a binder, emulsifier, stabilizer, and thickener in the pharmaceutical, cosmetic, and food industries (Mijinyawa et al. 2018; Andrade et al. 2020). Andrade et al. (2020) attributed the emulsifying action of the taro mucilage to arabinogalactan protein (AGP) glycoprotein, a macromolecule primarily formed by arabinogalactans connected to proteins.

Due to its distinct composition and functional properties, Taro starch is a unique and valuable ingredient in the food industry. It contains protein (5.6%), ash (0.8%), and phosphorus (0.4%) but a lower lipid concentration compared to other tropical roots like tiger nut and sweet potato. Furthermore, its low amylose content, high swelling capacity, and excellent water and oil retention stability make it suitable for food processing applications. The small granule size (1–5 mm diameter) and high digestibility further enhance its potential in creating functional foods like infant formula, bread, noodles, and filler ingredients for edible films (Kaushal et al. 2015; Singla et al. 2020). Taro-resistant starch offers potential health benefits due to its ability to bind with bile acids. This binding helps lower cholesterol levels, making it a promising dietary component for managing cardiovascular health (Simsek and El 2012). Taro starch is highly digestible, hypoallergenic, and gluten-free (Shah et al. 2022). These properties of taro starch can benefit individuals with peptic ulcers, pancreatic illness, chronic liver issues, gall bladder, and inflammatory bowel disorder (Cambie and Ferguson 2003; Ribeiro Pereira et al. 2020). Water-soluble non-starch polysaccharide from taro (*Colocasia esculenta*) peel was extracted using ultrasound and pulse electric field with potential use as a natural ingredient with special properties of high thermal stability and low viscosity to develop new food products with novel functionalities (Anwar et al. 2021).

Taro's leaves and flowers are nutritious vegetables (Miyasaka et al. 2019) of which protein (2.7%), fiber (2.8%), carbohydrates (3.1), ash (2.5%), and an 88.5% moisture content for the leaves were reported (Odedeji et al. 2014). This rich profile of protein, carbohydrates, and fiber underpins the importance of these by-products to develop fermented and unfermented Taro-based food products such as alcoholic beverages, yogurt alternatives, and deserts including frozen taro cake (Kulolo, enyekenyek), sticky cake (dodol), cookies, ice cream, and jam (Ferdaus et al. 2023). Taro leaves are prepared in several dishes as sauces, soups, or stews thanks to the high fiber content, vitamins, minerals, and secondary metabolites (Otekunrin et al. 2021; Shah et al. 2022). Kapoor et al. (2022) emphasized that taro could be valorized as a zero-wastage orphan food crop for food and nutritional security.

4.2 Non-food Applications

Production of taro products such as taro chips, taro cakes, and taro pastries generates large amounts of peel waste, often burned or used as fertilizer. However, this waste presents a valuable opportunity. Ash and Elliott (1991) reported that taro leaves are suitable forage for ensilage production, especially when combined with additives rich in water-soluble carbohydrates (WSC) such as molasses, papaya pulp, and ripe banana, a forage of high nutritional value can be produced.

The high starch content of taro peels makes them an excellent source material for various useful products, including glucose, ethanol, food additives, and even a nutrient source for cultivating microorganisms. A study used taro waste for *Lactobacillus acidophilus* BCRC 14079 cultivation. The study successfully demonstrated the potential of taro waste as a cost-effective and sustainable alternative for cultivating this beneficial bacterium, paving the way to produce non-viable probiotic products (Hsieh et al. 2016). As a taro waste management strategy, taro waste was converted into compost using herbicides, mechanical cutting, manual removal, or a combination of methods used to dispose of it in landfills. The final compost products created were of equal or higher quality to current compost standards (Sembera et al. 2019). Nafees et al. (2022) studied the effects of dietary starch from different sources, including taro flour, on the growth, nutrient utilization, and liver histology of juvenile tinfoil barb (*Barbonymus schwanenfeldii*) and reported that the dietary nutrients in taro flour can increase the daily feed intake and feed conversion ratio in juvenile tinfoil barb. Another study by Karki and Philip (2024) investigated the removal and distribution of personal care products and several pharmaceutical compounds such as metronidazole, naproxen, and methylparaben by taro leaves, *C. esculenta* and *Chrysopogon zizanioides* (Vetiver) under hydroponic conditions. They reported an uptake potential for the pollutants (5.23 to 5.76 mg/g of biomass dry weight), with remaining pollutant accumulation less than 10% in the experimental plants and concluded on the taro promising candidacy for environmental pollutants removal for sustainable development and environment protection and human well-being. Taro peels have been reported as a source of cellulose nanofibers (Daza-Orsini et al. 2024). They developed nanocomposite films using taro starch and cellulose nanofibers extracted from the root's peel as well as glycerol and water. They reported that adding cellulose nanofibers increases the film thickness, opacity, UV-light barrier capacity, and water swelling properties.

5 Valorization of By-products for Non-health Purposes, i.e. Source of Bioenergy

Bioethanol is a fuel source processed from plants which can lead to lower CO_2 emissions. A study highlighted the potential of taro waste as a valuable resource for bioethanol production. An efficient process was developed using the heat-resistant yeast *Kluyveromyces marxianus* K21 to convert taro waste into bioethanol. When scaled up to a 5-liter bioreactor, the process achieved an impressive ethanol concentration of 43.78 g/L and a productivity of 2.19 g/L/h, corresponding to a remarkable 94.2% theoretical ethanol yield (Wu et al. 2016). In another study, taro root waste, a readily available source of starch, was considered for bioethanol production. The conversion of starch into ethanol was achieved by enzymatic hydrolysis followed by fermentation. The analysis of the chemical composition of bioethanol identified ethanol (41.14%) and glycerol (29.08%) as the active components (Rosalina and Permana 2022). This study also indicated that taro root waste is a great potential source of bioethanol, an eco-friendly alternative fuel.

Additionally, promising research suggests taro waste could be a cost-efficient substrate for cultivating beneficial bacteria. Its high-quality genome has been sequenced, shedding light on its genetic makeup and potential applications in agriculture and biotechnology (Yin et al. 2021). New strategies in taro cultivation are moving toward developing new genotypes with enhanced nutritional and low anti-nutritional composition for improved human nutrition and industrial purposes (Donkor et al. 2023). Hydrothermal carbonization was utilized by Mahmoud et al. (2024) to valorize taro peel wastes to produce hydrochars using machine learning models for the levofloxacin adsorption and removal potential. They applied models and reported effective predictions for levofloxacin removal under different operating conditions. Their work provides a sustainable management solution for food industry wastes and possible waste valorization options to formulate effective hydrochar composites for water treatment in poultry, hospitals, and pharmaceutical industries.

6 Conclusion

Taro (*Colocasia esculenta*), one of the world's oldest crops, has been recognized for its edible, medicinal, nutritional, and economic value. Taro flourishes even in poor soil conditions, serving as a vital food source for resource-limited farmers and strengthening food security in challenging environments. Beyond its primary role as a dietary staple, taro has many health benefits owing to its distinctive bioactive compounds. Abundant in antioxidants, taro leaves exhibit anti-inflammatory, anticancer, and immune-enhancing properties. Research highlights its potential in managing diabetes, hypertension, and digestive disorders.

Moreover, taro acts as a prebiotic, fostering the growth of beneficial gut flora. Notably, even the by-products of taro processing offer significant value. Studies have successfully converted taro peels into bioethanol, offering a sustainable and environmentally friendly fuel alternative. Taro, an often overlooked but highly adaptable tuber crop, is poised for a bright future, thanks to the scientific input to explore its full potential. Overall, taro emerges as a multi-faceted crop with the potential to contribute to food security, human health, and a more sustainable future. Further research is warranted to unlock the potential of this remarkable tuber fully.

References

Alcantara RM, Hurtada WA, Dizon EI (2013) The nutritional value and phytochemical components of taro [*Colocasia esculenta* (L.) Schott] powder and its selected processed foods. J Nutr Food Sci 3(3):1–7

Andrade LA, de Oliveira Silva DA, Nunes CA, Pereira J (2020) Experimental techniques for the extraction of taro mucilage with enhanced emulsifier properties using chemical characterization. Food Chem 327:127095

Antonio-Estrada C, Bello-Pérez LA, Martínez-Sánchez CE, Montañez-Soto JL, Jiménez-Hernández J, Vivar-Vera MA (2009) Enzymatic production of maltodextrins from taro (*Colocasia esculenta*) starch. CyTA-J Food 7(3):233–241

Anwar M, Babu G, Bekhit AED (2021) Utilization of ultrasound and pulse electric field for the extraction of water-soluble non-starch polysaccharide from taro (*Colocasia esculenta*) peel. Innov Food Sci Emerg Technol 70:102691

Ash AJ, Elliott R (1991) Tropical crop and crop by-product additives can improve the quality of taro leaf (*Colocasia esculenta*) silage. J Agric Sci 117(2):233–239

Baro MR, Das M, Kalita A, Das B, Sarma K (2023) Exploring the anti-inflammatory potential of *Colocasia esculenta* root extract in in-vitro and in-vivo models of inflammation. J Ethnopharmacol 303:116021

Biren NS, Nayak BS, Bhatt SP, Jalalpure SS, Seth AK (2007) The anti-inflammatory activity of the leaves of *Colocasia esculenta*. Saudi Pharm J 15(3):228–232

Cambie RC, Ferguson LR (2003) Potential functional foods in the traditional Maori diet. Mutat Res/Fundam Mol Mech Mutagen 523:109–117

Champagne A, Hilbert G, Legendre L, Lebot V (2011) Diversity of anthocyanins and other phenolic compounds among tropical root crops from Vanuatu, South Pacific. J Food Compos Anal 24(3):315–325

Daza-Orsini SM, Medina-Jaramillo C, Caicedo-Chacon WD, Ayala-Valencia G, López-Córdoba A (2024) Isolation of taro peel cellulose nanofibers and its application in improving functional properties of taro starch nanocomposites films. Int J Biol Macromol 273:132951

Del Rosarlo M, Vinas A, Lorenz K (1999) Pasta products containing taro (*Colocasia esculenta*) and chaya (*Cnidoscolus chavamansa*). J Food Process Preserv 23(1):1–20

Donkor EF, Nyadanu D, Akromah R, Osei K, Odoom DA (2023) Evaluation and phenotypic plasticity of taro [*Colocasia esculenta* (l.) Schott.] genotypes for nutrient and anti-nutrient composition. PLoS One 18(9):e0291358

Dwivedi P, Dwivedi J, Patel D, Desai S, Meshram D (2016) Phytochemical analysis and assessment of in vitro urolithiatic activity of colocasia leaves. J Med Plant 4(6):43–47

Ehiobu JM, Ogu GI (2018) Phytochemical content and in vitro anti-mycelial efficacy of *Colocasia esculenta* L, *Manihot esculenta* Crantz and *Dioscorea rotundata* Poir leaf extracts on *Aspergillus niger* and *Botryodiplodia theobromae*. J Hortic Plant Res 1:9–18

Emmanuel-Ikpeme CA, Eneji CA, Essiet U (2007) Storage stability and sensory evaluation of taro chips fried in palm oil, palm olein oil, groundnut oil, soybean oil and their blends. Pak J Nutr 6(6):570–575

Ferdaus MJ, Chukwu-Munsen E, Foguel A, da Silva RC (2023) Taro roots: an underexploited root crop. Nutrients 15(15):3337

Gupta K, Kumar A, Tomer V, Kumar V, Saini M (2019) Potential of Colocasia leaves in human nutrition: review on nutritional and phytochemical properties. J Food Biochem 43(7):e12878

Hsieh SC, Liu JM, Pua XH, Ting Y, Hsu RJ, Cheng KC (2016) Optimization of *Lactobacillus acidophilus* cultivation using taro waste and evaluation of its biological activity. Appl Microbiol Biotechnol 100:2629–2639

Ikpeme-Emmanuel CA, Okoi J, Osuchukwu NC (2009) Functional, anti-nutritional and sensory acceptability of taro and soybean based weaning food. Afr J Food Sci 3(11):372–377

Iwashina T, Konishi T, Takayama A, Fukada M, Ootani S (1999) Isolation and identification of the flavonoids in the leaves of taro. Ann Tsukuba Bot Gard 18:71–74

Jane J, Shen L, Chen J, Lim S, Kasemsuwan T, Nip W (1992) Physical and chemical studies of taro starches and flours. Cereal Chem 69(5):528–535

Kapoor B, Singh S, Kumar P (2022) Taro (*Colocasia esculenta*): zero wastage orphan food crop for food and nutritional security. S Afr J Bot 145:157–169

Karki BK, Philip L (2024) Fate of pharmaceuticals and personal care products like metronidazole, naproxen, and methylparaben and their effect on the physiological characteristics of two wetland plants. Chem Eng J 483:149180

Kaur M, Kaushal P, Sandhu KS (2013) Studies on physicochemical and pasting properties of Taro (*Colocasia esculenta* L.) flour in comparison with a cereal, tuber and legume flour. J Food Sci Technol 50:94–100

Kaushal P, Kumar V, Sharma HK (2015) Utilization of taro (*Colocasia esculenta*): a review. J Food Sci Technol 52:27–40

Kubde MS, Khadabadi SS, Farooqui IA, Deore SL (2010) In-vitro anthelmintic activity of *Colocasia esculenta*. Pharm Lett 2(2):82–85

Kumar V, Sharma HK, Singh K (2023) Physico-chemical, functional and anti-nutritional properties of taro (Colocasia esculenta) flour as affected by cooking and drying methods. Afr. J. Food Sci. 17(11):241–255. https://doi.org/10.5897/AJFS2015.1370

Kumawat NS, Chaudhari SP, Wani NS, Deshmukh TA, Patil VR (2010) Antidiabetic activity of ethanol extract of *Colocasia esculenta* leaves in alloxan induced diabetic rats. Int J Pharmtech Res 2(2):1246–1249

Lai HM, Jeng ST, Lii CY (1998) 17O NMR and DSC for studying quality of taro paste as affected by processing and storage. LWT-Food Sci Technol 31(1):57–63

Legesse T, Bekele T (2021) Evaluation of improved taro (*Colocasia esculenta* (L.) Schott) genotypes on growth and yield performance in North-Bench woreda of Bench-Sheko zone, South-Western Ethiopia. Heliyon 7(12):e08630

Lu TJ, Chuang CW, Chang YH (2002) Sensory and physicochemical analyses on commercial taro ice products. J Food Drug Anal 10(1):55–63

Maga JA, Liu MB, Rey T (1993) Taro (*Colocasia esculenta*) extrusion. Carbohydr Polym 21(2–3):177–178

Mahmoud AED, Ali R, Fawzy M (2024) Insights into levofloxacin adsorption with machine learning models using nano-composite hydrochars. Chemosphere 355:141746

Mengane SK (2015) Antifungal activity of the crude extracts of *Colocasia esculenta* leaves in vitro on plant pathogenic fungi. Int Res J Pharm 6(10):713–714

Mijinyawa AH, Durga G, Mishra A (2018) Isolation, characterization, and microwave assisted surface modification of *Colocasia esculenta* (L.) Schott mucilage by grafting polylactide. Int J Biol Macromol 119:1090–1097

Mitharwal S, Kumar A, Chauhan K, Taneja NK (2022) Nutritional, phytochemical composition and potential health benefits of taro (*Colocasia esculenta* L.) leaves: A review. Food Chem 383:132406

Miyasaka SC, Bellinger MR, Kantar MB, Helmkampf M, Wolfgruber T, Paudel R, Shintaku M (2019) Genetic diversity in Taro (*Colocasia esculenta*). In: Nandwani D (ed) Genetic diversity in horticultural plants; sustainable development and biodiversity. Springer International Publishing, Cham, Switzerland, pp 191–215

Muller WS, Allen AL, Sikes A, Racicot K, Senecal A (2005) Development of fermented taro as a food preservative ingredient in intermediate moisture products. Technical report NATICK/TR-06/2005 (PDF Url: ADA439715, US Army Research, Development and Engineering Command Natick Soldier Center, Natick, Massachusetts)

Nafees MSM, Kamarudin MS, Karim M, Hassan MZ, de Cruz CR (2022) Effects of dietary starch sources on growth, nutrient utilization and liver histology of juvenile tinfoil barb (*Barbonymus schwanenfeldii*, Bleeker 1853). Aquacult Rep 23:101069

Nip WK, Whitaker CS, Vargo D (1994) Application of taro flour in cookie formulations. Int J Food Sci Technol 29(4):463–468

Njintang NY, Parker ML, Moates GK, Faulds CB, Smith AC, Waldron KW et al (2008) Microstructure and creep-recovery characteristics of achu (a taro-based paste) made from freeze dried taro chips as affected by moisture content and variety. J Food Eng 87(2):172–180

Njintang NY, Boudjeko T, Tatsadjieu LN, Nguema-Ona E, Scher J, Mbofung CM (2014) Compositional, spectroscopic and rheological analyses of mucilage isolated from taro (*Colocasia esculenta* L. Schott) corms. J Food Sci Technol 51:900–907

Odedeji JO, Oyeleke GO, Ayinde LA, Azeez LA (2014) Nutritional, antinutritional compositions and organoleptic analyses of raw and blanched cocoyam (*Colocasia esculenta*) leaves. J Environ Sci Toxicol Food Technol 8:45–48

Onsay EA, Baltar KC, Galicia ER, Pesino IRC (2022) The dynamics of Taro (*Colocasia esculenta*) through value chain analysis and crop accounting in Partido District, Camarines Sur, The Philippines. In: *Sustainable rural development perspective and global challenges*. IntechOpen

Onwueme IC, Charles WB (1994) Tropical root and tuber crops: production, perspectives and future prospects (No. 126). Food & Agriculture Organization, Rome

Otekunrin OA, Sawicka B, Adeyonu AG, Otekunrin OA, Rachoń L (2021) Cocoyam [*Colocasia esculenta* (L.) Schott]: exploring the production, health and trade potentials in sub-Saharan Africa. Sustain For 13:4483

Patil BR, Ageely HM (2011) Anti-lipid peroxidative activity of *Colocasia esculenta* leaf juice against CCl_4 and acetaminophen mediated cell damage. Int J Pharm Appl 2(3):141–149

Peluola-Adeyemi OA, Idowu MA (2014) Effect of extrusion parameters on the physical and functional properties of cocoyam (*Colocasia esculenta*) flour. J Environ Sci Toxicol Food Technol 8:29–34

Pereira PR, Corrêa ACNTF, Vericimo MA, Paschoalin VMF (2018) Tarin, a potential immunomodulator and COX-inhibitor lectin found in taro (*Colocasia esculenta*). Compr Rev Food Sci Food Saf 17(4):878–891

Ribeiro Pereira P, Bertozzi de Aquino Mattos E, Nitzsche Teixeira Fernandes Correa AC, Afonso Vericimo M, Margaret Flosi Paschoalin V (2020) Anticancer and immunomodulatory benefits of taro (*Colocasia esculenta*) corms, an underexploited tuber crop. Int J Mol Sci 22(1):265

Rodríguez-Miranda J, Ruiz-López II, Herman-Lara E, Martínez-Sánchez CE, Delgado-Licon E, Vivar-Vera MA (2011) Development of extruded snacks using taro (*Colocasia esculenta*) and nixtamalized maize (*Zea mays*) flour blends. LWT-Food Sci Technol 44(3):673–680

Rosalina AM, Permana AH (2022) Bioethanol production using taro roots waste (*Colocasia esculenta*) from Bogor Indonesia and analysis of chemical compounds. Rasayan J Chem 15(2):870–874

Saikia A, Das M, Saikia Q (2018) Hepatoprotective potential of the methanolic extract of *Colocasia esculenta* on iron overload induced mice. Med Plants – Int J Phytomed Relat Ind 10(3):243–251

Saxby S, Li Y, Lee C, Kim YS (2019) Assessing the prebiotic potential of Taro (*Colocasia esculenta*) with probiotic *Lactobacillus* species in an in vitro human digestion system (P20-022-19). Curr Dev Nutr 3:nzz040-P20-022-19

Sembera JA, Waliczek TM, Meier EJ (2019) Composting as an alternative management strategy for wild taro waste. HortTechnology 29(2):205–209

Shah YA, Saeed F, Afzaal M, Waris N, Ahmad S, Shoukat N, Ateeq H (2022) Industrial applications of taro (*Colocasia esculenta*) as a novel food ingredient: a review. J Food Process Preserv 46:e16951

Simsek S, El SN (2012) Production of resistant starch from taro (*Colocasia esculenta* L. Schott) corm and determination of its effects on health by in vitro methods. Carbohydr Polym 90(3):1204–1209

Singh B, Namrata N, Lokendra Kumar LK, Dwivedi SC (2011) Antibacterial and antifungal activity of *Colocasia esculenta* aqueous extract: an edible plant. J Pharm Res 4(5):1459–1460

Singla D, Singh A, Dhull SB, Kumar P, Malik T, Kumar P (2020) Taro starch: isolation, morphology, modification and novel applications concern-a review. Int J Biol Macromol 163:1283–1290

Soudy ID, Delatour P, Grancher D (2010) Effects of traditional soaking on the nutritional profile of taro flour (*Colocasia esculenta* L. Schott) produced in Chad. Rev Med Vet 1:37–42

Sulistiani RP, Afifah DN, Pemayun TGD, Widyastiti NS, Anjani G, Kurniawati DMA (2020) The effects of *Colocasia esculenta* leaf extract in inhibition of erythrocyte aldose reductase activity and increase of Haemoglobin in experimental rats. J Nutr Sci Vitaminol 66(Supplement):S320–S323

Taro Market Size & Share Analysis – Growth Trends & Forecasts (2024–2029). Available online: https://www.mordorintelligence.com/industry-reports/taro-market. Accessed 30 May 2024

Temesgen M, Retta N (2015) Nutritional potential, health and food security benefits of taro *Colocasia esculenta* (L.): a review. Food Sci Qual Manag 36:23–31

Waheed A, Rasool G, Asghar A (2010) Effect of interesterified palm and cottonseed oil blends on cookie quality. Agric Biol J N Am 1(3):402–406

Wills RB, Lim JS, Greenfield H, Bayliss-Smith T (1983) Nutrient composition of taro (*Colocasia esculenta*) cultivars from the Papua New Guinea highlands. J Sci Food Agric 34(10):1137–1142

Wu WH, Hung WC, Lo KY, Chen YH, Wan HP, Cheng KC (2016) Bioethanol production from taro waste using thermo-tolerant yeast *Kluyveromyces marxianus* K21. Bioresour Technol 201:27–32

Wudali SN, Barwad A, Banadka A, Shaikh A, Al-Khayri JM, Nagella P (2024) Bioactive compounds and biological activities of Taro (Colocasia esculenta (L.). Schott). In: Murthy HN, Paek KY, Park SY (eds) Bioactive compounds in the storage organs of plants. Reference series in Phytochemistry. Springer, Cham

Yin J, Jiang L, Wang L, Han X, Guo W, Li C, Zhou Y, Denton M, Zhang P (2021) A high-quality genome of taro (*Colocasia esculenta* (L.) Schott), one of the world's oldest crops. Mol Ecol Resour 21(1):68–77

Zubair MW, Imran A, Islam F, Afzaal M, Saeed F, Zahra SM, Akhtar MN, Noman M, Ateeq H, Aslam MA, Mehta S, Shah MA, Awuchi CG (2023) Functional profile and encapsulating properties of *Colocasia esculenta* (Taro). Food Sci Nutr 11(6):2440–2449

Bioactive Phytochemicals from Ginger (*Zingiber officinale*) By-products

Shumaila Ijaz, Javed Iqbal, Banzeer Ahsan Abbasi, Zakir Ullah, Siraj Uddin, Tabassum Yaseen, Farishta Zarshan, Habib Ali, Hassan Sher, Zahid Ullah, Ahmad Ali, Sobia Kanwal, and Tariq Mahmood

Contents

Shumaila Ijaz and Javed Iqbal contributed equally with all other contributors.

S. Ijaz
School of Biomedical Engineering, Shenzhen University Medical School, Shenzhen University, Shenzhen, China

J. Iqbal (✉) · T. Yaseen · F. Zarshan
Department of Botany, Bacha Khan University, Charsadda, Khyber Pakhtunkhwa, Pakistan
e-mail: tabassumyaseen@bkuc.edu.pk

B. A. Abbasi
Department of Botany, Rawalpindi Women University, Rawalpindi, Pakistan

Z. Ullah · S. Uddin · T. Mahmood
Department of Plant Sciences, Faculty of Biological Sciences, Quaid-i-Azam University, Islamabad, Pakistan
e-mail: zakirullah@bs.qau.edu.pk; tmahmood@qau.edu.pk

H. Ali
Department of Agricultural Engineering, Khwaja Fareed University of Engineering and Information Technology, Rahim Yar Khan, Pakistan

H. Sher · Z. Ullah · A. Ali
Center for Plant Sciences and Biodiversity, University of Swat, Charbagh, Pakistan
e-mail: zahidtaxon@uswat.edu.pk; draali@uswat.edu.pk

S. Kanwal
Department of Biology and Environmental Sciences, Allama Iqbal Open University, Islamabad, Pakistan
e-mail: sobia.kanwal@aiou.edu.pk

M. F. Ramadan (ed.), *Bioactive Phytochemicals in By-products from Leaf, Stem, Root and Tuber Vegetables*, https://doi.org/10.1007/978-3-031-80700-8_15

Abstract Ginger (*Zingiber officinale*) is a highly esteemed spice and medicinal plant of considerable economic significance on a global scale. Nevertheless, the processing of ginger results in substantial quantities of by-products, such as rhizome peels, leaves, and stems, which are frequently underused or disposed of as refuse. This exploration delves into the potential valorization of ginger by-products as bioactive compounds and bioenergy reservoirs. The preamble accentuates ginger's economic value and customary applications, underscoring the necessity for the sustainable exploitation of by-products. The chemical constitution of ginger by-products is deliberated, unveiling their abundance in bioactive compounds like gingerols, shogaols, paradols, zingerone, and phenolic compounds. These plant bioactive compounds have diverse biological and functional attributes, encompassing antioxidant, anti-inflammatory, antimicrobial, and anticancer properties. The analysis further scrutinizes the conceivable uses of ginger by-product extractions and bioactive compounds in the culinary sector as natural preservatives, flavoring agents, and functional components. Moreover, their roles in nutraceuticals, cosmeceuticals, and pharmaceuticals are examined, spotlighting their remedial potential. In addition to health-related uses, the valorization of ginger by-products for non-food objectives, such as bioenergy generation. The lignocellulosic biomass from ginger by-products can be a raw material for producing biofuels, biogas, and other sustainable energy resources, contributing to sustainable energy resolutions and circular bioeconomy objectives. There is a substantial promise of ginger by-products as valuable reservoirs of bioactive compounds and bioenergy, advocating for their sustainable exploitation and propelling economic, environmental, and health advantages.

Keywords Ginger · Ginger oleoresin · Gingerols · Shogaols · Bioactive compounds

1 Introduction

Ginger, known as *Zingiber officinale* Roscoe, pertains to the Zingiberaceae botanical family and is extensively farmed in the tropical regions of Asia as a perennial herb to gather its rhizome. Alongside its culinary application, ginger has been acknowledged since ancient eras for its numerous therapeutic advantages and has recently garnered significant attention in the realm of research. Diverse investigations have indicated abundant biologically active ginger elements that promote health. The utilization of ginger for addressing common maladies like colds, nausea, and headaches has been widely observed (Mao et al. 2019a, 2019b). The increasing interest in leveraging potential bioactive compounds from ginger has introduced a fresh perspective on extracting these natural substances (Kiyama 2020). Ijaz et al. (2023) have reported a number of both volatile (monoterpenes and sesquiterpenes) and non-volatile enriched oleoresin (gingerols, shogaols, paradols, and zingerone) from ginger that exhibit diverse bioactivities such as antioxidant properties (Abolaji et al. 2017), anti-inflammatory effects (Alsahli et al. 2021), antimicrobial characteristics (Akintobi et al. 2013, Beristain-Bauza et al. 2019), antiviral potential (Kaushik et al. 2020), and anticancer attributes (Mahomoodally et al. 2021), alongside other protective benefits (Choi et al. 2018), contributing to the maintenance of good health. However, the standard and volume of these bioactive compounds significantly rely on the protocol used for extraction, leading to the emergence of innovative technologies aimed at enhancing the efficiency of ginger oil extraction along with its compound preservation to its full extent within the extract.

Hydro-distillation and solvent extraction, the two traditional methods, remain prevalent in various industries due to their simplicity and lower initial investment. Nevertheless, significant drawbacks prevail in these approaches, including inefficiencies in yield and extract quality, environmentally unfriendly practices involving extensive organic solvent consumption, and elevated energy usage caused by prolonged extraction periods at high temperatures leading to the deterioration of thermolabile components. Alternatively, cutting-edge extraction techniques can be adopted based on the required criteria such as efficiency, sustainability, and innovation. These emerging advanced green extraction methods align with sustainable development objectives, emphasizing enhanced efficiency, economic viability, reduced energy consumption, and eco-friendliness (Mwaurah et al. 2020). As stated by Chemat et al. (2017), green extraction methodologies should adhere to six key principles: (i) minimal organic solvent usage, (ii) reduced energy consumption, (iii) short extraction durations, (iv) rational sourcing practices, (v) generation of by-products with high value, and (vi) production of safe and natural extracts. These principles have significantly enhanced production sustainability on an industrial scale by minimizing solvent usage, reducing reliance on fossil energy, and eliminating wastewater and hazardous outputs. Ongoing research into these cutting-edge technologies is imperative to identify gaps and opportunities for mitigating limitations within the field.

By-products generated from the processing of ginger, such as ginger peel, have the potential to be efficiently employed in the retrieval of significant constituents, such as polysaccharides and gingerol, as delineated in several scholarly articles. Various methodologies are used, including the ultramicro crushing of ginger peel to acquire submicron powder for the extraction of gingerol and polysaccharides (Chen et al. 2019), the production of ultramicro ginger powder for diverse applications such as seasoning and nutritional supplements (Zhang et al. 2020), and the development of ginger juice commodities with health advantages through the amalgamation of ginger with other elements like fruits and vegetables (Murthy and Gautam 2015). Additionally, the financial feasibility of transforming ginger by-products into herbal beverages like syrup "Sarabba" and spice syrup "Habbatussauda" has been validated, illustrating profitable results and augmented worth per kilogram of the final goods (Sabahannur 2022). Overall, these investigations underscore the prospects for the sustainable utilization of ginger by-products, curbing wastage and engendering valuable goods with economic and health merits.

2 Chemical Composition and Bioactive Compounds of Extracts from Ginger By-products

Ginger extracts derived from residual materials contain diverse bioactive components that offer notable health advantages. These components encompass monoterpenoids, sesquiterpenoids, phenolic compounds, aldehydes, ketones, alcohols, and esters (Dalsasso et al. 2023). Moreover, ginger is abundant in bioactive constituents like zingiberene, shogaols, gingerol, and curcumene, as depicted in Fig. 1, which play a pivotal role in its antioxidant, antimicrobial, anti-inflammatory, and anticancer attributes (Shalaby et al. 2023, Ijaz et al. 2024a, 2024b). The preservation of these compounds heavily relies on the extraction process, where techniques such as vacuum microwave drying in conjunction with ultrasound have shown potential in

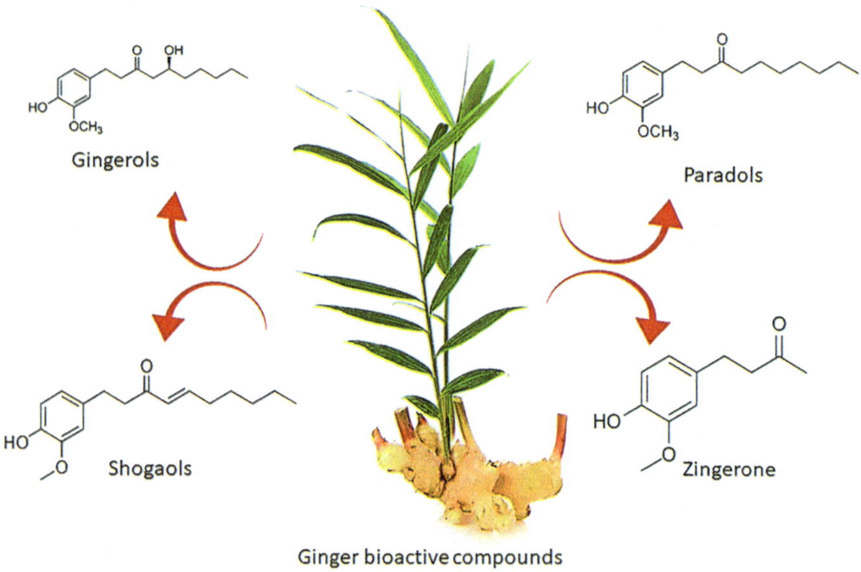

Fig. 1 Different kinds of bioactive compounds of ginger with their structures

upholding substantial levels of phenolic compounds and antioxidant efficacy while minimizing the impact on sesquiterpenes (Ashokkumar et al. 2020, Ijaz et al. 2024b). Encapsulation methodologies can offer additional protection to these bioactive compounds against degradation, volatilization, and oxidation, thereby augmenting their stability and utility across diverse food products.

2.1 *Gingerols*

Within ginger, gingerols are present as a pungent yellow liquid or in crystalline solid form with low melting points, which contributes to the simulation of taste buds spice receptors (Mao et al. 2019a, 2019b). With variable unbranched alkyl side chain, six homologs gingerols are identified, of which 6-gingerol is the most prevalent compound, with 4-, 7-, 8-, 10-, and 12-gingerol present in negligible quantities (Varakumar et al. 2017). Numerous research works have documented the anticancer properties of gingerols. For example, the impact of 6-gingerol on pain relief in rats with oral ulcerative mucositis (OUM) in squamous cell cancer was explored by Hitomi et al. (2017). The investigation revealed a notable analgesic effect attributed to inhibiting the Na + current–voltage activator associated with OUM-induced pain. Additionally, Zhang et al. (2017) highlighted the impact of Tongling white ginger extracted 6-gingerol on human cervical carcinoma (HeLa) cancer cells that hindered the proliferation by induction of structural alterations that triggered apoptosis in HeLa cells.

Moreover, multiple studies have unveiled the remarkable antibiofilm efficacy of gingerols. Kim et al. (2015) delved into the influence of 6-gingerol on pathogenic *Pseudomonas aeruginosa*. By interacting with the quorum sensing (QS) receptor regulator, biofilm formation is impeded by 6-gingerol. The QS-induced genes were suppressed with this interaction, reducing biofilm production and virulence factors. Furthermore, Choi et al. (2017) demonstrated that both 6-gingerol and 8-gingerol exhibited inhibitory effects on biofilm formation by *Pseudomonas aeruginosa*. Kim et al. (2015) signify the hydrogen-bonding interaction for biofilm inhibition. Additionally, Lee et al. (2018) emphasized the high efficacy of 6-gingerol homologs, compared to other gingerols, in reducing biofilm formation of *Candida albicans*. Furthermore, a lower virulence activity in a nematode model was noted for *Candida albicans*.

2.2 *Shogaols*

12-shogaol along homologs 6-, 8-, 10-, is notably pungent. The presence of high levels of shogaols that have undergone thermal processing is because of the gingerol conversion by a dehydration process triggered by heat, with 6-shogaol being the most abundant homolog in dried ginger, while fresh ginger contains only minimal traces of shogaols (Foudah et al. 2020, Ho and Su 2016, Kou et al. 2018). As the conversion of gingerols to shogaols is a heat-induced reaction, the type of heat (dry or moist) applied greatly influenced the process. Through a steaming process lasting 4 h, the highest conversion of 6-shagaol was achieved by Cheng et al. (2011), while Jung et al. (2018) generated (2890 mg/kg) of 6-shogaol through moist heat compared to dry heat, which produced only 1161 mg/kg of 6-shogaol at 130 °C lasting for 360 min. The authors also noted that the transformation of gingerols into shogaols was significantly impacted by temperatures, where higher temperatures accelerated the conversion rate, thereby shortening the duration of heat treatment (Jung et al. 2018; Zahra et al. 2021, 2023).

Several reports indicate superior pharmacological characteristics for shogaols compared to gingerols, particularly in terms of neuroprotective effects (Ohnishi et al. 2019), antioxidant properties (Hur et al. 2020), and anti-inflammatory activities (Ho and Su 2016). In the intestine, 6-shogaol inhibits TNF-a-induced barrier disruption as an anti-inflammatory agent (Luettig et al. 2016). 6-shogaol also suppresses 6-gingerol in its anti-adipogenic effects on 3T3-L1 cells. It impedes the adipogenesis process by suppressing key adipogenesis regulators and promoting lypolysis, thereby contributing to its anti-obesity properties (Suk et al. 2016). Additionally, 6-shogaol, combined with cisplatin and irradiation, has potent cytotoxic effects against head and neck squamous cell carcinoma (HNSCC) cell lines by enhancing apoptosis (Kotowski et al. 2018). Rampogu et al. (2018) investigated the photobiological response of enzyme 6-hydroxymethyl-7,8-dihydropterin pyrophospho-kinase (SaHPPK) to diverse ginger compounds, which confers resistance of *Staphylococcus aureus* to most antibiotics. In this regard, shogaols and gingerone-A, among the eight examined phytochemicals, displayed the highest molecular docking scores, indicating a strong, active site interaction, suggesting shogaols could act as SaHPPK inhibitors.

Moreover, 6-shogaol has also been linked with antibiofilm and antivirulence activities against *Candida albicans*. Out of six different gingerol and shogaol, Lee et al. (2018) revealed that 6-shogaol inhibits biofilm formation, hyphal growth, and fungus virulence. Due to α,β-unsaturated carbonyl group, shogaol possesses superior neuroprotective properties compared to gingerol by facilitating better interaction with Kelch-like ECH-associated protein (Keap)1, an essential heme oxygenase inducer crucial for heme degradation in cases of intracerebral hemorrhage (Ohnishi et al. 2019).

2.3 Paradols

Paradols are the non-pungent metabolites of shogaols (Sapkota et al. 2019), which have the potential to control oxidative stress and inflammatory reactions. They can potentially limit neuroinflammation of central nervous system (CNS) disorders, specifically considering the one linked with cerebral ischemia (Gaire et al. 2015). Nile and Park (2015) evaluated ginger extracts (6-gingerol, 6-shogaol, and 6-paradol) for different biological potencies, focusing on antioxidant potential. In vitro, assays for ginger components 6-gingerol, 6-shogaol, and 6-paradol extracted with ethyl acetate and water exhibited anti-inflammatory, potent antioxidant, and xanthine oxidase inhibitory properties. However, compared to these, 6-paradols displayed lower antioxidant capabilities.

For the antihyperglycemic effects, Wei et al. (2017) assessed ten active ginger components and homologs in the paradol, shogaol, gingerol, and zingerone series. Out of these assessed compounds, glucose utilization is stimulated by 6-paradol and 6-shogaol in 3T3-L1 adipocytes and C2C12 myotubes. Increased phosphorylation of 5′ adenosine monophosphate-activated protein kinase (AMPK) in 3 T3-L1 adipocytes is the major identified cause of these effects. In an in vivo experiment conducted on mice with a high-fat diet, 6-paradol, the primary metabolite of 6-shogaol, significantly lowered blood glucose, cholesterol levels, and body weight (Wei et al. 2017).

2.4 *Zingerone*

As a result of retro-aldol reaction of gingerol after drying or heat processing, gingerone, often referred to as vanillyl acetone, is created (Ahmad et al. 2015; Choi et al. 2018). In 2015, Hsiang et al. (2015) examined how zingerone and ginger affected inflammation. In a dose-dependent manner, ginger and zingerone suppressed lipopolysaccharide (LPS)-induced nuclear factor kappa-light-chain-enhancer of activated B cells (NF-κB) activities in cells; the greatest suppression was reported at 100 g/ml for ginger (84.5 ± 3.5%) and 96.2 ± 0.6%) for zingerone. Furthermore, LPS-induced NF-B-driven luminous intensities were decreased by ginger and zingerone in most organs, with the small intestine demonstrating the most suppression.

In their investigation of zingerone's protective effect against oxidative stress and inflammation in cisplatin-induced rat nephrotoxicity, Alibakhshi et al. (2018) found that zingerone significantly reduced tissue levels of malondialdehyde (MDA) while remarkably maintaining the activity of the enzymes catalase (CAT) and glutathione peroxidase (GPX) in kidney tissue when compared to cisplatin. Additionally, it was shown that gingerone inhibits the loss of glutathione (GSH) in kidney tissue and reduces tumor necrosis factor (TNF), which in turn reduces inflammation brought on by cisplatin. Gingerone enhanced glomerular diameters, RBC congestion, leukocyte infiltration, brush border loss, and vacuolation (fat deposit), among other histological alterations. In a rat model of nephrotoxicity using cisplatin, gingerone also displays nephroprotective benefits, mainly by suppressing inflammation and oxidative stress (Alibakhshi et al. 2018).

3 Traditional Techniques for Extraction

Traditional extraction methods, like Soxhlet and maceration, usually use a lot of hazardous solvents, take a long time, require much energy, time-consuming extraction causes thermal degradation of the compounds, and limited extraction rates

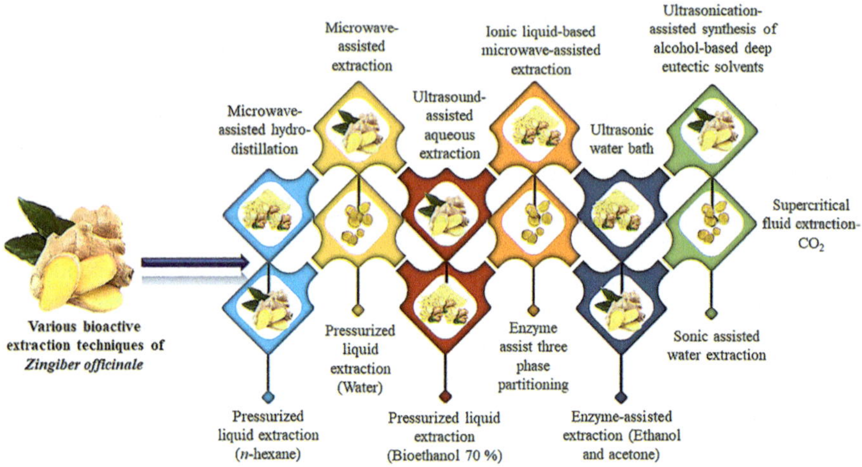

Fig. 2 The oleoresin, gingerols, shogaols, TPC (Total Phenolic Contents), TFC (Total Flavonoid Contents), and antioxidant activity of ginger were extracted using various extraction methods

(Essien et al. 2020; Giacometti et al. 2018, Ali et al. 2022). The list goes on. Non-conventional methods that align with sustainable development technologies have garnered significant interest as a means of overcoming these limits (Fig. 2).

4 Advanced Green Extraction Technologies

"Green technologies" refers to environmentally friendly technology that supports social and economic sustainability. Ginger's active compounds have been successfully recovered through the development of these state-of-the-art techniques, which include pressurized liquid extraction (PLE), sub- and supercritical fluid extraction (SFE), enzyme-assisted extraction (EAE), microwave-assisted extraction (MAE), and ultrasound-assisted extraction (UAE) (Fig. 3). The extraction medium is similar, but each method has a distinct mechanism, especially regarding cell disruption. Intensification methods include all aided extraction techniques, including UAE, MAE, and EAE (Zhang et al. 2022a, 2022b; Ali et al. 2023). As opposed to this, SFE technology makes use of green solvents, including ionic liquids (ILs), deep eutectic solvents (DESs), and natural deep eutectic solvents (NaDESs), in conjunction with aided extraction (Chemat et al. 2019). The cutting-edge green extraction methods are examined and contrasted in this article methodically, considering all relevant factors.

More than 400 bioactive chemical components are found in ginger. More often than not, bioactive ginger constituents include sulfonates, monoterpenoid glycosides, terpenes, diarylheptanoids, gingerol analogs, and phenylalkaloids. Numerous physiological and biochemical processes in the human body depend on ginger's

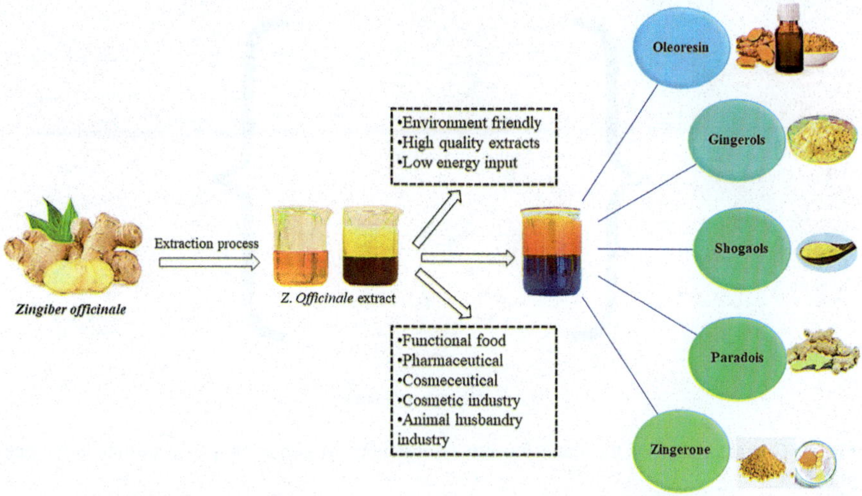

Fig. 3 Biological and functional properties of extracts and bioactive compounds from by-products

high concentrations of nutrients and chemical components. The chemical components of ginger exhibit promise in treating, preventing, and relieving illness. Ginger also contains high phenolic chemicals, lipids, terpenes, and carbs. Ginger's chemical makeup has led to its use as a nerve relaxant and treatment for brain illnesses such as ischemic stroke paralysis (Shukor et al. 2020). Ginger has been clinically effective in treating multiple ailments, including vascular disorders, cancer, emesis, bone disease, and metabolic problems.

Because of their bioactivity, several of ginger's chemical components have pharmacological actions and positive health impacts. Ginger's primary bioactive constituents include phenolic compounds, sesquiterpenes, and dialylheptanoids (Hoferl et al. 2015). Numerous research studies have demonstrated the unique health advantages and food value-boosting qualities of ginger and shogaols. Because of its chemical makeup, ginger is more popular among customers and is produced in the food business in an environmentally friendly manner. It is essential to use the great nutritional properties of ginger while creating new food items and pharmaceuticals. When blending or employing new items, ginger boosts food value and minimizes undesirable flavors. The anticarcinogenic, antioxidant, antibacterial, antidiabetic, anti-fatty, anti-gastric, anti-inflammatory, ulcer, and antiallergic properties of ginger's nutritional makeup have been utilized (Fig. 4) (Hoferl et al. 2015).

Ginger's chemical and nutritional makeup is linked to increased nutrition, including qualities that aid digestion, immunological stimulation, and growth promotion. It can also enhance the metabolism of proteins and fats (Shakya 2015). It possesses antiviral, antibacterial, antihyperglycemic, antioxidant, and antiparasitic qualities (Anzian et al. 2020). The Food and Drug Administration (FDA) states ground, crushed, and whole foods can benefit significantly from ginger's flavoring and

Fig. 4 Characteristics of the main bioactive substances in ginger by-products that may benefit health

nutritional value. Most Ethiopians use ginger with garlic when preparing meals at home or in restaurants. In tiny doses, they are frequently added to food as flavorings or preservatives, as well as to increase gastric juice flow and enhance food taste to stimulate hunger (Lawal et al. 2018). Ginger is regarded as safe for use as an herbal remedy, having few adverse effects on both people and animals (Da Silveira et al. 2019).

Certain cancers, degenerative illnesses including rheumatism, arthritis, and Alzheimer's disease, cardiovascular conditions like high blood pressure and atherosclerosis, vomiting, and diabetes mellitus can all be prevented with ginger (Azam et al. 2014; Lin et al. 2014). It has been observed that bioactive chemicals found in ginger have strong antibacterial, viral, fungal, and parasite disease-fighting properties (Chakraborty et al. 2020). Fiber, carbs, fats, sugars, protein, gum, ash, and essential oils are among the nutrients that are abundant in ginger. Because of its rich nutritional makeup and potent antioxidant and antibacterial qualities, it raises food's value. Foods infused with ginger can provide flavors, scents, and sights, boosting customer approval. Ginger has been used to extract a cheap, environmentally friendly fiber to make translucent cellulose films. In addition to its nutritional advantages, agricultural waste-derived ginger fiber possesses antibacterial potency that can be utilized for food safety and packaging purposes (Abral et al. 2020; Gul et al. 2022).

5 Bioactivities of Ginger By-products

The utilization of ginger to cure diverse groups of ailments or medical conditions, including gastrointestinal disorders, respiratory disorders, atherosclerosis, migraines, stomach ulcers, cholesterol, and much more, has been reported since ancient times (Puvaca 2018). Its anti-inflammatory and antioxidant properties help relieve rheumatoid arthritis discomfort. Because of their bioactive qualities, the primary ingredients of ginger shogaole, paradole, and gingerols can help the body both during and after cooking by enhancing the digestive and metabolic processes (Lei et al. 2017). The bioactivity of many phytochemicals found in ginger has been investigated. The amounts and processing characteristics of gingerols, shogaols, phenylbutenoids, diarylheptanoids, flavonoids, diterpenoids, and sesquiterpenoids determine how beneficial ginger is to health (Lin et al. 2012; Sivakrishnan 2019, Badshah et al. 2022). For cold relief, ginger rhizomes crushed with brown sugar and ginger powder on scrambled eggs have been consumed for cough relief.

In medical procedures, ginger essential oils are employed as medicinal agents and bioactive substances. Consequently, it is essential to the growth of scientific data supporting its possible pharmacological and nutritional value augmentation. Its importance in cooking, food processing, preservation, pharmacy, and antibiotics has not always been highlighted. Ginger has been researched for its bioactive potential in treating pregnancy-related motion nausea, morning sickness, and heart spasms.

The many bioactivities of ginger are caused by its phenolic compounds, which include paradols, shogaols, and gingerols. The bioactive components of ginger, known as gingerols, shogaol, gingerols, ingenol, paradols, and zingerones, have therapeutic and nutraceutical uses. Of them, the most pharmacologically active constituents are found in gingerol and shogaols, composed of 6-, 8-, and 10-structural analogs (Herve et al. 2019; Abdullahi et al. 2020). Ginger is utilized as an antibacterial, antidiabetic, and anticancer antioxidant due to its possible health advantages. There have been reports of anti-neuro-inflammatory, chemotherapy-induced reductions in nausea and vomiting (Nile and Park 2015; Li et al. 2019). It has been demonstrated that ginger can guard against the toxicity of medications like carbon tetrachloride (Singh and Avupati 2012; Naseer et al. 2022).

By preventing its hydrolysis, the aqueous ginger extract may prevent dietary fat from being absorbed through the gut wall. Ginger increases energy expenditure and fat beta-oxidation while decreasing fat absorption, indicating activity in the liver enzymes. Ginger may, therefore, offer protection against several illnesses, including heart disease, cancer, and atherosclerosis. According to rat experiments, Ginger shields DNA from oxidative damage caused by lipopolysaccharides (Herve et al. 2019). In addition, ginger possesses antiplatelet, hepatoprotective, neuroprotective, gastroprotective, and antihypertensive/cholesterol-lowering qualities. The bioactive components of ginger include flavonoids, phenolic compounds, compounds containing sulphur, tannins, alkaloids, phenolic diterpenes, and vitamins (Srivastava et al. 2019). Ginger's main purpose is to provide cuisine flavor, fragrance, texture, and color.

Atherosclerosis, cancer, diabetes, obesity, inflammation, arthritic joint, immunological deficiency, impacts of free radicals, bacteria, aging, and mental health problems have all been treated with ginger and several other phytochemicals (Batool et al. 2019; Batool et al. 2020, Shahbaz et al. 2021). The possible health advantages and benefits of bioactive substances derived from ginger may augment human cognitive functions. Ginger's polyphenols are utilized in various meals for human health and can support food processing in a healthy way. One viable method for treating and preventing COVID-19 infection is to use naturally occurring bioactive chemicals found in ginger to inhibit the SARS-CoV2 spike protein's attachment to ACE2.

6 Food and Non-food Applications of Extracts and Bioactive Compounds from By-products

Zingiber officinale, or ginger, is a plant that has several uses in the culinary and non-food sectors. It contains a variety of extracts and bioactive components. When incorporated into nanoliposomes, extracts such as ginger essential oil exhibit promise as intelligent sustained-release systems in simulated digestion, providing advantages in antioxidants and decreased cytotoxicity (Ekrami et al. 2023). *Z. officinale* extracts' antibacterial and antibiofilm properties, especially against multidrug-resistant *Helicobacter pylori*, demonstrate their promise in the fight against infections and improvement of antibiotic effectiveness (Elbestawy et al. 2023). Additionally, Z. *officinale* essential oils have the potential to be useful for culinary, cosmetic, and pharmaceutical applications due to their anti-inflammatory, antibacterial, and antioxidant qualities against a variety of strains (Zhang et al. 2020).

Aqueous-ethanolic and subcritical water extracts of ginger also show antibiofilm action against pathogenic bacteria, indicating their potential application as natural antimicrobial agents in the food sector (Akbari et al. 2023; Abbasi et al. 2019). *Zingiber officinale* by-product bioactive chemicals provide valuable advantages for various uses.

Since ancient times, ginger has been utilized worldwide as a traditional medicinal and culinary enhancer (Moawad et al. 2015). When food is prepared for consumption, ginger adds flavor and scent. The growing appeal of this product among customers can be attributed to its capacity to enhance the flavor of prepared dishes. Ginger's health-promoting qualities result from its high antioxidant content, volatile oils, shogaols, and gingerols (Jelled et al. 2015). A well-liked blooming plant with a solitary rhizome, ginger is frequently utilized in cooking and traditional medicine (Dosoky and Setzer 2018). The world now knows that ginger has many advantages, and this is mainly because fractionation and detection technologies in the field of ginger studies have advanced quickly.

Shogaols, zingerones, and gingerols are only a few of the useful components of ginger that have diverse therapeutic and nutritional uses (Liu et al. 2017. Ginger's ability to boost food value and serve as a medicinal herb is attributed to its phytochemical content, which includes phenylpropanoids, terpenes, flavonoids, and anthocyanins (Sajilata and Singhal 2012). Ginger is widely used in culinary products, medications, and cosmetics due to its bioactive components. Its bioactive components reduce free radicals, including phenols, terpenoids, flavonoids, gingerol, paradol, shogaols, zingerone, and cerumbone (sesquiterpene compound) (Ahmed et al. 2016). Numerous research studies have been conducted on ginger's bioactivities and antioxidant qualities. Ginger is also used in confectionary sectors globally because of its scent and flavor possibilities. Products including jam, pickles, chutney, liqueurs, ginger beer, and ginger wine all contain ginger as an ingredient (Zhu et al. 2018; Iqbal et al. 2017).

Ginger is frequently used in Ethiopian cuisine as a spice, a key component of major meals, and a component of traditional treatments. In addition to being used as a spice in foods like shiro stew, spiced chili pastes, ginger tea, and fiery red peppers known as berbere, ginger is also used for food processing together with other spices and herbs (Dessalegn 2020). Ethiopian food is well-known outside because ginger adds flavor (Hordofa and Tolossa 2020). Ginger adds flavor and scent and helps make nutrients more accessible. Natural food preservatives pique customers' interest visually and increase appetite (Opara and Chohan 2014). Ginger's active ingredients have the biological ability to facilitate nutrition absorption. As a result, it enhances the biological value of food and has several health advantages (Ali et al. 2021). When creating novel food items, ginger improves the final product's organoleptic qualities and aids in nutrient absorption.

One of the most prized and well-liked spices ginger is also recognized to improve the flavor of meals in human cultures worldwide (Dissanayake et al. 2020). For farmers, consumers, and other stakeholders, ginger is an extremely exciting and commercially significant crop because it increases the value of food items that include it. Consumers today want goods enhanced with naturally occurring spices, like ginger, in favor of artificial preservatives, which may pose health risks. While ginger in all forms boosts the nutritional value of food by supporting recommended nutrient intakes and stimulating the development of new food products, most ginger-growing regions do not have many technological applications that use ginger for biological and nutritional benefits.

Ginger from easily available home gardens is being used as appetizers by increasing groups. Transparency Market Research reports that ginger is one of the vegetable raw materials with the highest economic value (Amit et al. 2017). Further research is necessary, particularly in nations that produce ginger, on the biological application and usage of ginger in the food sector, medicine, cosmetics, fragrance, and dietary supplements. There is not enough information on ginger's rising value at this time. Additional research is required to evaluate the safety, effectiveness, and quality of food based on the use of ginger in traditional treatments and supplements. In addition to its recognized bioactive components, ginger may include antibacterial qualities and other biological activities. These molecules include gingerol, paradol, shogaol, and zingerone. The ethanolic ginger extract has been recognized to have antibacterial properties against microorganisms (Gurav et al. 2022; Tariku et al. 2016). This review was conducted to support the use of ginger in food systems and highlight its potential for bioactivity while boosting food value and health benefits. With more studies, it is possible to forecast the development of innovative foods and medications enhanced with ginger in the future.

7 Non-food Applications of Ginger By-products

Ginger rhizome's versatile qualities and bioactive substances have uses outside of the culinary sector. Ginger's main potential uses outside of food include producing metal nanoparticles, inhibiting metal corrosion, tenderizing meat products, preserving food packaging systems, and biodiesel from oxidation (Vichakshana et al. 2022). Furthermore, ginger has been shown to have medical benefits for improving metabolism and digestion and treating several illnesses, including colds, coughs, nausea, vomiting, and arthritis (Akhlaghi and Najafpour-Darzi 2023). Additionally, ginger's bioactive constituents, such as 6-gingerol, have antimicrobial, antioxidant, and anticancer qualities, which makes ginger a viable option for adjuvant treatment and chemotherapeutic roles in several diseases (Kubra and Rao 2012). Ginger's numerous uses outside the food industry demonstrate its adaptability and promise across various sectors.

The research articles emphasize ginger's wide range of medical uses and by-products. Ginger is well-known for its anti-inflammatory, anti-tumor, antibacterial, anti-emetic, hepatoprotective, and neuroprotective qualities. It contains chemicals like gingerol and shogaol (Liu et al. 2022). According to Menon et al. (2021), these constituents are essential for boosting immunity, functioning as antioxidants, and demonstrating anticancer properties. Furthermore, rheumatic illnesses, colds, nausea, vomiting, and other ailments have historically been treated with ginger and its by-products (Iqbal et al. 2018a, 2018b, 2018c; Verma and Bisen 2022; Nasori et al. 2022). Transient receptor potential (TRP) ion channel and signaling pathway regulation have been connected to the therapeutic benefits of ginger and its processed products against diseases such as primary dysmenorrhea (Dissanayake et al. 2020; Iqbal et al. 2019). Moreover, ginger's antiviral, anti-inflammatory, anticancer, and antioxidant properties are ethnomedically valuable and consistent with Ayurvedic and contemporary medical theories (Fig. 5).

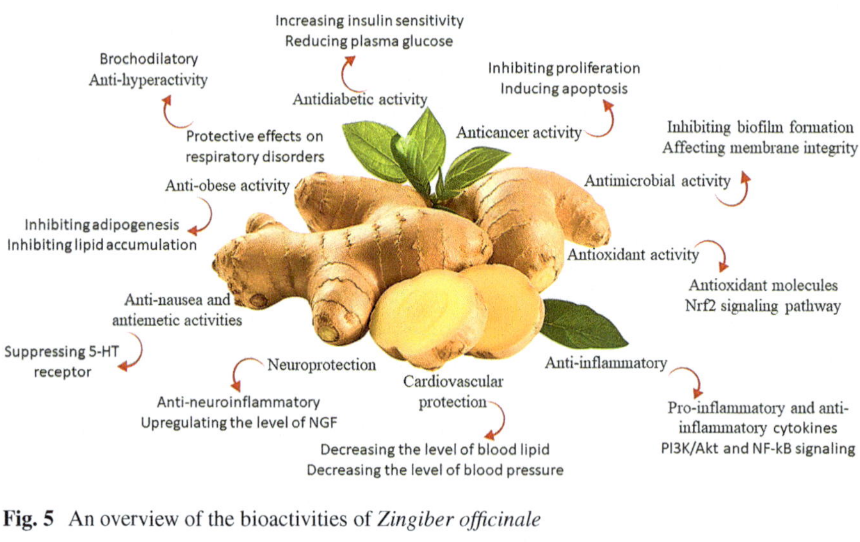

Fig. 5 An overview of the bioactivities of *Zingiber officinale*

8 Valorization of By-products for Non-health Purposes, i.e. Source of Bioenergy

A developing trend in many businesses is the value-adding of by-products for unhealthy objectives like resource optimization and environmental sustainability. The food processing, brewing, distilling, and gasification industries all produce by-products that may be turned into valuable materials such as biogas, biofuels, functional materials, food and aquaculture feed ingredients, and adsorption processes (Abbasi et al. 2018; He et al. 2023; Umego and Barry-Ryan 2023). The projects above seek to mitigate waste creation, improve the dynamics of the circular economy, and advance sustainable industrial methods. To successfully value-added various by-products, however, issues including scalability, integrated processing, and environmental impact assessment must be resolved (Caponio et al. 2022). Given the possibility of contaminants in food by-products, safety assessments are essential to guaranteeing the quality and acceptability of these valorized materials for human and animal use (Assis et al. 2023).

As a by-product of processing ginger, ginger waste has the potential for value addition in non-health industries, including bioenergy manufacturing. Research has indicated increased ginger waste, such as rhizomes, stems, and leaves, due to the rise in ginger consumption and industrial uses worldwide. Antibiotics have been successfully removed from water matrices using waste-derived materials like activated ginger biochar (AGB), which has excellent adsorption capabilities for various antibiotics (Meghani et al. 2023). Moreover, by reducing waste and increasing resource efficiency, the value-adding of food sector by-products, such as ginger waste, is essential for sustainable development (Inthalaeng et al. 2023). The waste-to-wealth strategy may be improved and support a circular economy and environmental sustainability by integrating ginger waste into biofuel production processes, such as bioethanol or biogas fermentation (Chandrasekaran 2012; Iqbal et al. 2019).

A key component of waste reduction and sustainable food processing is the value-adding of ginger by-products. Leftover ginger, such as activated ginger charcoal and ginger hydrosol, allows for extracting and repurposing essential chemicals for various uses. According to studies, adding ginger hydrosol to foot sanitizers may strengthen their antibacterial qualities (Meghani et al. 2023). Furthermore, eliminating antibiotics from water matrices using activated ginger biochar has shown strong adsorption capabilities, indicating the biochar's potential for environmental decontamination (Chandrasekaran 2012). Repurposing agro-food waste to reduce environmental pollution and improve resource efficiency aligns with the value-adding of ginger by-products, which supports a more sustainable food sector (Meghani et al. 2023).

9 Conclusion

Zingiber officinale, often known as ginger, is a popular seasoning and medicinal herb with significant economic value. Processing ginger yields significant by-products, including peels from the rhizome, leaves, and stems, which are often wasted or underutilized. These ginger by-products offer a priceless source of bioactive substances with potential uses across various industries. The chemical study of ginger by-products has discovered numerous bioactive chemicals, including gingerols, shogaols, paradols, zingerone, and other phenolic molecules. These phytochemicals' prominent biological and functional characteristics include antioxidative, anti-inflammatory, antibacterial, and anticancer effects. The food sector has shown promise in using ginger by-product extracts and their bioactive components as flavor enhancers, functional additives, and natural preservatives.

Furthermore, because of their therapeutic potential, they promise integration into medicines, cosmeceuticals, and nutraceuticals. Ginger by-products can be valued for uses other than food, such as the synthesis of bioenergy, in addition to uses in the medical field. The leftover lignocellulosic biomass from ginger may be used as a starting point to produce biofuels, biogas, and other renewable energy sources, which will help provide environmentally beneficial energy options. Valorizing by-products made from ginger is a practical way to make the most of a vital agricultural resource. The ginger business has the potential to contribute to waste reduction while also yielding benefits for the environment, economy, and public health by exploring the uses of bioactive chemicals in many industries and exploiting them. Additional research and development efforts are necessary to improve extraction techniques, identify bioactive substances, and create scalable technologies for the value-adding of ginger by-products.

References

Abbasi BA, Iqbal J, Mahmood T, Khalil AT, Ali B, Kanwal S, Ahmad R (2018) Role of dietary phytochemicals in modulation of miRNA expression: natural swords combating breast cancer. Asian Pac J Trop Med 11(9):501–509

Abbasi BA, Iqbal J, Ahmad R, Bibi S, Mahmood T, Kanwal S, Hameed S (2019) Potential phytochemicals in the prevention and treatment of esophagus cancer: A green therapeutic approach. Pharmacol Rep 71(4):644–652

Abdullahi A, Khairulmazmi A, Yasmeen S, Ismail IS, Norhayu A, Sulaiman MR, Ahmed OH, Ismail MR (2020) Phytochemical profiling and antimicrobial activity of Ginger (*Zingiber officinale*) essential oils against important Phytopathogens. Arab J Chem 13(11):8012–8025. https://doi.org/10.1016/j.arabjc.2020.09.031

Abolaji AO, Ojo M, Afolabi TT, Arowoogun MD, Nwawolor D, Farombi EO (2017) Protective properties of 6-gingerol-rich fraction from *Zingiber officinale* (Ginger) on chlorpyrifos-induced oxidative damage and inflammation in the brain, ovary and uterus of rats. Chem Biol Interact 270:15–23

Abral H, Ariksa J, Mahardika M, Handayani D, Aminah I, Sandrawati N, Sugiarti E, Muslimin AN, Rosanti SD (2020) Effect of heat treatment on thermal resistance, transparency and antimicrobial activity of sonicated Ginger cellulose film. Carbohydr Polym 240(2):375–379. https://doi.org/10.1016/j.carbpol.2020.116287

Ahmad B, Rehman MU, Amin I, Arif A, Rasool S, Bhat SA et al (2015) A review on pharmacological properties of zingerone (4-(4-Hydroxy-3-methoxyphenyl)-2-butanone). Sci World J 2015:816364

Ahmed MARH, Mustafa AI, Hussan HAR, Elfaki AE (2016) Proximate analysis, protein and starch digestibility, specific volume, and sensory evaluation of (gluten-free) tin bread. OALib J 03(06):1–8. https://doi.org/10.4236/oalib.1103208

Akbari S, Didar Z, Vazifedoost M, Hajirostamloo B, Mohtashami M (2023) Antibiofilm activity of ginger (*Zingiber officinale*) extracts *in vitro* and food model. J Food Process Preserv 2023:5134332

Akhlaghi N, Najafpour-Darzi G (2023) Potential applications of Ginger rhizomes as a green biomaterial: a review. Int J Eng 36(2):372–383

Akintobi OA, Onoh CC, Ogele JO, Idowu AA, Ojo OV, Okonko IO (2013) Antimicrobial activity of *Zingiber officinale* (ginger) extract against some selected pathogenic bacteria. Nat Sci 11(1):7–15

Ali AS, Saied EA, Theeban SM, Humayran MM, Shaker AA, Almuhayawi MS, Alfadil A (2021) Phytochemicals against COVID -19, pharmacological perspectives: A systematic review. J Pharm Res Int 33(53):34–55. https://doi.org/10.9734/jpri/2021/v33i53B33678

Ali S, Khan MR, Iqbal J, Shah SA, Abbasi BA, Yaseen T, Kazi M (2022) Chemical characterization and evaluation of the nephroprotective potential of *Parrotiopsis jacquemontiana* (Decne) Rehder and Periploca hydaspidis Falc crude extract in CCl_4-induced male Sprague-Dawley rats. Saudi J Biol Sci 29(2):702–712

Ali S, Sayed AS, Saeed RF, Iqbal J, Ijaz S, Munazir M (2023) Ethnomedicinal plant use value in lower swat, Pakistan. Ethnobot Res Appl 25:1–22

Alibakhshi T, Khodayar MJ, Khorsandi L, Rashno M, Zeidooni L (2018) Protective effects of zingerone on oxidative stress and inflammation in cisplatin-induced rat nephrotoxicity. Biomed Pharmacother 105:225–232

Alsahli MA, Almatroodi SA, Almatroudi A, Khan AA, Anwar S, Almutary AG et al (2021) 6-Gingerol, a major ingredient of ginger attenuates diethylnitrosamine-induced liver injury in rats through the modulation of oxidative stress and anti-inflammatory activity. Mediat Inflamm 2021:1–17

Amit SK, Uddin M, Rahman R, Islam SM, Khan MS (2017) A review on mechanisms and commercial aspects of food preservation and processing. Agric Food Secur 6(1):1–22. https://doi.org/10.1186/s40066-017-0130-8

Anzian A, Muhialdin BJ, Mohammed NK, Kadum H, Marzlan AA, Sukor R, Meor Hussin AS (2020) Antibacterial activity and metabolomics profiling of torch Ginger (*Etlingera Elatior* Jack) flower oil extracted using subcritical carbon dioxide (CO_2). Evid Based Complement Alternat Med 2020:4373401. https://doi.org/10.1155/2020/4373401

Ashokkumar K, Murugan M, Dhanya MK, Sathyan T, Raj S, Mathews N (2020) Traditional uses, phytochemistry, and pharmacological properties of *Zingiber officinale* essential oil and extracts. In: Ethnopharmacological investigation of Indian spices. IGI Global, pp 62–84

Assis AC, Panizio R, Calado L, Brito P, Mourão P (2023) Use of by-products from gasification and carbonization from polymeric residues and biomass for application in liquid phase adsorption. Environments 10(5):74

Azam F, Amer AM, Abulifa AR, Elzwawi MM (2014) Ginger components as new leads for the design and development of novel multi-targeted anti-Alzheimer's drugs: a computational investigation. Drug Des Devel Ther 8:2045–2059. https://doi.org/10.2147/DDDT.S67778

Badshah H, Mumtaz AS, Hussain I, Ali B, Iqbal J, Ali S (2022) Wild morels in Pakistan: environmental and trading statues. In: Biodiversity, conservation and sustainability in Asia: volume 2: prospects and challenges in south and middle Asia. Springer International Publishing, Cham, pp 691–702

Batool R, Aziz E, Salahuddin H, Iqbal J, Tabassum S, Mahmood T (2019) Rumex dentatus could be a potent alternative to treatment of microbial infections and of breast cancer. J Tradit Chin Med 39(6):772–779

Batool R, Aziz E, Iqbal J, Salahuddin H, Tan BKH, Tabassum S, Mahmood T (2020) In vitro antioxidant and anti-cancer activities and phytochemical analysis of *Commelina benghalensis* L. root extracts. Asian Pac J Trop Biomed 10(9):417–425

Beristain-Bauza SDC, Hernández-Carranza P, Cid-Pérez TS, Ávila-Sosa R, Ruiz-López II, Ochoa-Velasco CE (2019) Antimicrobial activity of ginger (*Zingiber officinale*) and its application in food products. Food Rev Intl 35(5):407–426

Caponio F, Piga A, Poiana M (2022) Valorization of food processing by-products. Foods 11(20):3246

Chakraborty M, Afrin T, Munshi SK (2020) Microbiological quality and antimicrobial potential of extracts of different spices. Food Res 4(2):375–379. https://doi.org/10.26656/fr.2017.4(2).303

Chandrasekaran M (ed) (2012) Valorization of food processing by-products. CRC Press

Chemat F, Rombaut N, Sicaire AG, Meullemiestre A, Fabiano-Tixier AS, Abert-Vian M (2017) Ultrasound assisted extraction of food and natural products. Mechanisms, techniques, combinations, protocols and applications. A review. Ultrason Sonochem 34:540–560

Chemat F, Abert Vian M, Ravi HK, Khadhraoui B, Hilali S, Perino S, Fabiano Tixier AS (2019) Review of alternative solvents for green extraction of food and natural products: panorama, principles, applications and prospects. Molecules 24(16):3007

Chen GT, Yuan B, Wang HX, Qi GH, Cheng SJ (2019) Characterization and antioxidant activity of polysaccharides obtained from ginger pomace using two different extraction processes. Int J Biol Macromol 139:801–809

Cheng XL, Liu Q, Peng YB, Qi LW, Li P (2011) Steamed ginger (*Zingiber officinale*): changed chemical profile and increased anticancer potential. Food Chem 129(4):1785–1792

Choi JinGyu CJ, Kim SunYeou KS, Jeong MinSun JM, Oh MyungSook OM (2018) Pharmacotherapeutic potential of ginger and its compounds in age-related neurological disorders. Pharmacol Ther 182:56–69

Choi JiWon CJ, Park HoYoung PH, Oh MyungSook OM, Yoo HyeHyun YH, Lee SangHoon LS, Ha SangKeun HS (2017) Neuroprotective effect of 6-paradol enriched ginger extract by fermentation using Schizosaccharomyces pombe. J Funct Foods 31:304–310

Choi JG, Kim SY, Jeong M, Oh MS (2018) Pharmacotherapeutic potential of ginger and its compounds in age-related neurological disorders. Pharmacol Ther 182:56–69

Da Silveira Vasconcelos M, Mota EF, Gomes-Rochette NF, Nunes-Pinheiro DCS, Nabavi SM, de Melo DF (2019) Chapter 3.18- Ginger (Zingiber officinale Roscoe). In: Nabavi SM, Silva AS (eds) Nonvitamin and nonmineral nutritional supplements. Academic Press, pp 235–239. https://doi.org/10.1016/B978-0-12-812491-8.00034-5

Dalsasso RR, Valencia GA, Monteiro AR (2023) Improving ginger's bioactive composition by combining innovative drying and extraction technologies. Plant Foods Hum Nutr 78(4):755–761

Dessalegn E (2020) Effect of extraction solvents on antioxidant and α-amylase inhibition activities of spiced green chili paste: an in vitro study. Adv Life Sci Technol 77:5–13

Dissanayake KGC, Waliwita WALC, Liyanage RP (2020) A review on medicinal uses of *Zingiber officinale* (ginger). Int J Health Sci Res 10(6):142–148

Dosoky NS, Setzer WN (2018) Chemical composition and biological activities of essential oils of curcuma species. Nutrients 10(9):10–17. https://doi.org/10.3390/nu10091196

Ekrami M, Babaei M, Fathi M, Abbaszadeh S, Nobakht M (2023) Ginger essential oil (*Zingiber officinale*) encapsulated in nanoliposome as innovative antioxidant and antipathogenic smart sustained-release system. Food Biosci 53:102796

Elbestawy MK, El-sherbiny GM, Moghannem SM (2023) Antibacterial, antibiofilm, and anti-inflammatory activities of *Zingiber officinale* extract against helicobacter pylori. Microbiol Res 14(3):1124–1138

Essien SO, Young B, Baroutian S (2020) Recent advances in subcritical water and supercritical carbon dioxide extraction of bioactive compounds from plant materials. Trends Food Sci Technol 97:156–169

Foudah AI, Shakeel F, Yusufoglu HS, Ross SA, Alam P (2020) Simultaneous determination of 6-shogaol and 6-gingerol in various ginger (*Zingiber officinale* Roscoe) extracts and commercial formulations using a green RP-HPTLC-densitometry method. Food Secur 9(8):1136

Gaire BP, Kwon OW, Park SH, Chun KH, Kim SY, Shin DY, Choi JW (2015) Neuroprotective effect of 6-paradol in focal cerebral ischemia involves the attenuation of neuroinflammatory responses in activated microglia. PLoS One 10(3):e0120203

Giacometti J, Kovačević DB, Putnik P, Gabrić D, Bilušić T, Krešić G et al (2018) Extraction of bioactive compounds and essential oils from Mediterranean herbs by conventional and green innovative techniques: a review. Food Res Int 113:245–262

Gul F, Khan I, Iqbal J, Abbasi BA, Shahbaz A, Capasso R, Mahmood T (2022) Phytochemistry, biological activities and in silico molecular docking studies of *Oxalis pes-caprae* L. compounds against SARS-CoV-2. J King Saud Univ-Sci 34(6):102136

Gurav TP, Dholakia BB, Giri AP (2022) A glance at the Chemodiversity of Ocimum species: trends, implications, and strategies for the quality and yield improvement of essential oil. Phytochem Rev 21(3):879–913. https://doi.org/10.1007/s11101-021-09767-z

He Y, Allen J, Huang H (2023) Food by-products valorization technologies: Brewer's spent grain

Herve T, Raphaël KJ, Ferdinand N, Victor Herman N, Willy Marvel NM, D'alex C, Laurine Vitrice TF (2019) Effects of ginger (*Zingiber officinale*, Roscoe) essential oil on growth and laying performances, serum metabolites, and egg yolk antioxidant and cholesterol status in laying Japanese quail. J Vet Med 2019:7857504. https://doi.org/10.1155/2019/7857504

Hitomi S, Ono K, Terawaki K, Matsumoto C, Mizuno K, Yamaguchi K et al (2017) (6)-gingerol and (6)-shogaol, active ingredients of the traditional Japanese medicine hangeshashinto, relief oral ulcerative mucositis-induced pain via action on Na+ channels. Pharmacol Res 117:288–302

Ho SC, Su MS (2016) Optimized heat treatment enhances the anti-inflammatory capacity of ginger. Int J Food Prop 19(8):1884–1898

Höferl M, Stoilova I, Wanner J, Schmidt E, Jirovetz L, Trifonova D, Stanchev V, Krastanov A (2015) Composition and comprehensive antioxidant activity of Ginger (*Zingiber officinale*) essential oil from Ecuador. Nat Prod Commun 10(6):1085–1090. https://doi.org/10.1177/1934578X1501000672

Hordofa TS, Tolossa TT (2020) Cultivation and postharvest handling practices affecting yield and quality of major spices crops in Ethiopia: a review. Cogent Food Agric 6(1):1. https://doi.org/10.1080/23311932.2020.1788896

Hsiang CY, Cheng HM, Lo HY, Li CC, Chou PC, Lee YC, Ho TY (2015) Ginger and zingerone ameliorate lipopolysaccharide-induced acute systemic inflammation in mice, assessed by nuclear factor-κB bioluminescent imaging. J Agric Food Chem 63(26):6051–6058

Hur J, Lee Y, Lee CJ, Park HY, Choi SY (2020) 6-shogaol suppresses oxidative damage in L6 muscle cells. Appl Biol Chem 63:1–6

Ijaz S, Iqbal J, Abbasi BA, Ullah Z, Yaseen T, Kanwal S, Cho WC (2023) Rosmarinic acid and its derivatives: current insights on anticancer potential and other biomedical applications. Biomed Pharmacother 162:114687

Ijaz S, Iqbal J, Abbasi BA, Tufail A, Yaseen T, Uddin S, Sharifi-Rad J (2024a) Current stage of preclinical and clinical development of guggulsterone in cancers: challenges and promises. Cell Biol Int 48(2):128–142

Ijaz S, Iqbal J, Abbasi BA, Kanwal S, Tavafoghi M, Ahmed MZ, Mahmood T (2024b) Investigation of bioactive constituents and evaluation of different *in vitro* antimicrobial, antioxidant, and cytotoxicity potentials of different *Portulacaria afra* extracts. J King Saud Univ-Sci 36(2):103033

Inthalaeng N, Gao Y, Remón J, Dugmore TI, Ozel MZ, Sulaeman A, Matharu AS (2023) Ginger waste as a potential feedstock for a zero-waste ginger biorefinery: a review. RSC Sustain 1(2):213–223

Iqbal J, Abbasi BA, Mahmood T, Kanwal S, Ali B, Shah SA, Khalil AT (2017) Plant-derived anticancer agents: A green anticancer approach. Asian Pac J Trop Biomed 7(12):1129–1150

Iqbal J, Abbasi BA, Batool R, Mahmood T, Ali B, Khalil AT, Ahmad R (2018a) Potential phytocompounds for developing breast cancer therapeutics: nature's healing touch. Eur J Pharmacol 827:125–148

Iqbal J, Abbasi BA, Ahmad R, Mahmood T, Kanwal S, Ali B, Badshah H (2018b) Ursolic acid a promising candidate in the therapeutics of breast cancer: current status and future implications. Biomed Pharmacother 108:752–756

Iqbal J, Abbasi BA, Khalil AT, Ali B, Mahmood T, Kanwal S, Ali W (2018c) Dietary isoflavones, the modulator of breast carcinogenesis: current landscape and future perspectives. Asian Pac J Trop Med 11(3):186–193

Iqbal J, Abbasi BA, Ahmad R, Batool R, Mahmood T, Ali B, Munir A (2019) Potential phytochemicals in the fight against skin cancer: current landscape and future perspectives. Biomed Pharmacother 109:1381–1393

Jelled A, Fernandes Â, Barros L, Chahdoura H, Achour L, Ferreira IC, Cheikh HB (2015) Chemical and antioxidant parameters of dried forms of Ginger rhizomes. Ind Crop Prod 77:30–35. https://doi.org/10.1016/j.indcrop.2015.08.052

Jung MY, Lee MK, Park HJ, Oh EB, Shin JY, Park JS et al (2018) Heat-induced conversion of gingerols to shogaols in ginger as affected by heat type (dry or moist heat), sample type (fresh or dried), temperature and time. Food Sci Biotechnol 27:687–693

Kaushik S, Jangra G, Kundu V, Yadav JP, Kaushik S (2020) Anti-viral activity of *Zingiber officinale* (Ginger) ingredients against the Chikungunya virus. Virus 31:270–276

Kim HS, Lee SH, Byun Y, Park HD (2015) 6-Gingerol reduces Pseudomonas aeruginosa biofilm formation and virulence via quorum sensing inhibition. Sci Rep 5(1):8656

Kiyama R (2020) Nutritional implications of ginger: chemistry, biological activities and signaling pathways. J Nutr Biochem 86:108486

Kotowski U, Kadletz L, Schneider S, Foki E, Schmid R, Seemann R et al (2018) 6-shogaol induces apoptosis and enhances radiosensitivity in head and neck squamous cell carcinoma cell lines. Phytother Res 32(2):340–347

Kou X, Ke Y, Wang X, Rahman MRT, Xie Y, Chen S, Wang H (2018) Simultaneous extraction of hydrophobic and hydrophilic bioactive compounds from ginger (*Zingiber officinale* Roscoe). Food Chem 257:223–229

Kubra IR, Rao LJM (2012) An impression on current developments in the technology, chemistry, and biological activities of ginger (*Zingiber officinale* Roscoe). Crit Rev Food Sci Nutr 52(8):651–688

Lawal AR, Olayinka BU, Murtadha RA, Ayinla A, Etejere EO (2018) Comparative analysis of phytochemical and proximate composition of *Allium sativum* L. and *Zingiber officinale* Rosc. Nig J Basic Appl Sci 26(2):82–87. https://doi.org/10.5455/NJBAS.265778

Lee JH, Kim YG, Choi P, Ham J, Park JG, Lee J (2018) Antibiofilm and antivirulence activities of 6-gingerol and 6-shogaol against Candida albicans due to hyphal inhibition. Front Cell Infect Microbiol 8:299

Lei H, Wei Q, Wang Q, Su A, Xue M, Liu Q, Hu Q (2017) Characterization of ginger essential oil/Palygorskite composite (GEO-PGS) and its anti-bacteria activity. Mater Sci Eng C 73:381–387. https://doi.org/10.1016/j.msec.2016.12.093

Li H, Xie L, Ma Y, Zhang M, Zhao Y, Zhao X (2019) Effects of drying methods on drying characteristics, physicochemical properties and antioxidant capacity of okra. LWT 101:630–638. https://doi.org/10.1016/j.lwt.2018.11.076

Lin Z, Wang H, Xu Y, Dong J, Hashi Y, Chen S (2012) Identification of antioxidants in Fructus Aurantii and its quality evaluation using a new on-line combination of analytical techniques. Food Chem 134(2):1181–1191. https://doi.org/10.1016/j.foodchem.2012.02.161

Lin RJ, Chen CY, Lu CM, Ma YH, Chung LY, Wang JJ, Lee JD, Yen CM (2014) Anthelmintic constituents from Ginger (*Zingiber officinale*) against Hymenolepis Nana. Acta Trop 140:50–60. https://doi.org/10.1016/j.actatropica.2014.07.009

Liu Q, Meng X, Li Y, Zhao C-N, Tang G-Y, Li H-B (2017) Antibacterial and antifungal activities of spices. Int J Mol Sci 18(6):1283. https://doi.org/10.3390/ijms18061283

Liu X, Meng X, Su X, Ren K, Ning C, Qi X, Zhang S (2022) The mechanism of ginger and its processed products in the treatment of estradiol valerate coupled with oxytocin-induced dysmenorrhea in mice *via* regulating the TRP ion channel-mediated ERK 1/2/NF-κB signaling pathway. Food Funct 13(21):11236–11248

Luettig J, Rosenthal R, Lee IFM, Krug SM, Schulzke JD (2016) The ginger component 6-shogaol prevents TNF-α-induced barrier loss via inhibition of PI3K/Akt and NF-κB signaling. Mol Nutr Food Res 60(12):2576–2586

Mahomoodally MF, Aumeeruddy MZ, Rengasamy KR, Roshan S, Hammad S, Pandohee J et al (2021) Ginger and its active compounds in cancer therapy: from folk uses to nano-therapeutic applications. In: Seminars in cancer biology, vol 69. Academic Press, pp 140–149

Mao QQ, Xu XY, Cao SY, Gan RY, Corke H, Beta T, Li HB (2019a) Bioactive compounds and bioactivities of ginger (*Zingiber officinale* Roscoe). Food Secur 8(6):185

Mao QQ, Xu XY, Cao SY, Gan RY, Corke H, Beta T, Li HB (2019b) Bioactive compounds and bioactivities of ginger (*Zingiber officinale* Roscoe). Foods 8(6):185

Meghani R, Lahane V, Kotian SY, Lata S, Tripathi S, Ansari KM, Yadav AK (2023) Valorization of Ginger waste-derived biochar for *simultaneous multiclass* antibiotics remediation in aqueous medium. ACS Omega 8(12):11065–11075

Menon V, Elgharib M, El-awady R, Saleh E (2021) Ginger: from serving table to salient therapy. Food Biosci 41:100934

Moawad SA, El-Ghorab AH, Hassan M, Nour-Eldin H, El-Gharabli MM (2015) Chemical and microbiological characterization of Egyptian cultivars for some spices and herbs commonly exported abroad. Food Nutr Sci 6(07):643–659. https://doi.org/10.4236/fns.2015.67068

Murthy PS, Gautam R (2015) Ginger oleoresin chemical composition, bioactivity and application as bio-preservatives. J Food Process Preserv 39(6):1905–1912

Mwaurah PW, Kumar S, Kumar N, Attkan AK, Panghal A, Singh VK, Garg MK (2020) Novel oil extraction technologies: process conditions, quality parameters, and optimization. Compr Rev Food Sci Food Saf 19(1):3–20

Naseer S, Iqbal J, Naseer A, Kanwal S, Hussain I, Tan Y, Mahmood T (2022) Deciphering chemical profiling, pharmacological responses and potential bioactive constituents of *Saussurea lappa* Decne. Extracts through in vitro approaches. Saudi J Biol Sci 29(3):1355–1366

Nasori AS, Wiguna B, Sulaswaty A, Atmaji P, Mardliyati E, Purwanto W et al (2022, December) Medicinal plant extraction of *Zingiber officinale* rhizome using response surface methodology (RSM) and characterization of the product. In: IOP conference series: earth and environmental science, vol 1116, No. 1. IOP Publishing, p 012059

Nile SH, Park S (2015) Chromatographic analysis, antioxidant, anti-inflammatory, and xanthine oxidase inhibitory activities of Ginger extracts and its reference compounds. Ind Crop Prod 70:238–244. https://doi.org/10.1016/j.indcrop.2015.03.033

Ohnishi M, Ohshita M, Tamaki H, Marutani Y, Nakayama Y, Akagi M et al (2019) Shogaol but not gingerol has a neuroprotective effect on hemorrhagic brain injury: contribution of the α, β-unsaturated carbonyl to heme oxygenase-1 expression. Eur J Pharmacol 842:33–39

Opara EI, Chohan M (2014) Culinary herbs and spices: their bioactive properties, the contribution of polyphenols and the challenges in deducing their true health benefits. Int J Mol Sci 15(10):19183–19202. https://doi.org/10.3390/ijms151019183

Puvača N (2018) Bioactive compounds in selected hot spices and medicinal plants. Technol Eng Manag 1(1):8–17

Rampogu S, Baek A, Gajula RG, Zeb A, Bavi RS, Kumar R et al (2018) Ginger (*Zingiber officinale*) phytochemicals-gingerenone-A and shogaol inhibit SaHPPK: molecular docking, molecular dynamics simulations and in vitro approaches. Ann Clin Microbiol Antimicrob 17:1–15

Sabahannur S (2022) Ease of commercialization and downstream processing of Ginger-processed products. Jurnal Manajemen Bisnis 9(1):88–93

Sajilata MG, Singhal RS (2012) Quality indices for spice essential oils. In: Peter KV (ed) *Handbook of herbs and spices*, 2nd. Woodhead Publishing Ltd, New York Washington, DC, pp 42–54

Sapkota A, Park SJ, Choi JW (2019) Neuroprotective effects of 6-shogaol and its metabolite, 6-paradol, in a mouse model of multiple sclerosis. Biomol Ther 27(2):152

Shahbaz A, Abbasi BA, Iqbal J, Fatima I, Zahra SA, Kanwal S, Mahmood T (2021) Chemical composition of *Gastrocotyle hispida* (Forssk.) bunge and *Heliotropium crispum* Desf. and evaluation of their multiple in vitro biological potentials. Saudi J Biol Sci 28(11):6086–6096

Shakya SR (2015) Medicinal uses of Ginger (*Zingiber Officinale* Roscoe) improves growth and enhances immunity in aquaculture. Int J Chem Stud 3:83–87

Shalaby EA, Shanab SM, Hafez RM, El-Ansary AE (2023) Chemical constituents and biological activities of different extracts from ginger plant (*Zingiber officinale*). Chem Biol Technol Agric 10(1):14

Shukor NIA, Misran A, Ahmad SH, Xue YT, Mahmud TMM, Saari N (2020) The changes of chemical quality of Ginger during postharvest storage at chilling temperature. Food Res 4(5):1653–1662. https://doi.org/10.26656/fr.2017.4(5).109

Singh A, Avupati VR (2012) Development and validation of UV-spectrophotometric method for the estimation of curcumin in standardized polyherbal formulations. J Young Pharm 9(4):491–495. https://doi.org/10.5530/jyp.2017.9.96

Sivakrishnan S (2019) Traditional herbal medicines—a review. Int J Res Anal Rev 5:611–614

Srivastava N, Singh S, Gupta AC, Shanker K, Bawankule DU, Luqman S (2019) Aromatic Ginger (*Kaempferia galanga* L.) extracts with ameliorative and protective potential as a functional food, beyond its flavor and nutritional benefits. Toxicol Rep 6:521–528. https://doi.org/10.1016/j.toxrep.2019.05.014

Suk S, Seo SG, Yu JG, Yang H, Jeong E, Jang YJ et al (2016) A bioactive constituent of ginger, 6-shogaol, prevents adipogenesis and stimulates lipolysis in 3 T 3-L 1 adipocytes. J Food Biochem 40(1):84–90

Tariku H, Kassahun S, Gezahegne G (2016) First report of Ginger (*Zingiber officinale*) bacterial wilt disease in Ethiopia. Res J Agric For Sci 4(4):5–9

Umego EC, Barry-Ryan C (2023) Review of the valorization initiatives of brewing and distilling by-products. Crit Rev Food Sci Nutr 64:1–17

Varakumar S, Umesh KV, Singhal RS (2017) Enhanced extraction of oleoresin from ginger (*Zingiber officinale*) rhizome powder using enzyme-assisted three phase partitioning. Food Chem 216:27–36

Verma R, Bisen PS (2022) Ginger-a potential source of therapeutic and pharmaceutical compounds. J Food Bioact 18

Vichakshana GAD, Young DJ, Choo WS (2022) Extraction, purification, food applications, and recent advances for enhancing the bioavailability of 6-gingerol from ginger–a review. Qual Assur Saf Crops Foods 14(4):67–83

Wei CK, Tsai YH, Korinek M, Hung PH, El-Shazly M, Cheng YB et al (2017) 6-Paradol and 6-shogaol, the pungent compounds of ginger, promote glucose utilization in adipocytes and myotubes, and 6-paradol reduces blood glucose in high-fat diet-fed mice. Int J Mol Sci 18(1):168

Zahra SA, Iqbal J, Abbasi BA, Shahbaz A, Kanwal S, Shah SL, Mahmood T (2021) Antimicrobial, cytotoxic, antioxidants, enzyme inhibition activities, and scanning electron microscopy of *Lactuca orientalis* (Boiss.) Boiss seeds. Microsc Res Tech 84(6):1284–1295

Zahra N, Iqbal J, Arif M, Abbasi BA, Sher H, Nawaz AF, Calina D (2023) A comprehensive review on traditional uses, phytochemistry and pharmacological properties of *Paeonia emodi* wall. Ex Royle: current landscape and future perspectives. Chin Med 18(1):23

Zhang F, Zhang JG, Qu J, Zhang Q, Prasad C, Wei ZJ (2017) Assessment of anti-cancerous potential of 6-gingerol (Tongling white Ginger) and its synergy with drugs on human cervical adenocarcinoma cells. Food Chem Toxicol 109:910–922

Zhang J, Dong Y, Nisar T, Fang Z, Wang ZC, Guo Y (2020) Effect of superfine-grinding on the physicochemical and antioxidant properties of *Lycium ruthenicum* Murray powders. Powder Technol 372:68–75

Zhang F, Zhu F, Chen B, Su E, Chen Y, Cao F (2022a) Composition, bioactive substances, extraction technologies and the influences on characteristics of *Camellia oleifera* oil: a review. Food Res Int 156:111159

Zhang S, Zhang L, Yu M, Luo D, Chen S, Liu W et al (2022b) Essential oils of *Zingiber officinale*: chemical composition, in vivo alleviation effects on TPA induced ear swelling in mice and in vitro bioactivities. Front Nutr 9:1043175

Zhu J, Chen H, Song Z, Wang X, Sun Z (2018) Effects of Ginger (Zingiber Officinale Roscoe) on type 2 diabetes mellitus and components of the metabolic syndrome—a systematic review and meta-analysis of randomized controlled trials. Evid-Based Complement Altern Med 5692962:1–11. https://doi.org/10.1155/2018/5692962

Part V
Future Trends and Conclusion

Future Trends and Conclusion on Bioactive Phytochemicals in By-products from Leaf, Stem, Root, and Tuber Vegetables

Mohamed Fawzy Ramadan

Contents

Abstract The potential utilization of leaf, stem, root, and tuber vegetable by-products as natural sources of bioactive compounds has attracted interest among researchers in recent years. Leaf, stem, root, and tuber vegetables are quotidian constituents of the food cycle worldwide, generating copious quantities of by-products. These by-products are often discarded without any treatment, leading to concerns about sustainable utilization. Hence, the assessment of potentiality among these by-products has become obligatory. Many edible leaf, stem, root, and tuber vegetable by-products are already being utilized in various forms for human consumption. Recent research endeavors have been focused on exploring the potential

M. F. Ramadan (✉)
Department of Clinical Nutrition, Faculty of Applied Medical Sciences, Umm Al-Qura University, Makkah, Saudi Arabia
e-mail: mfhassanien@uqu.edu.sa

M. F. Ramadan (ed.), *Bioactive Phytochemicals in By-products from Leaf, Stem, Root and Tuber Vegetables*, https://doi.org/10.1007/978-3-031-80700-8_16

of bioprocessing of the by-products into value-added bioproducts for food, feed, nutraceutical, pharmaceutical, biotechnological, and bioenergetic applications. Besides daily consumption, vegetables have emerged as major sources of bioactive compounds that contribute to the reduction of several chronic diseases and health maintenance. This work highlights the current situation and valorization potential of leaf, stem, root, and tuber vegetable by-products.

Keywords Circle economy · Environment · Biowaste · Green chemistry · Phytochemicals · Food industry · Feed

1 Bioactivity of Phytochemicals

Phytochemicals, especially phenolics, flavonoids, and other polyphenols, including phenolic acids, flavonols, flavones, and isoflavones, are reported to have hepatoprotective, diuretic, chemopreventive, and antioxidant traits. Phenolics donate hydrogen and stabilize free radicals through electron transfer reactions. The biological effects of vegetable phytochemicals are affected by the type of vegetable, methods of extraction, analytical approaches used, etc. Moreover, flavonoids have general antioxidant effects by acting as scavengers and having metal-chelating properties. Bioactive phytochemicals and end metabolites recover free oxygen radicals and may ultimately protect cells from oxidative stress. Essential oils obtained from various by-products could be used as natural preservatives in food products due to their antibacterial, antifungal, and antioxidant traits. Essential oils showed strong antimicrobial, antioxidant, skin-whitening, and oil-reducing effects. Meanwhile, crude extracts protect against oxidative stress and play a crucial role in cell signaling and gene regulation. Phytochemicals are phytogenic bioactive compounds, generally low molecular weight substances of non-nutritional nature, that occur widely in edible plants (Caballero et al. 2021; Santacroce et al. 2024; Reguengo et al. 2022; Sorrenti et al. 2023; Teruel-Andreu et al. 2021; Akter and Rabeta 2021; Garcia-Alonso et al. 2023; Rojas-García et al. 2023; Favela-González et al. 2020; López-Salas et al. 2021; Elimam et al. 2022).

Inflammation is a complex biological response that involves four key components, namely heat, redness, swelling, and pain. Without inflammation, wounds would not heal, infections would not be cleared, and cancer cells would have an uncontrolled rate of growth. Inflammation is a fundamental protective response that is essential to maintain homeostasis. The objective of many plant products has focused on the inhibition of the early events in inflammation, such as changes in prostaglandin metabolism (cox activity), stabilization of lysosomal membranes, and inhibition of histamine release from mast cells. Plant extracts and their bioactive phytochemicals have shown anti-inflammatory traits by suppressing production, signaling, and actions of pro-inflammatory mediators (Meizlish et al. 2021; Mollaei et al. 2020; Silvagno et al. 2020; Murakami et al. 2020; Chitimus et al. 2020; Moreira Lopes et al. 2020; Margraf et al. 2020).

1.1 Antioxidant Activity

Studies have been evaluated on the bioactive phytochemicals present in the by-products of leaf, stem, root, and tuber vegetables and their antioxidant activity. Phytochemicals identified in considerably lesser quantities and novel bioactive compounds have been documented. Commercially available extracts from the by-products of already established commodities have also been listed. The effect of different processing techniques on the recovery of bioactive compounds and antioxidant activities has also been presented. Bioactivity-guided isolation of target compounds with antioxidant activity, their bioaccessibility, and toxicological studies have been reviewed. Reports have been compiled to explore potential food, feed, pharmaceutical, nutraceutical, or cosmeceutical applications for commercial utilization. Innovative bioprocessing technologies for generating value-added bioproducts from these by-products have been discussed (Trigo et al. 2022; Reguengo et al. 2022; Mohd Basri et al. 2021; Karimi et al. 2021; Patra et al. 2022; 'Aqilah et al. 2023; Zaky et al. 2024; Núñez-Gómez et al. 2023; Meral et al. 2022).

1.2 Anti-Inflammatory Activity

Vegetables are important sources of bioactive phytochemicals, especially those in the Labiatae, and Cruciferae families. The anti-inflammatory impacts of extracts from these by-products were investigated in RAW 264.7 macrophage cells stimulated with lipopolysaccharide (LPS). The results showed that the inhibition of nitric oxide production was significantly greater in the root, tuber, and stem extracts compared with control by-products. However, inhibition was not observed for leaf by-products. Tuber and root by-products had the most potent suppressive effects on the production of pro-inflammatory cytokines, interleukin-6, and tumor necrosis factor-α in LPS-activated macrophages. Generally, by-products from root, tuber, and stem vegetables contained higher total phenolic contents than leaf vegetables. By-products obtained from the tuber and stem vegetables are particularly suitable candidates for the development of functional products based on higher extraction yields and bioactive contents (Yimer et al. 2020; Yesmin et al. 2020; Amo-Mensah et al. 2020; Borquaye et al. 2020; Anokwah et al. 2022; Sahukari et al. 2021).

Vegetable by-products are the edible remnants of vegetables consumed for their foliant, stem, root, or tuber. Lactic acid fermented, leaf and stem vegetable by-products obtained from lettuce and bok choy, respectively, contained quercetin, which has been reported for their anti-inflammatory and anti-cancer properties. Phenolic acids have been known to provide both anti-inflammatory and anti-oxidative activity within tissues. Chlorogenic acid is a caffeoylquinic acid that occurs in a large number of plants and is the predominant phenolic acid in tuber and stem vegetables. Its derivatives have been reported to be beneficial in the prevention of chronic inflammatory diseases. However, anti-inflammatory activity on the hot water extracts of leaf and stem vegetable by-products has not previously been described. By-products have been described as a natural source of bioactive substances with protective effects against oxidative stress and inflammation in human cells. The accumulation of chronic inflammation in tissues has been demonstrated to be a high-risk factor for metabolic syndrome, including obesity, type-2 diabetes, and hyperlipidemia (Joly et al. 2020; Cebulak et al. 2022; Makori et al. 2020; Krochmal-Marczak et al. 2020; Alves Filho et al. 2020; Mangiapelo et al. 2023; Musilová et al. 2024; Yang et al. 2021).

2 Applications of Biowaste-Derived Phytochemicals

Bioactive phytochemicals with a diverse range of bioactivities have been extracted and utilized from by-products of vegetables, fruits, cereal, and legumes. The recent advances in extraction methods and technologies, especially those involving non-thermal methods, combined with the increased focus on functional foods, smart preservatives, and eco-sustainable packaging solutions, have greatly enhanced the interest in vegetable by-products as phytochemical sources. There is a growing emphasis on the need to find novel applications for these by-products, especially in the food, nutraceutical, and pharmaceutical industries. The applications of by-product-derived phytochemicals are highlighted with specific industrial examples, particularly within the food and pharmaceutical industries. Emerging technologies or products developed in the past decade are emphasized to give readers an idea of the current possibilities and opportunities within the relevant sectors (Patra et al. 2022; Reguengo et al. 2022; ‘Aqilah et al. 2023; Coman et al. 2020; Galali et al. 2020; Cano-Lamadrid and Artes-Hernandez 2021; Delgado et al. 2022; Ramadan Hassanien 2023; Marcillo-Parra et al. 2021).

2.1 Food Industry

By-products created from the preparation, processing, and storage of various products are generated. Vegetable by-products, while being considerably larger in quantity, remain predominantly underutilized. They can be employed to extract bioactive phytochemicals, with potential applications in food, medicine, and nutraceuticals; nevertheless, very few have actually been exploited. The exploitation of such bioactive phytochemicals would aid in reducing waste and pollution from the food processing industry, creating new avenues for the food industry and trade. It would also meet public demand for health-promoting, functional, and nutraceutical foods. This work presents an additional approach to utilizing by-products from vegetables, moving toward waste-free by-product utilization. Vegetable by-products have been established as a substantial source of bioactive phytochemicals and their potential application in the food industry. Waste management of vegetable by-products utilizing biodegradation techniques and fermentation processes should be considered for future research.

Food industries generate large quantities of by-products or waste that remain poorly utilized. These by-products are actually an important source of bioactive phytochemicals, such as phenolics, carotenoids, vitamin C, dietary fibers, and antioxidants, with the potential of being utilized in the food industry. A large quantity of by-products are produced during the preparation, processing, cooking, and storage of various food products. Vegetables are predominantly cultivated, consumed, and wasted by around 30% worldwide. Vegetable by-products consist of discarded parts, such as skins, after vegetable consumption or processing and unevaluated residues/growth after vegetable harvesting (leaves, stems, and roots). Tuber and rhizome vegetable by-products consist of waste bulbs, shoots, and non-edible parts after harvesting. Generally, by-products derived from root, stem, leaf, and tuber-based vegetable processing are a potential source of natural bioactive phytochemicals. However, due to a lack of a proper by-product waste management scheme and consumer preference for only edible parts, these by-products remain poorly utilized. Moreover, the majority of vegetable by-products are free from pesticides and herbicides due to non-commercial use (Reguengo et al. 2022; Trigo et al. 2022; Coman et al. 2020; ‘Aqilah et al. 2023; Marcillo-Parra et al. 2021; Galali et al. 2020; Majumder and Annegowda 2021; Sorrenti et al. 2023; Ramadan Hassanien 2023).

2.2 Pharmaceutical Industry

The amalgamation of chemical biology with advances in genomics, proteomics, and metabolomics has brought renewed interest in phytochemicals as potential drug candidates. Any natural compound derived from plants with protective effects on human health can be categorized under phytochemicals. The diverse phytochemical space of 200,000 natural compounds makes plants a readily available source for drug discovery. Many agents used in anti-cancer therapy (e.g., vinca alkaloids and paclitaxel) were originally obtained through chemotherapeutic screening of plants. The bioactive nature of natural compounds is often attributed to a complex set of functions ranging from antioxidant, anti-inflammatory, and antibacterial to affecting glucose and lipid metabolism. Plants rely on a plethora of phytochemicals as their primary defense mechanism against disease and herbivory. Currently, over 25% of all modern medications are produced from higher plants. The most widely utilized drugs in the early twenty-first century (e.g., salicylic acid, quinine, morphine, taxol, and vincristine) were shark-vetted from plants. Phytochemicals are continuously being tested in preclinical models to understand their mechanisms of action against disease. Compounds from rice bran, curcumin, curcuminoids, soy isoflavones, garlic derivatives, wheat bran, green tea polyphenols, and orange juice have demonstrated promise as potential chemopreventive agents by modulating multiple targets involving microflora, mutagenesis, signaling pathways, survival, proliferation, apoptosis, epigenetic factors, cell adhesion, and inflammation.

In many developing countries, medicinal plants are the primary source of remedies for controlling illness and disease. These countries possess the greatest biodiversity of plants, many of which reflect a unique source of active principles used in traditional medicine for centuries and a well-established knowledge of herbal treatments. The growing reliance on medicinal plants is due to the perception that they are (i) less expensive than synthetic drugs, (ii) more effective for treating chronic illness, (iii) have fewer side effects, (iv) are more compatible with cultural beliefs, and (v) have been safety-tested by time (Varijakzhan et al. 2020; Gorlenko et al. 2020; Babich et al. 2020; Porras et al. 2020; Dzobo 2022; Nasim et al. 2022; Parveen et al. 2020; Jugran et al. 2021; Süntar 2020; Najmi et al. 2022).

3 Regulatory Aspects and Safety Considerations

3.1 Regulatory Guidelines

Bioactive phytochemicals have been traditionally regarded as secondary metabolites present in fruits and vegetables, which are not directly involved in normal plant growth or development. However, this perception has changed over the years. Studies have uncovered bioactive properties in a multitude of these compounds, extending their functionality to areas such as health improvement, fertility stimulation in plants, and plant growth regulations. With increasing awareness of the health issues associated with synthetic fertilizers and pesticides, research has shifted toward developing eco-friendly alternatives derived from bioactive natural compounds. Several national and international organizations have devised rules and regulations governing the commercialization of these bioactive compounds and bioproducts. Guidelines for gathering phytochemicals of industrial importance from plants and seeds have been issued by WHO. Their plant-genome product toxicity studies recommendation includes the evaluation of individual and cumulative toxicities, teratogenicity, mutagenicity, etc. For industrial applicability, comparative genomics studies at full scale of the phytochemical targeted on bioactive compounds from plant tissues, commonly stem, leaf, tuber, and seed, are needed. The bioactivity studies, their targets, the mechanism of action of phytochemical compounds, and the probable molecular scale tracking of different bioactivity may be commercially patentable (Wink 2022; Halder and Jha 2023; Ahmed et al. 2024; Pinto et al. 2021; Behl et al. 2021a, 2021b; Mohan et al. 2020; Haruna and Yahaya 2021; Roy et al. 2022).

3.2 *Toxicological Studies*

The organisms exposed to soil phytochemicals, mainly the earthworms, aquatic plants, or dorsal fins of fishes, were selected to conduct the toxicity study. The bioactivity of crude extracts, fractions, or pure individual compounds has been documented for lipid peroxidation, pavement cell rupturing, necrotic lesions, and germination inhibition effects. Although the toxicity on plants, nematodes, fungi, and bacteria has been tested, the toxicity effect among the bio-compounds has been largely neglected until now. The toxic bioactive compounds with potent weed or insect-killing properties have the potential to invoke bioactive targeted classification. The comprehensive understanding of toxicity variation among the selected biochemicals against a particular pest can rule out secondary metabolites from the same family of compounds. The toxicity study on the potent bioactive at different concentrations has opened new avenues for resistance mechanism studies and organism-target studies for the broad-spectrum insecticidal or herbicidal bi-selection (Dawurung et al. 2021; Erhabor et al. 2021; Mogana et al. 2020; Oldoni et al. 2021; Angeloni et al. 2021; Mahnashi et al. 2021; Abubakar and Haque 2020; Diniz do Nascimento et al. 2020; Aljubiri et al. 2021).

4 Future Perspectives and Research Directions

There is a growing interest in the bioactive phytochemicals of vegetable by-products, as they are significant sources of health-promoting compounds. Their sustainable exploitation is a priority for the scientific community. However, research on the bioactive compounds of by-products from understudied food sources and their application in human health-promoting products is still lagging.

Future research should focus on the following: (i) widening the spectrum of food by-products from various plant sources with potential health benefits that are not broadly studied, (ii) the sustainable extraction of bioactive compounds from food by-products by exploring green extraction technologies, (iii) investigation of the influence of technological/processing conditions of by-products on the bioactivities of their bioactive components, and (iv) in-depth investigation of the safety, metabolism, absorption, and viability of in vitro and in vivo studies of bioactive compounds or by-product fractions exhibiting potential health benefits (Teshome et al. 2023; Trigo et al. 2022; Calleja-Gómez et al. 2024; Santacroce et al. 2024; Tlais et al. 2020; Manthei et al. 2023; Szabo et al. 2022; Sorrenti et al. 2023).

4.1 Emerging Trends

The need for sustainable food manufacturing, coupled with a growing awareness of the health benefits of consuming vegetables, is inducing the modernization and innovation of food-processing systems. Novel conservation technologies such as enzyme application, edible packaging, non-thermal treatments, and bioprocessing systems are being researched and studied to retain the quality of the edible parts of food while rendering by-products useful. Several countries have issued government policies to improve the competitiveness of by-product exploitation in the valorization of edible food and improve the economic viability of by-product exploitation. The pursuit of a sustainable environment and the establishment of public organizations to control waste is also on the rise.

4.2 Areas for Further Investigation

Research regarding bioactive phytochemicals in by-products from vegetables remains limited. Although scientific publications focus on commonly known compounds, such as phenolics, carotenoids, alkaloids, and phytosterols, information on other less-studied compounds is scarce and needs to be completed. There is a lack of knowledge regarding the modeling of phytochemicals, particularly with respect to their structure and activity relationship. Phytochemicals are a complex mixture of compounds containing multiple hydroxyl groups, different substituents, and stereogenic centers, which makes prediction challenging. This issue becomes even more complicated when investigating food extracts as a source of bioactive phytochemicals, as the mixture profile and composition could vary largely with the extraction approach. There is also a lack of knowledge regarding the digestibility and bioavailability of by-product phytochemicals. Research has mainly focused on the biological activity and chemical composition of extracts.

Further studies are required to clarify the metabolic fate of phytochemicals in the in vivo gastrointestinal system. Additionally, processing effects, particularly food processing modes such as heating and fermentation, are largely lacking in the area of by-products. Approaches such as in vitro screening of biological activity can be useful tools in the early stages of such investigation. Due consideration should also be given to the side effects of anti-nutritional compounds often occurring in vegetation and plant extracts, which could affect long-term consumption of extracts. As much of the vegetable by-product and food waste research has focused on safety aspects, prior work regarding food safety investigations could be combined with application approaches. Such investigation areas could lead to improved knowledge and a wider understanding of vegetable by-products and extracts as a source of bioactive phytochemicals and nutrients (Cano-Lamadrid and Artes-Hernandez 2021; Sorrenti et al. 2023; Reguengo et al. 2022; Majumder and Annegowda 2021; Kainat et al. 2022).

References

'Aqilah NMN, Rovina K, Felicia WXL, Vonnie JM (2023) A review on the potential bioactive components in fruits and vegetable wastes as value-added products in the food industry. Molecules 28(6):2631

Abubakar AR, Haque M (2020) Preparation of medicinal plants: basic extraction and fractionation procedures for experimental purposes. J Pharm Bioallied Sci 12(1):1–10

Ahmed MH, Karkush SI, Ali SA, Mohammed AA (2024) Phytochemicals: a new arsenal in drug discovery. Int J Med Sci Dent Health 10(01):29–44

Akter B, Rabeta MS (2021) Synbiotic and antioxidant activity of fruit by-products and their effect on human health. Food Res 5(1):24–35

Aljubiri SM, Mahgoub SA, Almansour AI, Shaaban M, Shaker KH (2021) Isolation of diverse bioactive compounds from *Euphorbia balsamifera*: cytotoxicity and antibacterial activity studies. Saudi J Biol Sci 28(1):417–426

Alves Filho EG, Sousa VM, Rodrigues S, de Brito ES, Fernandes FA (2020) Green ultrasound-assisted extraction of chlorogenic acids from sweet potato peels and sonochemical hydrolysis of caffeoylquinic acids derivatives. Ultrason Sonochem 63:104911

Amo-Mensah J, Darko G, Borquaye LS (2020) Anti-inflammatory and antioxidant activities of the root and bark extracts of *Vitex grandifolia* (Verbanaceae). Sci Afr 10:e00586

Angeloni S, Spinozzi E, Maggi F, Sagratini G, Caprioli G, Borsetta G et al (2021) Phytochemical profile and biological activities of crude and purified *Leonurus cardiaca* extracts. Plants 10(2):195

Anokwah D, Kwatia EA, Amponsah IK, Jibira Y, Harley BK, Ameyaw EO et al (2022) Evaluation of the anti-inflammatory and antioxidant potential of the stem bark extract and some constituents of *Aidia genipiflora* (DC.) dandy (Rubiaceae). Heliyon 8(8):e10082

Babich O, Sukhikh S, Pungin A, Ivanova S, Asyakina L, Prosekov A (2020) Modern trends in the in vitro production and use of callus, suspension cells and root cultures of medicinal plants. Molecules 25(24):5805

Behl T, Kumar K, Brisc C, Rus M, Nistor-Cseppento DC, Bustea C et al (2021a) Exploring the multifocal role of phytochemicals as immunomodulators. Biomed Pharmacother 133:110959

Behl T, Rocchetti G, Chadha S, Zengin G, Bungau S, Kumar A et al (2021b) Phytochemicals from plant foods as potential source of antiviral agents: an overview. Pharmaceuticals 14(4):381

Borquaye LS, Doetse MS, Baah SO, Mensah JA (2020) Anti-inflammatory and anti-oxidant activities of ethanolic extracts of *Tamarindus indica* L.(Fabaceae). Cogent Chem 6(1):1743403

Caballero S, Mereles L, Burgos-Edwards A, Alvarenga N, Coronel E, Villalba R, Heinichen O (2021) Nutritional and bioactive characterization of *Sicana odorifera* naudim vell. Seeds by-products and its potential hepatoprotective properties in Swiss albino mice. Biology 10(12):1351

Calleja-Gómez M, Roig P, Pateiro M, Domínguez R, Lorenzo JM, Fernández-López J et al (2024) Health promoting benefits of plant-based by-products extracts obtained by innovative technologies. Curr Opin Food Sci 57:101161

Cano-Lamadrid M, Artes-Hernandez F (2021) By-products revalorization with non-thermal treatments to enhance phytochemical compounds of fruit and vegetables derived products: A review. Foods 11(1):59

Cebulak T, Krochmal-Marczak B, Stryjecka M, Krzysztofik B, Sawicka B, Danilčenko H, Jarienè E (2022) Phenolic acid content and antioxidant properties of edible potato (*Solanum tuberosum* L.) with various tuber flesh colours. Foods 12(1):100

Chitimus DM, Popescu MR, Voiculescu SE, Panaitescu AM, Pavel B, Zagrean L, Zagrean AM (2020) Melatonin's impact on antioxidative and anti-inflammatory reprogramming in homeostasis and disease. Biomolecules 10(9):1211

Coman V, Teleky BE, Mitrea L, Martău GA, Szabo K, Călinoiu LF, Vodnar DC (2020) Bioactive potential of fruit and vegetable wastes. Adv Food Nutr Res 91:157–225

Dawurung CJ, Nguyen MT, Pengon J, Dokladda K, Bunyong R, Rattanajak R et al (2021) Isolation of bioactive compounds from medicinal plants used in traditional medicine: Rautandiol B, a potential lead compound against *Plasmodium falciparum*. BMC Complement Med Ther 21:1–12

Delgado A, Chammem N, Issaoui M, Ammar E (2022) Bioactive phytochemicals from olive (*Olea europaea*) processing by-products. In: Bioactive phytochemicals from vegetable oil and oilseed processing by-products. Springer International Publishing, Cham, pp 1–37

Diniz do Nascimento L, Barbosa de Moraes AA, Santana da Costa K, Pereira Galúcio JM, Taube PS, Leal Costa CM et al (2020) Bioactive natural compounds and antioxidant activity of essential oils from spice plants: new findings and potential applications. Biomolecules 10(7):988

Dzobo K (2022) The role of natural products as sources of therapeutic agents for innovative drug discovery. In: Comprehensive pharmacology, pp 408–422

Elimam DM, Ramadan MF, Elshazly AM, Farag MA (2022) Introduction to Mediterranean fruits bio-wastes: chemistry, functionality and techno-applications. In: Ramadan MF, Farag MA (eds) Mediterranean fruits bio-wastes. Springer, Cham. https://doi.org/10.1007/978-3-030-84436-3_1

Erhabor RC, Aderogba MA, Erhabor JO, Nkadimeng SM, McGaw LJ (2021) In vitro bioactivity of the fractions and isolated compound from *Combretum elaeagnoides* leaf extract against selected foodborne pathogens. J Ethnopharmacol 273:113981

Favela-González KM, Hernández-Almanza AY, De la Fuente-Salcido NM (2020) The value of bioactive compounds of cruciferous vegetables (Brassica) as antimicrobials and antioxidants: A review. J Food Biochem 44(10):e13414

Galali Y, Omar ZA, Sajadi SM (2020) Biologically active components in by-products of food processing. Food Sci Nutr 8(7):3004–3022

Garcia-Alonso A, Sanchez-Paniagua Lopez M, Manzanares-Palenzuela CL, Redondo-Cuenca A, López-Ruíz B (2023) Edible plant by-products as source of polyphenols: prebiotic effect and analytical methods. Crit Rev Food Sci Nutr 63(31):10814–10835

Gorlenko CL, Kiselev HY, Budanova EV, Zamyatnin Jr AA, Ikryannikova LN (2020) Plant secondary metabolites in the battle of drugs and drug-resistant bacteria: new heroes or worse clones of antibiotics? Antibiotics 9(4):170

Halder M, Jha S (2023) Medicinal plants and bioactive phytochemical diversity: a fountainhead of potential drugs against human diseases. In: Medicinal plants: biodiversity, biotechnology and conservation. Singapore, Springer Nature Singapore, pp 39–93

Haruna A, Yahaya SM (2021) Recent advances in the chemistry of bioactive compounds from plants and soil microbes: a review. Chem Afr 4(2):231–248

Joly N, Souidi K, Depraetere D, Wils D, Martin P (2020) Potato by-products as a source of natural chlorogenic acids and phenolic compounds: extraction, characterization, and antioxidant capacity. Molecules 26(1):177

Jugran AK, Rawat S, Devkota HP, Bhatt ID, Rawal RS (2021) Diabetes and plant-derived natural products: from ethnopharmacological approaches to their potential for modern drug discovery and development. Phytother Res 35(1):223–245

Kainat S, Arshad MS, Khalid W, Zubair Khalid M, Koraqi H, Afzal MF et al (2022) Sustainable novel extraction of bioactive compounds from fruits and vegetables waste for functional foods: a review. Int J Food Prop 25(1):2457–2476

Karimi A, Kazemi M, Samani SA, Simal-Gandara J (2021) Bioactive compounds from by-products of eggplant: functional properties, potential applications and advances in valorization methods. Trends Food Sci Technol 112:518–531

Krochmal-Marczak B, Cebulak T, Kapusta I, Oszmiański J, Kaszuba J, Żurek N (2020) The content of phenolic acids and flavonols in the leaves of nine varieties of sweet potatoes (*Ipomoea batatas* L.) depending on their development, grown in Central Europe. Molecules 25(15):3473

López-Salas L, Borrás-Linares I, Quintin D, García-Gomez P, Giménez-Martínez R, Segura-Carretero A, Lozano-Sánchez J (2021) Artichoke by-products as natural source of phenolic food ingredient. Appl Sci 11(9):3788

Mahnashi MH, Alqahtani YS, Alqarni AO, Alyami BA, Jan MS, Ayaz M et al (2021) Crude extract and isolated bioactive compounds from *Notholirion thomsonianum* (Royale) Stapf as multitargets antidiabetic agents: in-vitro and molecular docking approaches. BMC Complement Med Ther 21:1–13

Majumder P, Annegowda HV (2021) Fruit and vegetable by-products: novel ingredients for a sustainable society. In: Valorization of Agri-food wastes and by-products. Academic Press, pp 133–156

Makori SI, Mu TH, Sun HN (2020) Total polyphenol content, antioxidant activity, and individual phenolic composition of different edible parts of 4 sweet potato cultivars. Nat Prod Commun 15(7):1934578X20936931

Mangiapelo L, Blasi F, Ianni F, Barola C, Galarini R, Abualzulof GW et al (2023) Optimization of ultrasound-assisted extraction of chlorogenic acid from potato sprout waste and enhancement of the in vitro total antioxidant capacity. Antioxidants 12(2):348

Manthei A, López-Gámez G, Martín-Belloso O, Elez-Martínez P, Soliva-Fortuny R (2023) Relationship between physicochemical, techno-functional and health-promoting properties of fiber-rich fruit and vegetable by-products and their enhancement by emerging technologies. Foods 12(20):3720

Marcillo-Parra V, Tupuna-Yerovi DS, González Z, Ruales J (2021) Encapsulation of bioactive compounds from fruit and vegetable by-products for food application–A review. Trends Food Sci Technol 116:11–23

Margraf A, Ludwig N, Zarbock A, Rossaint J (2020) Systemic inflammatory response syndrome after surgery: mechanisms and protection. Anesth Analg 131(6):1693–1707

Meizlish ML, Franklin RA, Zhou X, Medzhitov R (2021) Tissue homeostasis and inflammation. Annu Rev Immunol 39(1):557–581

Meral R, Kose YE, Ceylan Z, Cavidoglu İ (2022) The potential use of agro-industrial by-products as sources of bioactive compounds: A nanotechnological approach. Stud Nat Prod Chem 73:435–466

Mogana R, Adhikari A, Tzar MN, Ramliza R, Wiart CJBCM (2020) Antibacterial activities of the extracts, fractions and isolated compounds from Canarium patentinervium Miq. Against bacterial clinical isolates. BMC Complement Med Ther 20:1–11

Mohan S, Elhassan Taha MM, Makeen HA, Alhazmi HA, Al Bratty M, Sultana S et al (2020) Bioactive natural antivirals: an updated review of the available plants and isolated molecules. Molecules 25(21):4878

Mohd Basri MS, Abdul Karim Shah NN, Sulaiman A, Mohamed Amin Tawakkal IS, Mohd Nor MZ, Ariffin SH et al (2021) Progress in the valorization of fruit and vegetable wastes: active packaging, biocomposites, by-products, and innovative technologies used for bioactive compound extraction. Polymers 13(20):3503

Mollaei M, Abbasi A, Hassan ZM, Pakravan N (2020) The intrinsic and extrinsic elements regulating inflammation. Life Sci 260:118258

Moreira Lopes TC, Mosser DM, Gonçalves R (2020) Macrophage polarization in intestinal inflammation and gut homeostasis. Inflamm Res 69(12):1163–1172

Murakami Y, Ishikawa K, Nakao S, Sonoda KH (2020) Innate immune response in retinal homeostasis and inflammatory disorders. Prog Retin Eye Res 74:100778

Musilová J, Franková H, Fedorková S, Lidiková J, Vollmannová A, Sulírová K et al (2024) Comparison of polyphenols, phenolic acids, and antioxidant activity in sweet potato (*Ipomoea batatas* L.) tubers after heat treatments. J Agric Food Res 18:101271

Najmi A, Javed SA, Al Bratty M, Alhazmi HA (2022) Modern approaches in the discovery and development of plant-based natural products and their analogues as potential therapeutic agents. Molecules 27(2):349

Nasim N, Sandeep IS, Mohanty S (2022) Plant-derived natural products for drug discovery: current approaches and prospects. The Nucleus 65(3):399–411

Núñez-Gómez V, González-Barrio R, Periago MJ (2023) Interaction between dietary fibre and bioactive compounds in plant by-products: impact on bioaccessibility and bioavailability. Antioxidants 12(4):976

Oldoni TLC, Merlin N, Bicas TC, Prasniewski A, Carpes ST, Ascari J et al (2021) Antihyperglycemic activity of crude extract and isolation of phenolic compounds with antioxidant activity from *Moringa oleifera* Lam. Leaves grown in Southern Brazil. Food Res Int 141:110082

Parveen B, Parveen A, Parveen R, Ahmad S, Ahmad M, Iqbal M (2020) Challenges and opportunities for traditional herbal medicine today, with special reference to its status in India. Ann Phytomed 9(2):97–112

Patra A, Abdullah S, Pradhan RC (2022) Review on the extraction of bioactive compounds and characterization of fruit industry by-products. Bioresour Bioprocess 9(1):14

Pinto T, Aires A, Cosme F, Bacelar E, Morais MC, Oliveira I et al (2021) Bioactive (poly) phenols, volatile compounds from vegetables, medicinal and aromatic plants. Foods 10(1):106

Porras G, Chassagne F, Lyles JT, Marquez L, Dettweiler M, Salam AM et al (2020) Ethnobotany and the role of plant natural products in antibiotic drug discovery. Chem Rev 121(6):3495–3560

Ramadan Hassanien MF (2023) Introduction to bioactive phytochemicals from vegetable oil and oilseed processing by-products. In: Bioactive phytochemicals from vegetable oil and oilseed processing by-products. Springer International Publishing, Cham, pp 3–10

Reguengo LM, Salgaço MK, Sivieri K, Júnior MRM (2022) Agro-industrial by-products: valuable sources of bioactive compounds. Food Res Int 152:110871

Rojas-García A, Fernández-Ochoa Á, Cádiz-Gurrea MDLL, Arráez-Román D, Segura-Carretero A (2023) Neuroprotective effects of agri-food by-products rich in phenolic compounds. Nutrients 15(2):449

Roy A, Datta S, Bhatia KS, Jha P, Prasad R (2022) Role of plant derived bioactive compounds against cancer. S Afr J Bot 149:1017–1028

Sahukari R, Punabaka J, Bhasha S, Ganjikunta VS, Kondeti Ramudu S, Kesireddy SR et al (2021) Phytochemical profile, free radical scavenging and anti-inflammatory properties of Acalypha Indica root extract: evidence from in vitro and in vivo studies. Molecules 26(20):6251

Santacroce L, Bottalico L, Charitos IA, Castellaneta F, Gaxhja E, Topi S et al (2024) Exploitation of natural by-products for the promotion of healthy outcomes in humans: special focus on antioxidant and anti-inflammatory mechanisms and modulation of the gut microbiota. Antioxidants 13(7):796

Silvagno F, Vernone A, Pescarmona GP (2020) The role of glutathione in protecting against the severe inflammatory response triggered by COVID-19. Antioxidants 9(7):624

Sorrenti V, Burò I, Consoli V, Vanella L (2023) Recent advances in health benefits of bioactive compounds from food wastes and by-products: biochemical aspects. Int J Mol Sci 24(3):2019

Süntar I (2020) Importance of ethnopharmacological studies in drug discovery: role of medicinal plants. Phytochem Rev 19(5):1199–1209

Szabo K, Mitrea L, Călinoiu LF, Teleky BE, Martău GA, Plamada D et al (2022) Natural polyphenol recovery from apple-, cereal-, and tomato-processing by-products and related health-promoting properties. Molecules 27(22):7977

Teruel-Andreu C, Andreu-Coll L, López-Lluch D, Sendra E, Hernández F, Cano-Lamadrid M (2021) Ficus carica fruits, by-products and based products as potential sources of bioactive compounds: a review. Agronomy 11(9):1834

Teshome E, Teka TA, Nandasiri R, Rout JR, Harouna DV, Astatkie T, Urugo MM (2023) Fruit by-products and their industrial applications for nutritional benefits and health promotion: a comprehensive review. Sustainability 15(10):7840

Tlais AZA, Fiorino GM, Polo A, Filannino P, Di Cagno R (2020) High-value compounds in fruit, vegetable and cereal byproducts: an overview of potential sustainable reuse and exploitation. Molecules 25(13):2987

Trigo JP, Alexandre EM, Saraiva JA, Pintado ME (2022) High value-added compounds from fruit and vegetable by-products–characterization, bioactivities, and application in the development of novel food products. Crit Rev Food Sci Nutr 60(8):1388–1416

Varijakzhan D, Chong CM, Abushelaibi A, Lai KS, Lim SHE (2020) Middle eastern plant extracts: an alternative to modern medicine problems. Molecules 25(5):1126

Wink M (2022) Current understanding of modes of action of multicomponent bioactive phytochemicals: potential for nutraceuticals and antimicrobials. Annu Rev Food Sci Technol 13(1):337–359

Yang H, Liao Q, Ma L, Luo W, Xiong X, Luo Y et al (2021) Features and genetic basis of chlorogenic acid formation in diploid potatoes. Food Chem: Mol Sci 3:100039

Yesmin S, Paul A, Naz T, Rahman AA, Akhter SF, Wahed MII et al (2020) Membrane stabilization as a mechanism of the anti-inflammatory activity of ethanolic root extract of Choi (*Piper chaba*). Clin Phytoscience 6:1–10

Yimer T, Birru EM, Adugna M, Geta M, Emiru YK (2020) Evaluation of analgesic and anti-inflammatory activities of 80% methanol root extract of *Echinops kebericho* M.(Asteraceae). J Inflamm Res 13:647–658

Zaky AA, Akram MU, Rybak K, Witrowa-Rajchert D, Nowacka M (2024) Bioactive compounds from plants and by-products: novel extraction methods, applications, and limitations. AIMS Mol Sci 11(2):150–188

The manufacturer's authorised representative in the EU is Springer Nature Customer Service Centre GmbH, Europaplatz 3, 69115 Heidelberg, Germany. If you have any concerns regarding our products, please contact ProductSafety@springernature.com

Printed and bound by CPI Group (UK) Ltd, Croydon, CR0 4YY
30/09/2026
02225563-0008